Learning Materials in Biosciences

Learning Materials in Biosciences textbooks compactly and concisely discuss a specific biological, biomedical, biochemical, bioengineering or cell biologic topic. The textbooks in this series are based on lectures for upper-level undergraduates, master's and graduate students, presented and written by authoritative figures in the field at leading universities around the globe.

The titles are organized to guide the reader to a deeper understanding of the concepts covered. Each textbook provides readers with fundamental insights into the subject and prepares them to independently pursue further thinking and research on the topic. Colored figures, step-by-step protocols and take-home messages offer an accessible approach to learning and understanding. In addition to being designed to benefit students, Learning Materials textbooks represent a valuable tool for lecturers and teachers, helping them to prepare their own respective coursework.

More information about this series at http://www.springer.com/series/15430

Florian Rüker • Gordana Wozniak-Knopp
Editors

Introduction to Antibody Engineering

 Springer

Editors
Florian Rüker
Institute of Molecular Biotechnology,
Department of Biotechnology
University of Natural Resources and
Life Sciences, Vienna (BOKU)
Vienna, Austria

Gordana Wozniak-Knopp
Christian Doppler Laboratory for Innovative
Immunotherapeutics, Institute of Molecular
Biotechnology, Department of Biotechnology
University of Natural Resources and
Life Sciences, Vienna (BOKU)
Vienna, Austria

ISSN 2509-6125   ISSN 2509-6133   (electronic)
Learning Materials in Biosciences
ISBN 978-3-030-54629-8   ISBN 978-3-030-54630-4   (eBook)
https://doi.org/10.1007/978-3-030-54630-4

This Springer imprint is published by the registered company Springer Nature Switzerland AG.
The registered company address is: Gewerbestrasse 11, 6330 Cham, Switzerland

Preface

The textbook *Introduction to Antibody Engineering* intended for students of natural sciences at the master's level, but written for all fans of the antibody molecule, is a stroll through early days of antibody discovery and milestone findings on their molecular properties toward their most recent groundbreaking applications. Antibodies are today one of the most important classes of therapeutic molecules, saving lives and improving quality of lives of millions of patients suffering from cancer and autoimmune diseases. It is fascinating how these molecules, produced by the immune system in a wide variety and innate superb stability, were discovered to be amenable to further biotechnological and biomedical developments. The fact that their activity alone can account for the recovery from a diseased state and that it can often even be attributed to a single therapeutic agent incited the ambition of many scientists to analyze the activity of their molecular building blocks and their organization. Obliging to fit the broad spectrum of requirements of their end use, such as graded degrees of specific antigen affinity, tailored half-life, and variable engagement of effector cells, this class of molecules has yielded suitable candidates for various therapeutic applications. Their amenability to molecular evolution has led to the development of sophisticated display and selection platforms based on phage, yeast, insect, and mammalian cells. Their modular composition enabled the discovery of miniaturized antibodies and antibody-like formats, which still mediate antibody-like antigen capturing, but can exhibit improved biodistribution, bioavailability, and even stunningly improved potency in their envisioned use. Incredible new biology that can result from a combination of two or more antigen specificities in a single molecule has prompted engineering efforts to produce bispecific antibody-based agents, which have in the last ten years entered the stage of clinical testing in large numbers. One unique turn in antibody development was the attachment of a viciously poisonous toxin, which is however only active when released from the antibody moiety, and thereby the exquisite specificity and the deleterious effect on the cell were combined to optimize targeted toxicity. For the production of therapeutic agents, it is mandatory to secure a source of high-level expression of a homogenous product that fulfills highly structured quality control criteria, and stability engineering may be decisive for their further development. Apart from their therapeutic value, antibodies are the core components of diagnostic tests in clinical and research diagnostics.

It would not be possible to cover this complex field without the contributions of experts from academia and industry, who are not only eminent researchers in the antibody field, but also specialists in the particular topics relevant to the ever-expanding sphere of antibody engineering. We wish to express sincere gratitude for sharing their knowledge and investing their valuable time and composing excellent chapters, which render the conceptual background, applicative aspects, and timely views on the particular subject a pleasant read. We thank Antony Rees for an enchanting overview of antibody history; Janina Klemm, Lukas Pekar, Simon Krah, and Stefan Zielonka for a comprehensive comparison of antibody display systems; Biao Ma and Michael Osborn for enlightening on the timely topic of transgenic animals as platforms for antibody discovery; Laura Rhiel and Stefan Becker for systematically summarizing applications of antibodies in therapy, diagnosis, and science; Stephan Dickgiesser, Marcel Rieker, and Nicolas Rasche for the thorough description of manifold aspects of the exciting class of antibody–drug conjugates; Doreen Könning and Jonas Schäfer for the well-structured overview of the colorful collection of alternative binding scaffolds; Astrid Holzinger and Hinrich Abken for the valuable insights into the therapeutic options with CAR-T cells; Neil Brewis for the inspirational ideas on the improvement of the key characteristics of antibodies; Björn Hock and Henri Kornmann for their highly application-oriented views on challenges in the manufacture of therapeutic antibodies; and Lina Heistinger, David Reinhart, Diethard Mattanovich, and Renate Kunert for the informative comparison of animal cell- and yeast-based platforms for production of antibodies and antibody fragments. We would like to acknowledge Sivachandran (Siva) Ravanan for excellent assistance in the editorial process, immense patience, and understanding. Finally, a word of appreciation for all our students past and present, whose genuine interest in antibody engineering was rewarding and motivating for us through many years of lecturing.

Vienna, Austria   Florian Rüker
                                                                       Gordana Wozniak-Knopp

Contents

Introduction by the Editors

1

Florian Rüker and Gordana Wozniak-Knopp

Abstract

Aimed at Master's and PhD students in biotechnology, molecular biology, and immunology, and all those who are interested in antibody engineering, the Textbook on Antibody Engineering provides a timely overview of the current state of the art of this important field in biotechnology and in medicine. Starting with a historical overview, the reader is introduced to the various ways of generating and expressing specifically binding antibodies and is familiarized with strategies to engineer their specific properties, in particular antigen binding, elicitation of effector functions, in vivo half-life, stability, and manufacturability. Further engineering of antibodies leads to miniaturization, bispecificity or to their use as carriers of payloads, e.g., in the form of antibody–drug conjugates. In addition, the use of antibodies, antibody fragments, and alternative binding scaffolds as targeting molecules, such as for the engineering of CAR-T cells, is covered in the book, as well as the important field of antibody validation.

F. Rüker (✉)
Institute of Molecular Biotechnology, Department of Biotechnology, University of Natural Resources and Life Sciences, Vienna (BOKU), Vienna, Austria
e-mail: florian.rueker@boku.ac.at

G. Wozniak-Knopp
Christian Doppler Laboratory for Innovative Immunotherapeutics, Institute of Molecular Biotechnology, Department of Biotechnology, University of Natural Resources and Life Sciences, Vienna (BOKU), Vienna, Austria
e-mail: gordana.wozniak@boku.ac.at

© Springer Nature Switzerland AG 2021
F. Rüker, G. Wozniak-Knopp (eds.), *Introduction to Antibody Engineering*,
Learning Materials in Biosciences,
https://doi.org/10.1007/978-3-030-54630-4_1

Antibodies have been unpreceded by any other class of molecules that served humanity to fight infectious, autoimmune, and cancerous disease. The first experiments harnessing their activity were based on transfer of serum of immunized animals to prevent illness in unprotected ones, and at that time their actual existence had not been even revealed. The structure of the antibody molecule was first described at the end of the nineteenth century, and with increasing insight in their molecular properties and accompanying technical advances in their recombinant expression and downstream processing, antibodies have grown to the presently most important class of therapeutics, with 6 antibodies among the 10 top-selling drugs in 2019.

What are the reasons behind this success story? Antibodies act naturally to counter a plethora of pathogens invading the host organism and demonstrate an immense level of diversity to cover the required spectrum of specificities. The scientific endeavors that explained the mechanisms underlying the creation of such large numbers of molecular species, highly specific and strongly affine to their target, rendered the "one gene-one protein" principle oblivious in this case and revealed the power of combinatorial diversity, arising from several unique modes of combinations of diverse immunoglobulin gene regions, as well as affinity maturation as a consequence of somatic hypermutation. These were the sparks that ignited the engineering mind: how can we transfer these intricate protocols from nature into a laboratory setting and apply them to design efficient therapeutic agents? With the use of recombinant DNA techniques as well as the development of powerful display and selection platforms including transgenic animals, the ground principles underlying antibody molecule formation could be put into practice in controlled laboratory procedures. As a result, an ever-increasing number of valuable therapeutic antibody-based products, improved in their functionality, safety, and convenience for the patient, could be documented in the last years. The success of the scientific community in this respect was corroborated with the Nobel Prize for Chemistry in 2018, awarded in the recognition of importance of the translational value of directed evolution of biological molecules.

A particular property of the antibody molecule is its ability to exert a multitude of functions via simultaneous interaction with its cognate antigen, mediated by the antibody's variable regions, and several other ligands that bind to specialized regions within the Fc fragment of the antibody. Antibodies of the IgG1 class, the most common subset of immunoglobulins in the serum, can activate natural killer cells to attack the targeted cells or trigger the complement cascade to induce their potent lysis. Yet another feature of many antibodies makes them particularly successful therapeutics: their engagement with the neonatal Fc receptor, expressed on endothelial cells, rescues them from lysosomal degradation, and they return into bloodstream, which extends their serum half-life in vivo up to several weeks. Several engineering efforts have been oriented into optimization of these functionalities. Of particular note is a very successful class of therapeutic reagents that includes fusion proteins, composed of receptor molecules and an Fc fragment, joining properties of specific targeting and favorable pharmacokinetics profile.

The last decade has witnessed the spectrum of functionalities of an antibody molecule be even further extended. Engineered molecular architectures that allow bi- or multi-specific antigen recognition within a single antibody molecule led to discovery of novel biological phenomena that now form the basis of promising therapeutic approaches. Further, in antibody-drug conjugates, the exquisite specificity of the antibody is coupled with the ferocious toxicity of molecules that cause DNA damage or inhibit the formation of the mitotic spindle in the targeted cell. These two novel classes of antibody-based therapeutics are currently the most fiercely developing ones, with several compounds that have already reached the stage of clinical approval.

Antibodies are great, but for certain applications, their size is relatively large: at 150 kDa, efficient tissue penetration is difficult to achieve. The early observation that specific antigen binding can be achieved even with isolated variable domains of the antibody incited the interest in the in-depth study of the organization of immunoglobulin domains and optimization of their stability, to endow this simple structural unit termed Fv with antibody-like antigen recognition properties. The concept of an intrinsically stably folded protein core that holds in place more variable patches on its surface, which mediate antigen interaction, was embodied in several successful attempts to engineer specific recognition into a number of proteinaceous scaffolds, many of which have entered the pathway of clinical testing.

While we are being surrounded with the immense variety of antigen recognition molecules that the advanced techniques of molecular biology have made available, there are further fields of biotechnological science that have contributed to the rise of antibody-based therapeutics: enhanced recombinant expression, efficient downstream processing, and refined analytical procedures that allow precise characterization of the active compound. Currently, therapeutic antibodies are engineered not only for antigen binding and effector functions but also for developability, which adds to their accessibility and safety and finally to the benefit for the patient.

With this, we wish for the readers to discover the momentum of joint features that has prompted the fascinating antibody molecule onto today's pedestal: its diversity that harbors the ability to find that perfect matching one, its stability that leverages the room for improvement, and its very unique virtue of persistent defense.

Antibodies: A History of Their Discovery and Properties

2

Anthony Rees

Contents

Keywords

Early vaccines · Anti-toxin discovery · Antibodies as proteins · Haptens and antigens · Antibody chains and primary structure · Antibody 3D structure · The genetics of antibody formation · Antibody diversity · Antibody effector functions

A. Rees (✉)
ReesConsulting, Uppsala, Sweden
e-mail: rees@reesconsultingab.com

© Springer Nature Switzerland AG 2021
F. Rüker, G. Wozniak-Knopp (eds.), *Introduction to Antibody Engineering*,
Learning Materials in Biosciences,
https://doi.org/10.1007/978-3-030-54630-4_2

What You Will Learn in This Chapter

This is the story of how the antibody molecules of the immune system were discovered. The early beginnings on immunity were phenomenological, preaching the dogma of miasmatic theory to explain the origin of infection and disease, even in the face of the proof that vaccines could act as a protective therapy. It was not until the beginning of the twentieth century that antibodies as independent entities began to be identified. Even then, the stranglehold on the nature of proteins by the dogmatic school of chemistry held back an understanding of the real nature of antibodies, and it would take until the second half of the twentieth century before the polypeptide nature of protein structure and the role of DNA as the genetic material would enable the full picture to emerge where antibodies would be shown to be dynamic molecular entities. You will learn how the development of physical methods in the 1960s and 1970s catapulted forward our understanding of the solution behavior of antibodies and the role of x-ray crystallography in explaining the structural basis of antigen recognition. The acceleration in the 1980s and 1990s made possible by the introduction of protein engineering techniques and more sophisticated molecular biology methods gave rise to the birth of a new age of antibody-based therapeutics.

2.1   Introduction

The history of antibodies can be partitioned into several physical and temporal phases. The early discoveries of the eighteenth and nineteenth centuries were more observations of protective effects observed in response to direct inoculation of low levels of virulent disease vectors (bacteria, viruses), later also experimenting with primitive vaccines intended to cure or prevent the disease (e.g., Jenner's cowpox). Such protective or curative effects observed would have been mediated by antibody responses in concert with T-cell responses, but at the time explanations were more phenomenological than mechanistic since knowledge of protein molecules and protective immune cells was well beyond the science of the day. As the chemistry of proteins began to be uncovered in the early twentieth century, antibodies took on a more specific form as large molecules, although their composition and the means by which they exerted their protective effects were poorly understood. As an understanding of proteins began to emerge and new techniques for protein characterization became available in the 1920s–1940s, antibody size, crude shape, recognition, and neutralization properties began to be defined along with the distinctive differences between different antibody types or "classes." At this time the detailed structural basis of antibody-antigen interactions was also beginning to be proposed although even the most prominent scientific leaders of the time were sometimes led down blind alleys in their search for mechanisms that, for example, explained how similar antibody structures could recognize a multitude of different foreign antigens. By the middle of the

twentieth century, clarity began to surface leading to an understanding of antibody structure in the early 1970s and culminating in the 1980s with an unravelling of the genetic
blueprint by which antibodies were assembled from multiple gene segments in the B-
lymphocyte. These genetic breakthroughs led to an explanation of how a diverse repertoire
of antibody structures could be produced by the immune system. We will now explore each
of these temporal periods in a little more detail.

2.2   Early Beginnings

Edward Jenner, credited with the discovery of vaccination (against smallpox), was born on
May 17, 1749, in the village of Berkeley in Gloucestershire, England. He studied surgery as
an apprentice in a small English village (Chipping Sodbury) under the surgeon apothecary
Daniel Ludlow and later the country surgeon, George Hardwicke. He then took a position
at St. George's Hospital in London under the wing of the well-known "scientific surgeon,"
John Hunter. In 1772 Jenner returned home to Berkeley where he became practitioner and
surgeon to the local community. He was only 23 years of age but had amassed 9 years of
knowledge and experience in dissection and investigation, an extraordinary erudition for
one so young. Jenner's interest in the smallpox disease may have been sparked during his
earlier apprenticeship with Ludlow and Hardwicke where it is said (not everyone agrees)
that he came across the dairy maid folklore "If I am exposed to and succumb to cowpox, I
will never get smallpox." A well-known seventeenth-century nursery rhyme, combined
with some poetical exegesis, supports perhaps this folklore assertion that suggested dairy
maids exposed to cowpox never had smallpox scars on their faces "Where are you going to
my pretty maid? I'm going a milking sir, she said. . . What is your fortune my pretty maid?
My face is my fortune sir, she said."

During the eighteenth century, smallpox claimed around 400,000 lives per year in
Europe. Of the survivors a third became blind. The method of variolation (Lat.
varius = spotted or stained), in which fluid from live smallpox sores was injected
subcutaneously into individuals as a protective procedure, had been known for some
hundreds of years. Its adoption in Europe however only began after a series of procedures
had been carried out initially on members of the English aristocracy and then in a more
controlled set of trials on inmates at Newgate Prison in London and later orphaned children,
in both instances with some moderate success.

Jenner had a different approach. He reported 23 case studies involving individuals or
groups of individuals who had (1) been exposed naturally to cowpox and then smallpox,
(2) been naturally exposed to cowpox and were then variolated, and (3) had experienced
neither infection and were then inoculated with cowpox followed by smallpox [1]. The first
of the case studies appears to have occurred in 1743, but the most famous of these was the
case of Sarah Nelms, a milkmaid on a farm near to Berkeley. Reported as case XVI Jenner
observed:

SARAH NELMES, a dairymaid at a Farmer's near this place, was infected with the Cow Pox from her master's cows in May, 1796. She received the infection on a part of the hand . . .. A large pustulous sore and the usual symptoms accompanying the disease were produced in consequence. . ..

Here was the opportunity to test the hypothesis that exposure to cowpox protected the individual against the more dangerous smallpox. Jenner identified a young village boy, James Phipps, who had not yet been variolated, and injected the secretions from the Sarah Nelms sores under Phipp's skin as two superficial incisions, each about half an inch long. This was in May 1796 and was the first example of a human-to-human vaccination (derived from the Latin *vacca* = cow). Two months later Jenner followed up the cowpox vaccination by variolation of Phipps with smallpox by the usual procedure, and extraordinarily the smallpox exposure caused no infection. Jenner's attempt to get his results published by the Royal Society of London failed for "lack of sufficient experimental evidence." His conviction that this was an enormously important medical advance led him to finance the publication himself. This was not enough however to convince the entire medical community. Skepticism abounded until several renowned London physicians supported Jenner's procedure. While many countries recognized the power of Jenner's vaccination procedure, including Napoleon, who required his army to be vaccinated, and forward-looking physicians in America, in some parts of Europe including Britain, skepticism was rife, and it took until the late nineteenth century for the British medical establishment to finally recognize the power of cowpox vaccination. One of the reasons for the reluctance of many physicians to adopt vaccination was their almost fanatical adherence to miasmatic (Gk *miasmatos*; Latin *miasma* = noxious vapors) theories of infection where bad air and other environmental influences were thought to be the origin of disease.

The reasons for the effectiveness of cowpox vaccination were not understood by Jenner or any of the medical establishment since the notion of a protective immunological "system" in the body was well outside the body of medical knowledge of the time. That was to change extremely slowly and not without considerable skepticism from all scientific quarters. In 1860 Pasteur working first in Lille, France, and then Paris proposed his "germ theory," initially to explain how specific organisms (yeast) were necessary for alcohol during fermentation and were also responsible for souring (bacteria). These critical observations led to the germ theory of disease. Neither were fully accepted for decades afterward due to the strong adherence of the scientific community to the notion of "spontaneous generation" or in the case of fermentation "spontaneous transformation." Along with Robert Koch's parallel studies in Berlin, the microbiological origins of infectious diseases began to be established, despite continuing skepticism from the "chemical school of thought" that considered all diseases to be due to chemical imbalances in the body, caused by either internal or external influences.

The discovery that an inactivated or "attenuated" pathogen (e.g., by chemical treatment or oxidation) could protect against infection by the virulent version of the pathogen was to be the coup de grâce for the miasmatic and purely chemical theories of infection. The first practical demonstration of this occurred at Pouilly-le-Fort (France) where in 1881 the

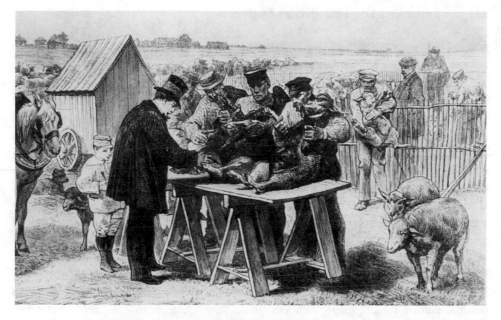

Fig. 2.1 The Pouilly-le-Fort vaccination in 1881. Reproduced with permission from SPL and TT Bild AB, Sweden

veterinarian Rossignol challenged Pasteur to publicly prove his vaccination approach by preventing anthrax infection in sheep (Fig. 2.1).

The experiment was a total success and validated the attenuation approach for many other infectious diseases (see [2] for further details). But this was in animals, and, while impressive, it had not been carried out with humans. In fact, human vaccination had not been attempted since Jenner, despite Pasteur's claim that Pouilly-le-Fort was the first demonstration of an "experimental vaccine," rather than something simply taken from a cowpox-infected "blister" as the inoculant [2]. Four years later Pasteur set the record straight with an experiment that would never be permitted today, this time with rabies. On July 6, 1885, three people bitten by a rabid dog presented themselves at Pasteur's laboratory in Paris. The youngest, a 9-year-old boy, Joseph Meister, was considered the most in danger. Pasteur had been working with rabies in rabbits and knew it attacked the nervous system. By exposing the spinal cords of infected rabbits to air over a period of 1–15 days, his experiments had shown that the rabies virus [3] gradually became attenuated over the exposure time period until it was non-infectious. He implanted rabbit cord material over a 10-day period in the boy's upper abdomen, starting with the longest, 15 days, exposure and gradually working down to the more virulent 1-day exposure. Meister survived the process, and 3 months later, Pasteur presented the case with great excitement to the French Academy of Sciences [4].

At this time medical science knew the "what," but they didn't understand the "how." How did vaccination cause protection and what was happening in the body that caused it? That was about to change, but not without many bumps in the road.

2.3   Serum as the Protection Matrix

In Robert Koch's Berlin Institute in the late 1880s, Emil von Behring and the visiting Japanese researcher Shibasaburō Kitasato had been working on attenuation of tetanus bacilli in an attempt to find a vaccine. They had treated the bacilli with iodoform (CHI_3, an oxidizing agent) in what they called a "disinfectant" preparation and then inoculated rabbits with the treated bacilli. The animals were completely protected when challenged with the virulent form of tetanus. von Behring believed that somehow the blood had become altered, and together he and Kitasato took serum from the previously exposed rabbits and injected it into mice whereupon the mice were completely immune to challenge with untreated tetanus bacilli. A further critical advance came from the work of Emile Roux at the Pasteur Institute in Paris and Kitasato in Berlin who were able to isolate the toxic entity from the diphtheria and tetanus bacilli, respectively, by filtration though special ceramic filters (e.g., the Chamberland) with miniscule natural pores that would block the passage of bacteria and other microorganisms but allow molecular aggregates through. Kitasato working with von Behring showed that when either the immune rabbit serum was pre-incubated with the tetanus toxin and the mixture injected into mice or the mice received the antiserum after challenge with the toxin, they again were completely protected. Following up on these observations, von Behring, whose main interest was diphtheria, repeated the procedure with the filtered diphtheria toxin and obtained identical results. These were the first examples of "prophylactic serum therapy," von Behring believing the serum acted as an "antitoxin." Since diphtheria was a more important problem in Germany (and the rest of Europe) than tetanus, it quickly became a commercially driven and highly successful clinical treatment where horses were used to produce large amounts of anti-diphtheria toxin antiserum. In 1901 von Behring received the Nobel Prize for his diphtheria antiserum discovery (Fig. 2.2) [5].

Also working in Koch's laboratory was Paul Ehrlich who had been working on chemical redox dyes as sensors for the oxygen status of cells. During the development of production batches of diphtheria toxin by the Hoechst company, used to generate the horse antiserum, Ehrlich was asked to address a problem of batch-to-batch variation in efficacy, a particularly acute problem during storage. He discovered that the toxin seemed to contain several components, the virulent toxin itself and various derivatives (what he called *toxoids*), that retained binding to the serum antitoxin component with varying affinities but which had lost their toxicity over time. His famous *giftspektrum*, reflecting different measured affinities of the various toxoids, showed stepwise binding which he interpreted as a reflection of irreversible covalent bonding between toxoids and antitoxin, a conclusion consistent with his earlier views that the bonding of oxygen species to his redox dyes was also irreversible. So, binding of antitoxins (antibodies) to their targets was covalent!

Meanwhile, Thorvald Madsen in Copenhagen and Svante Arrhenius in Uppsala, who had carried out similar studies with tetanus toxin, obtained results that questioned Ehrlich's covalent bonding hypothesis. Their results were in agreement with the reversible equilibrium binding model of Guldberg and Waage (Oslo) since the binding curve was a

Fig. 2.2 Emil von Behring. Reproduced courtesy of the Wellcome Collection, London. CC BY

continuous not a step curve. Despite this new evidence, even the great physical chemist Walter Nernst commented in 1904 that the concept of a reversible process collided with the facts. Sometimes great chemists are not always great biologists!

While continuing to develop his theory of antitoxin-toxin interactions, Ehrlich's ideas were heavily influenced by the early nineteenth-century rather primitive understanding of proteins, what they were chemically, and what their role was in the body. The notion of a toxin, virus, or bacterium that caused infection revolved around the idea that pathogens, when entering the body, were toxic because they sequestered nutrients that were essential for the body by cross-reacting (essentially competing) with "nutrient receptors" on the cell surfaces of various tissues. The cellular response to this challenge was then to over-produce the receptors as soluble entities which were the antitoxins that bound the relevant toxin and toxoids. According to this theory, if as a result of toxin overload nutrient supply to the tissues and organs became acutely reduced, the body could succumb to serious disease and even death.

Ehrlich's model of the antitoxin process was illustrated in his Croonian lecture to the Royal Society of London on March 22, 1900 [6] (see Fig. 2.3).

While his biology was somewhat off-center, the parallel with the antigen stimulation of B cells to produce soluble antibodies was, unknowingly, not far away. But there was another property of antitoxins Ehrlich's model needed to explain. Antitoxins were known to mediate agglutination and/or hemolysis (e.g., of red cells) which his side-chain hypothesis in its initial iteration found difficulty in explaining. Ehrlich then "expanded" his model to include the addition of a lysis-inducing moiety as an integral part of each antitoxin. He

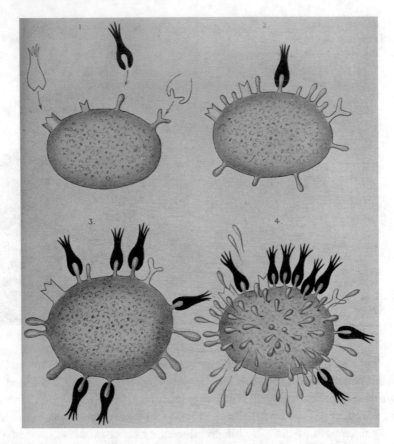

Fig. 2.3 Ehrlich's representation of the union of toxin molecules with side-chains of the "cell protoplasm" (the antitoxins) by means of their "haptophore groups." Reproduced from Plate 6, Ehrlich, P., 'On Immunity with Special Reference to Cell Life', Royal Society Croonian Lecture 1900, in Roy. Soc. Proc. 1900, 66, pp. 424–448, Copyright © The Royal Society, with permission from the Royal Society, London, UK

called this complex the "amboreceptor" (Fig. 2.4), and it worked as a dual recognition antitoxin—the antitoxin itself and the lytic entity which Ehrlich named "complement." While creative, it had a significant economy problem. Since each amboreceptor was required to be specific for a particular antigen, every different antitoxin would also require a different complement entity. Creative but biologically enormously wasteful.

In parallel the Belgian immunologist Jules Bordet proposed another way out of the hemolysis problem [7]. He also postulated a lytic substance (he called it "alexine") which was present as a separate entity in the blood. When the cell (e.g., a red cell) had been prepared by a toxin attaching to the antitoxin on the cell surface, the cell would become somehow sensitized, and alexine would then directly lyse the cell. In Bordet's model, however, there was no direct interaction of alexine with the antibody. Two wrongs don't normally make a right, but marrying the two approaches of Ehrlich and Bordet together

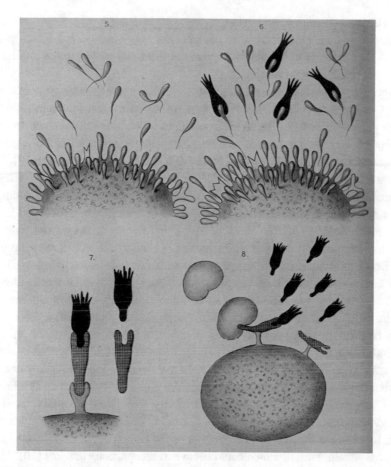

Fig. 2.4 Ehrlich's "amboreceptor" concept in which antitoxin binds to both cellular side-chains and a "complement" molecule. Reproduced from Plate 7, Ehrlich, P., 'On Immunity with Special Reference to Cell Life', Royal Society Croonian Lecture 1900, in Roy. Soc. Proc. 1900, 66, pp. 424–448, Copyright © The Royal Society, with permission from the Royal Society, London, UK

now (at the time they were passionately opposed to each other's ideas), the mechanism came passably close to how the complement system works, at least in principle.

2.4   Specificity and the Nature of Proteins

It would have been much simpler for Ehrlich, Bordet, and others if the nature of proteins as long chain amino acids, folded into three-dimensional shapes that depended on the amino acid sequences, was known. It was not. In fact, the powerful chemistry lobby around Europe, particularly strong in Vienna, believed that proteins were "colloids" consisting of mixtures of small molecules bound together with the mediation of essential surface charges. The great chemist, Emil Fischer, stated in 1906 that natural proteins were mixtures of substances such as peptides whose composition was much simpler than previously thought. A year later he noted that there was not the slightest guarantee that the natural proteins were homogeneous substances [8].

A further question that exercised the minds of Ehrlich and others was one of specificity. Friedrich Obermeyer and Ernst Peter Pick, working in Vienna, had shown in 1906 that if an antigen (e.g., bovine serum albumin) was modified by various chemical treatments (e.g., iodination) and then used to immunize rabbits, the serum generated only reacted with the modified albumin and not the native protein (still thought to be a colloid) [9]. This was arguably the earliest demonstration of "chemical specificity" of antitoxins. They further showed that the rabbit serum reacted with iodo-albumin from any species (including plants) but not with the native albumin from those species or with albumin modified by another chemical modification. Their conclusion, over-simplified as it turned out, was that the modification had destroyed the albumin species specificity. It would take the tenacious work of Karl Landsteiner, a Viennese physician with a significant knowledge of chemistry from his time working with Emil Fischer in Wurzburg, Max von Gruber in Vienna, and others, to explain what was happening. Likely influenced by Picks mammoth review of the subject in 1912, Landsteiner developed a much clearer picture of specificity during antitoxin binding through his meticulous attention to experimental detail allied with a smart mind. From 1912 to well after his Nobel Prize in 1930, Landsteiner addressed the fundamental question embedded in the dogma of the day, that proteins were the sole species that could act as antigens. So how could a simple chemical modification such as Pick had suggested bypass that requirement? By generating modified proteins that introduced "haptens" (small molecule modifiers of proteins) using chemically gentle methods, Landsteiner showed that recognition of the hapten could be separated from that of the native protein, both of which could behave as recognition elements [10]. This was a blow for Ehrlich's "toxin cross-reactivity with preexisting antitoxins (nutrient receptors)" theory since many of the modifications introduced by Landsteiner would never have been seen by the immunized animals so how could they have such a variety of specific, preexisting antitoxins!

Landsteiner's most important discovery was made when he extended the observations of Ehrlich and Morgenroth who showed that agglutinating antiserum raised against red blood cells from closely related species could be distinguished on the basis of antiserum titer and adsorption experiments. Landsteiner's question was: can the red cell antigens be distinguished *within* an individual? As early as 1901, he had shown that normal serum from one individual animal could agglutinate red blood cells from another animal of the same species. Further, he had also shown that injection of red blood cells could generate agglutinating antisera that differentiated members of the same species. The chemical basis of this took a decade to work out, but from these carefully executed agglutination experiments came the definition by Landsteiner of the blood group antigens, A, B, O, and AB [11]. He would receive a Nobel Prize in 1930 for this important work. A further important notion of antigenic species that properly buried the "protein as the sole antigen" theory was carried out in the mid-1920s at the Rockefeller Institute in New York (Karl Landsteiner has moved from Europe to the Rockefeller in 1922) by Oswald Avery and Michael Heidelberger who were working on *Pneumococcus*. The key result of these experiments was that the carbohydrate moieties on the surface of *Pneumococcus* were able to behave as antigens in precipitation reactions with antibodies but required association with protein (e.g., as on the red blood cell surface) in order to generate a carbohydrate-specific immune response. From this point forward, the notion that proteins were the sole antigenic species was shown to be simply wrong.

But, despite these momentous discoveries, the exact nature of antitoxins, now more commonly called antibodies, remained inside the box. As Landsteiner noted in an extensive account of his most important experiments, *The Specificity of Serological Reactions*, published in 1933, revised in 1936 in German, and published posthumously in 1945 in English:

> "...no finished theory of antibody reactions has yet been attained...This is not too surprising in view of the fact that ... the chemical structures responsible for the specificity of one of the reactants—the antibodies—are still unknown." [12]

It is easy to comprehend the frustration great biological scientists such as Landsteiner must have felt when attempting to put together the mechanistic pieces of antibody recognition when the structure of proteins was unknown. The colloid theory remained dominant dogma, and even with the advent of x-ray diffraction, the picture was still blurred. Rudolph Brill was working in Berlin in the 1920s on the structure of the fibrous protein, silk fibroin. It was known that fibroin consisted mainly of glycine and alanine amino acids. In 1923 Brill reported that the unit cell (the molecular repeating unit) of the crystals of fibroin consisted of four parallel glycine-alanine dipeptide units. Since it was thought that a molecule could not be larger than the content of the unit cell, Brill could not distinguish between a polymer of repeating units of Gly-Ala and a non-covalent aggregate of diketopiperazine units, in fact, a colloid.

Even with results from the newly discovered ultracentrifuge by Theodor Svedberg, working in Uppsala, Sweden, the colloidal school tentacles influenced the interpretation of protein structure. During his Nobel Prize lecture in 1926, Svedberg described studies in which the molecular mass of hemoglobin was shown to be around 69,000 Daltons and that of ovalbumin ~34,000 Daltons. These mass numbers were obtained in the presence and absence of salt (NaCl), disproving at least the assumption that protein colloidal structures depended on the presence of salts. Despite debunking the requirement that salts were necessary to bind the colloidal elements together, Svedberg, other than commenting that the proteins were homogeneous in exhibiting an extremely narrow particle size distribution, made no further comment on their chemical properties. In 1937 after further ultracentrifuge experiments on many different proteins, including results by other laboratories on "serum globulin" (the collective name for antibodies and showing a mass of about 150,000 Daltons), Svedberg's conclusions were still influenced by the colloidal theory:

"...on the whole only a very limited number of masses seems to be possible. Probably the protein molecule is built up by successive aggregation of definite units, but that only a few aggregates are stable..." [13]

2.5   Theories of Protein Structure

During the 1930s, various models of protein structure were proposed, including Frank and Wrinche's cyclol model, the repeating amino acid unit model of Bergmann and Niemann, and other more exotic models (see [2], Chapter 4). In April 1939, Linus Pauling and Carl Niemann, the latter now working with Pauling at Caltech, unceremoniously dumped the colloidal theory of protein structure in their paper "The Structure of Proteins." In the first paragraph they stated:

It is our opinion that the polypeptide chain structure of proteins, with hydrogen bonds and other interatomic forces...acting between polypeptide chains, parts of chains, and side-chains, is compatible not only with the chemical and physical properties of proteins but also with the detailed information about molecular structure in general... [14]

This was a momentous advance in the understanding of protein structure. In writing this introductory statement, the authors gave credit to Emil Fischer who had suggested a polypeptide chain possibility for proteins some decades earlier, although in the end favoring the colloidal particle theory. Curiously, Pauling and Niemann failed to acknowledge the remarkable insight of the Chinese scientist Hsien Wu, educated at Harvard and MIT and then working in Peiping (now Beijing) who, in 1929, proposed not only a chain theory of proteins but also that such chains folded into three-dimensional compact structures (see [2], pp. 47–48).

Armed with the polypeptide chain theory and other experimental data, Pauling took on the problem of antibody recognition. Experimental work on the serum globulin fraction by Heidelberger and Pedersen in 1937 had shown using ultracentrifugation that immune and non-immune rabbit and horse serum globulin had approximately the same molecular mass, 160,000 Daltons. Elvin Kabat visiting Uppsala analyzed the serum globulin fractions using the new electrophoresis method recently developed by Tiselius and showed by ultracentrifuge studies that two of the fractions had masses of 160,000 (the "γ-band") and 900,000 (the "heavy band"—later named IgM). The more interesting observation was that in immune sera the γ-band had increased in intensity compared with non-immune sera. What was perhaps the *experimentum crucis* was the fact that after adsorption of the fractions with antigen, the γ-band had reduced in intensity [15].

Putting together all the published information available, in 1940 Pauling published his theory of antibody structure and how it explained antigen recognition [16]. The key elements of this theory were:

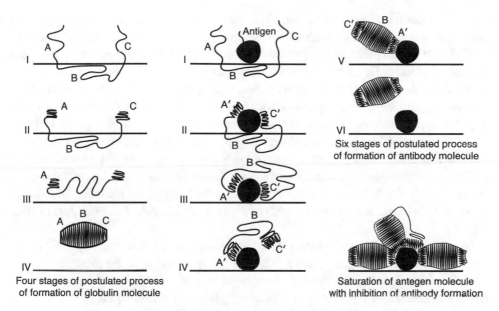

Fig. 2.5 Pauling's diagrams representing four stages in the process of formation of an antibody (left side of figure) and six stages in the process of formation of an antibody-antigen complex. The lower right cartoon shows an antigen molecule surrounded by bound antibody molecules or parts of molecules and thus prevented from further antibody formation

- Antibodies have complementarity to the antigen and are held together by "strong" forces.
- Antibodies have a molecular weight of ~160,000 and probably arise from the γ-fraction of serum.
- "Bivalency" is sufficient to explain all complexes formed during the precipitin reaction.
- *All* antibodies are *identical* in composition.

Despite his extraordinary molecular insight, to explain how antibodies could recognize many different antigens, Pauling had boxed himself in with the last of these properties. He then needed to develop a model that explained how a single antibody composition could recognize a multitude of antigens. His model is shown in Fig. 2.5.

Pauling called this his "instructive model." It required the same polypeptide chain sequence to fold into different conformations under direction of the antigen. To accomplish this, he postulated flexible regions (rich in glycine) at the antibody chain extremities that would fold around each antigen in different ways to form specific antibody-antigen complexes. Thus, non-immune globulin was no different to immune globulin except when antigen was bound. A further consequence was that since each antibody was bivalent, each of the recognition "arms" could in theory bind a different antigen. While some data supported this notion, Landsteiner and Scheer had shown in 1938 that when antibodies were raised against two different antigens simultaneously and then the single hapten-specific antibodies were adsorbed out, there was no evidence that any of the remaining antibodies were able to bind two different haptens. Pauling was quick to admit that there existed contradictory data and responded in an honorable scientific manner: "...with the kind cooperation of Dr Landsteiner we are continuing this investigation..." [17].

While Pauling's model explained how diverse antigens could be accommodated using specific, non-covalent, bonding and provided an acceptable model for precipitation and agglutination, it had two (at least) critical weaknesses. It could not explain the fact that "antigen-instructed" antibodies could persist in the circulation after disappearance of the antigen and neither could it explain how antibody affinity increased after secondary challenges with antigen. These were fundamental issues that the highly influential cellular immunologist, Frank Macfarlane Burnet, working in Melbourne, Australia, had drawn attention to in his influential monograph, *The Production of Antibodies*, published in 1941 [18].

The explanation of how antibody diversity was generated was caught in the net of contemporary genetic dogma where proteins were considered to be the genes. It was not until the game-changing experiments of Avery, MacLeod, and McCarthy on *Pneumococcus* in 1944 that DNA came out of the "interesting but not particularly important" material category and hit the limelight as the "transforming principle": at least, the limelight at the time for a few insightful scientists but still hotly debated by the "proteins are the genes" school. Even Avery and colleagues were cautious in their conclusions stating that the transformations they observed may have been caused by "other substances" in the preparation. The well-known Rockefeller molecular biologist Alfred Mirsky exemplified the prevailing uncertainty by commenting at a Cold Spring Harbor Symposium some 3 years later:

> In the present state of knowledge, it would be going beyond the experimental facts to assert that the specific agent transforming bacterial types is a deoxyribonucleic acid. [19]

What Mirsky was advising was that there needed to be more experimental evidence to establish whether protein or nucleic acid was the genetic material. This would come in the classic experiments of Hershey and Chase in 1952 in which differential radioactive labelling of protein and nucleic acid (S^{35} and P^{32}, respectively) demonstrated that DNA was the replication element in the T2 bacteriophage. By the mid-1950s, any lingering doubt that DNA was the transforming principle had disappeared.

2.6   Antibody Sequences: Homogeneous or Heterogeneous?

It would take the combined efforts of what we might call the "antibody galacticos," Rodney Porter in Fred Sanger's laboratory in Cambridge and later at the National Institute of Medical Research (NIMR) in London and St Mary's Hospital Medical School, London, Gerald Edelman at the Rockefeller Institute in New York, Edgar Haber at Harvard

University, and Charles Tanford at Duke University, to unravel the antibody primary structure and recognition mysteries, but not without cul-de-sacs along the way. Porter began his studies of non-immune and immune (anti-ovalbumin) rabbit globulin in 1948 while in Sanger's laboratory. After purification of the globulin fraction and N-terminal sequencing (Sanger's Edman method developed on insulin), he found no difference in the amino terminal residues of non-immune and immune globulins. In fact, all had the same sequence of the first five amino acids (Ala-Leu-Val-Asp-Glu?), and reduction of the disulfide bonds had no effect on ovalbumin binding. Here was the first cul-de-sac. Not knowing that the rabbit heavy chain had a blocked N-terminus and would thus not have been visible, he unknowingly had only sequenced the N-terminus of the light chain. With the knowledge that the globulin had a mass of 150,000 Daltons, he erroneously concluded that the antibody was a single continuous chain of ~1500 amino acids. Such a result was consistent with Pauling's single homogeneous sequence model, a theory Porter believed was supported by his results (published in 1950) [20]. Anyway, there was no alternative model, as Sanger noted in 1952, 6 years before he would get his first Nobel Prize in Chemistry:

> It is impossible with the small amount of experimental evidence at present to form any general theory of protein structure or to form any principles that govern the arrangement of amino acids in proteins. . . [21]

After a period in which he had learned advanced methods of chromatography at NIMR, Porter and John Humphrey showed, using different antigens, that the behavior of the immune γ-globulin fraction differed chromatographically from that of non-immune γ-globulin. Using the enzyme papain, Porter found that γ-globulin could be split into three fragments (I, II, and III), only two of which (I and II—now known as the Fab fragments) were able to bind antigen, while fragment III was less soluble and easily crystallizable (the Fc fragment). They published the results in 1959 [22]. He was still wedded to the Pauling model however and was forced to introduce some odd explanations of the fragmentation pattern to arrive at how two supposedly different antigen-binding arms could be joined to a non-binding region in a single linear chain! It would be Gerald Edelman working with Miroslav Poulik at the Rockefeller Institute who would put to bed the single chain globulin theory by some rather simple experiments. In their 1961 paper, they showed that γ-globulin could be separated into different chains when chemicals that reduced the disulfide bonds were added in the presence of denaturing agents. The smaller fragments produced were not consistent with a single chain model, and further, the fragments (heavy and light chains) were orthogonal in their identity with Porter's papain fragments. In the Discussion Edelman and Poulik commented:

> . . .γ-globulin molecules. . .consist of several polypeptide chains linked by disulfide bonds. Bivalent antibodies may contain two chains that are similar or identical in structure . . . may arise from various combinations of different chains as well as from differences in the sequence of amino acids within each type of chain. [23]

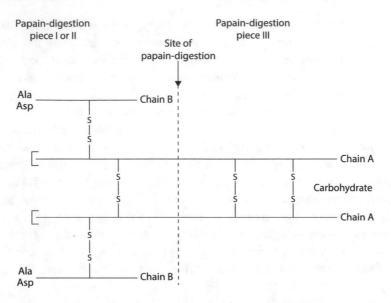

Fig. 2.6 The proposed four-chain structure of rabbit gamma globulin, redrawn from the original. Note the by then understood blocked ([) N-terminus of the heavy chain. Reproduced with permission from Fleischman, J.B., et al., The Arrangement of the Peptide Chains in γ-Globulin, Biochemical Journal, Volume 88, Part 2, pp. 220–228, Copyright © 1963 the Biochemical Society

Meanwhile Porter had not been idle. At a special symposium at Columbia University, New York, in March 1962, he proposed a chain structure for rabbit γ-globulin which explained both his and Edelman's results (see Fig. 2.6), published in full detail in 1963 [24]. The one remaining assumption that would only become clear a decade later was his belief that antigen binding resided solely in the A (heavy) chain.

Edelman then published his own model in 1964, and while different from Porter's, it still involved four chains for each antibody, any heavy chain of which was allowed to pair with any other light chain. It was different in that it showed the antibody as a solid "flat sausage" which had the heavy and light chains interacting with each other but no heavy chain-heavy chain interactions. He proposed that the antigen binding zones, comprising both heavy and light chain regions, were located at each end of the sausage, antigen specificity then being defined by amino acid differences in that region [25].

Two further sets of studies squared the circle on antibody specificity. Edgar Haber demonstrated in 1964 that, after disulfide reduction and denaturation of the γ-globulin chains, little residual structure remained in the molecule. The denatured form could then be reversed by oxidation and refolding (demonstrated for RNAse by Anfinsen in 1959) whereupon antigen binding was regained. His experiments also proved that both chains were required for binding and, critically, that binding specificity resided in the particular γ-globulin sequence, a serious blow for conformational/instructive theories of antigen recognition that postulated a common sequence for all antibodies. The coup de grâce was finally delivered as a result of a series of elegant experiments by Tanford in 1965. The

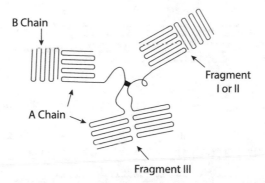

Fig. 2.7 Cartoon (redrawn) of IgG chain arrangement, as predicted by Tanford in 1965. A chain = heavy chain, B chain = light chain. Reproduced with permission from Noelken, M.E. et al., Gross conformation of rabbit 7S γ-immunoglobulin and its papain-cleaved fragments, The Journal of Biological Chemistry, Volume 240, Number 1, pp. 218–224, Copyright © 1965 by the American Society for Biochemistry and Molecular Biology

dinitrophenyl (DNP) hapten was attached to lysine and rabbit antibodies specific for DNP-lysine generated. When the γ-globulin was denatured and refolded in the absence of DNP-lysine, it recovered its ability to bind to the hapten. Further, if a non-specific γ-globulin was denatured and then refolded in the presence of DNP-lysine, it was unable to bind the hapten. Conclusion: antigen instruction did not exist. In the same year, Tanford, on the basis of sedimentation and viscosity experiments on rabbit γ-globulin and Porter's type I, II, and III papain fragments, produced a model of its heavy and light chain arrangement (Fig. 2.7). While the detail of the heavy chain and light chain variable domain quaternary relationship was not quite correct, it was much closer to the topology of an IgG than the simple "elliptical" model used at the time for most other protein shapes. Using optical rotatory dispersion (ORD) on γ-globulin and the various fragments, Tanford also correctly showed that antibodies possessed negligible α-helix content, hence its absence from his schematic diagram.

Definitive, extensive amino acid sequence data on antibody chains would arrive with the work of Norbert Hilschman and Lyman Craig, also at the Rockefeller Institute and contemporaries of Edelman. In 1962 Edelman and Joseph Gally had shown that Bence Jones proteins were similar to the light chains of γ-globulin produced in patients with multiple myeloma and to normal γ-globulin light chains. By 1965 Hilschman and Craig had carried out sequencing studies on two of such patients (famously known as Roy and Cummings) after isolating the overproduced proteins. Two important primary structure results emerged. First, the C-terminal sequences of the chains appeared to be almost identical between the two patients. Second, the two N-terminal regions varied considerably. Edelman and Gally speculated on the evolutionary mechanism that could generate such differences within a single chain, proposing that antibody diversity was created by the accumulation of tandem duplicated copies of the chains, containing differences in their variable regions, followed by crossing over to produce the different chain sequences (see

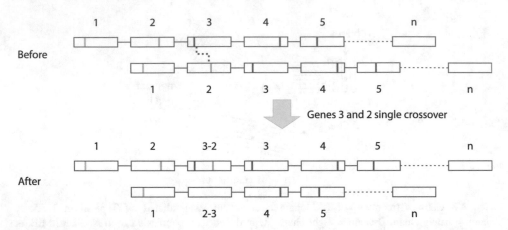

Fig. 2.8 Diagram indicating a single crossing-over (3 => 2) in a hypothetical double array of tandem duplicated genes (5 genes shown in the diagram). The colored lines indicate point mutations at different positions in each duplicated gene. (**a**) Before crossing-over; (**b**) after crossing-over between genes 3 and 2. Adapted from [26]

Fig. 2.8) [26]. An interesting hypothesis but at odds of course with the existing "germline theory," which had its drawbacks requiring as it did a different gene for every specificity required.

At the MRC in Cambridge, England, César Milstein also threw his hat into the "crossing-over" arena. At the same time, he made suggestions on the possible locations of varying portions of amino acid sequence that might be responsible for antigen recognition. His observation on light chains, that the more variable sequences were clustered in two regions of the V-region, the first around positions 30–32 and the second at 92–94 and 96, was the first clue that specific regions of hypervariability were present in variable regions. He had in fact identified the CDR1 and CDR3 regions of the light chain, but he also allowed for the possibility of other highly variable clusters. In addressing the possible genetic origins of antibody diversity in a 1967 paper, Milstein was uncertain about exactly which mechanism operated:

> Three or more structural genes seem to code for the N-terminal half, and so it seems necessary to assume some mechanism of chromosomal rearrangement or insertion involving "half genes". Unless…a very large number of genes is proposed in the germ line, an added mechanism of somatic hypermutation should operate on the variable section of the gene [27]

By 1969 Edelman with five other Rockefeller colleagues had sequenced both the heavy and light chain of the myeloma γG1 (IgG1) known as EU. Some of the conclusions of this study [28], a *tour de force* of protein chemistry and a watershed moment in our understanding of the structure of antibodies, were:

1. Understanding the contribution of the V-region to antigen binding requires structural information.
2. Tyrosine residues appear to be involved in antigen binding.
3. The presence of disulfide bridges between the VH and VL chains suggests a role in fixing the structures around these bonds.
4. The presence of "closely homologous" C_H1 and C_L constant regions may act as stabilizing domains to facilitate antigen recognition by the variable domains.
5. The C_H1 and C_L constant domains appear to be conserved across various animal species.
6. The Fc regions show close resemblance between human and rabbit gamma globulins suggesting they may have evolved to exhibit specific functions, such as activating the complement system.

2.7   γ-Globulin: Isotypes and Allotypes

The early 1960s picture of a single species of γ-globulin did not explain all of the serological analysis carried out in many laboratories, even when put together with the important sequence analysis of Porter, Edelman, and others. During 1964–1967 several groups began to uncover the different human light chain and heavy (γG) isotypes, although for the heavy chain some of the studies were on both normal and myeloma proteins and used different nomenclatures, complicating the interpretation. Initial studies using Bence Jones protein sequences to characterize light chains, which were shown to be identical to normal light chains, led to the isotypic kappa and lambda distinction, including the important fact that a single B cell could only make one or the other light chain isotype. The final piece of the light chain jigsaw was put in place by Frank Putnam in two publications in 1969 and 1970 where the complete sequences of a kappa and lambda Bence Jones light chain protein, respectively, were described. The sequencing confirmed the two-domain feature of the light chain, a variable and a constant sequence domain. For γG heavy chains, Henry Kunkel and other colleagues had evaluated the published data available and in 1966 proposed a nomenclature to the WHO that brought together the different labels for human γ-globulin heavy chain isotypes into a single system [29], summarized in Table 2.1 (see [2], Chapter 8, for a fuller account that also includes the discoveries of IgM, IgA, IgD, IgE, and their isotypes).

Seven years later (1973), Kunkel reviewed the extensive serological data available and published estimates of the relative levels of the different γ-chain isotypes present in circulating human serum antibodies: IgG1 66 ± 8%; IgG2 23 ± 8%; IgG3 7.3 ± 3.8%; and IgG4 4.2 ± 2.6%. The percentages clearly suggested roles that were immunologically distinct, complement activation being one that had long been studied and which Kunkel referred to as a possible role for the Fc region [30]. But in the absence of detailed sequence and in particular three-dimensional structures of the isotypes, the critically important differences would remain unclear, and until the family of Fc receptors and their functions

Table 2.1 WHO nomenclature of IgG subclasses. Adapted from [29]

Existing γ-globulin name	Occurrence as myelomas (%)	WHO proposed name	Heavy chain name
We, C, or γ2b	70–80	IgG1 or γG1	γ1
Ne or γ2a	13–18	IgG2 or γ2	γ2
Vi, Z, or γ2c	6–8	IgG3 or γ3	γ3
Ge or γ2d	3	IgG4 or γ4	γ4

had been characterized a decade and more later, the full functional role of Fc regions would remain obscure.

But there was another feature of Fc that would not have been visible to Kunkel just by examining the sequences, although from serology was known to be important. As far back as 1956, Jacques Oudin had reported to the French Academy of Sciences a phenomenon that would become an important aspect of the future clinical use of antibodies. When a non-immune rabbit was injected with antiserum, or antiserum-antigen precipitates, from another rabbit that had been immunized against various antigens, the serum from the non-immune rabbit showed precipitation behavior with the donor serum. Oudin's interpretation was that sera from the same species were not identical and further, it was not possible to immunize an animal of the same species with a protein of that species without generating an immune response. He called the reactive component antibodies in the serum *allotypes*. On the basis of extensive serology using 45 different rabbits, Oudin designated 7 different allotypes and speculated that, since it was likely to have a genetic origin, it would likely be seen in other species [31, 32]. At the same time, Rune Grubb, working in Lund, Sweden, established a similar pattern of allotypes in human sera. By 1964 Grubb and Kunkel had described a set of allotypic markers on the antibody γ-chain (called Gm) that defined the different allotypic behavior [33]. The potential importance of these observations was highlighted by Allen and Kunkel in 1963 when they reported the presence of anti-Gm agglutinating antibodies in 71% of the sera from 24 children who had received multiple blood transfusions. In 1965 the WHO proposed a new naming procedure for the human chain markers that included 14 gamma chain (γG1, 2, and 3 but none for γG4) markers (Gm 1–14) and 3 kappa chain markers (Km 1–3). By the 1980s it had been shown that these allotypic features are antigenic determinants (typically singe amino acid changes), within the Fc region for the heavy chain and the constant region of the light chain, that defined chain polymorphisms transmitted as dominant autosomal Mendelian traits and differentially expressed by Negroid, Caucasoid, and Mongoloid populations. Today, many more human allotypes are known, and most are associated with IgG1–3 subclasses with a small number of allotypes present on IgA2 (A2m) [34]. The importance of IgG1 Gm allotypes has been well established by their influence on the pharmacokinetics of clinically useful antibodies in vivo [35].

2.8   The Structural Basis of Antibody Recognition Revealed

By the close of 1967, a small number of light chain sequences and partial sequences had been determined. Some of these data were commented on in Milstein's *Nature* paper published on October 28, 1967 referred to earlier, in which he noted the sequence variability in two segments of the light chain variable region. Elvin Kabat may have had more data to examine and would have been aware of Milstein's analysis since Kabat's 1968 paper was not submitted until the end of November 1967. In his analysis of this still rather limited sequence information, Elvin Kabat made some key observations, some of which would turn out to be correct and others less so. He identified two sequence segments in the light chain variable regions that were less conserved than most other positions and speculated they were "possibly" involved in antigen recognition. He also noted a third region that he suggested might be involved in folding of the antibody. He was also persuaded that the high frequency of glycine residues could play a role in the requirement for antibodies to be "flexible," to better wrap around their antigens (shades of Pauling's instructive theory still present here).

A third section of sequence variability (around residues 50–55) was identified in Prague in 1969 by Frantisek Franěk working on porcine antibodies and believed to be important in antigen recognition [36]. As more antibody sequences became available, Kabat, now collaborating with the mathematician, Tai Te Wu, analyzed 77 light chain sequences and plotted the amino acid variability as a function of position, clearly showing the 3 hypervariable sequence segments and their likely role as complementarity-determining residues [37] (CDRs; Fig. 2.9). By suggesting these three segments were involved as CDRs, Kabat had cemented the sequence origin of specificity in antibodies, despite at that point having no idea what the relationship between them was in three dimensions. While this was an important advance in defining the antibody at the protein level, Kabat also speculated on the genetic mechanism by which the hypervariable sequence segments (CDR1 and CDR3) arrived in the variable region and concluded it was best explained if they were inserted as small segments of DNA or "episomes."

But all this was based on the light chain, while the heavy chain, for which sequence information was slow to arrive, was considered to be more important in antigen recognition. Donald Capra, at Mt. Sinai School of Medicine in New York, an ex-PhD student of Kunkel and rising star, sequenced a number of IgG heavy chains, and a year after, the Wu and Kabat paper revealed the same hypervariability segments Kabat had described for light chains. In his follow-on analysis of all the now more extensive sequence data, also published in 1971, Kabat concluded that at least the CDR1 and CDR3 segments in both chains were involved in forming the complementary binding surface but he was still uncertain about the role of CDR2. The answer would arrive 2 years later with the x-ray structure of the Fab fragment from myeloma antibody *New* by Roberto Poljak and in parallel a structure of the light chain dimer, Mcg, by Allen Edmundson. Poljak's structure clearly showed Kabat and Capra's CDRs, three from each chain, at one end of the Fab variable region "in close spatial proximity" and perfectly positioned to provide a

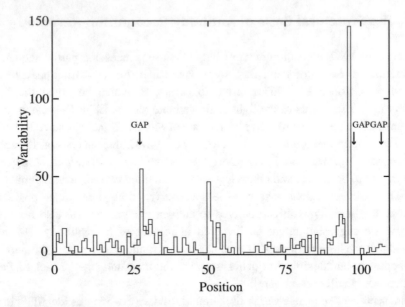

Fig. 2.9 Variability at different amino acid positions for the variable region of the light chains. GAP indicates positions at which episome insertions were proposed. Reproduced with permission from [37]

continuous surface for antigen binding (see Fig. 2.10a). What also became clear from both structures was that the beta-sheet "immunoglobulin fold" was orchestrated by the positioning of the glycine residues in tight beta turns (see Fig. 2.10b) at the ends of the arrowed beta strands that turn back on each other to form an anti-parallel beta-sheet. Kabat had incorrectly surmised the glycine residues were there to introduce chain flexibility necessary for antigen binding.

2.9   The Origins of Antibody Diversity

By 1970 various theories were circulating in an attempt to explain the thorny problem of how a different antibody for each of a myriad (it was supposed) of antigens could be encoded in the genome. By 1970 Lee Hood and David Talmage, two of the great early molecular biologists, carried out a calculation that went as follows:

- DNA content of human = 2.3×10^{-11} g (sperm cell (haploid))
- Molecular weight of one base pair = 6.2×10^2
- DNA content (per cell) = 3.7×10^9 base pairs
- One variable region = 107 amino acids
- One variable gene = 321 pairs
- 20,000 variable genes = 6.4×10^6 pairs
- 20,000 variable genes = 0.2% of genetic material of human haploid cells

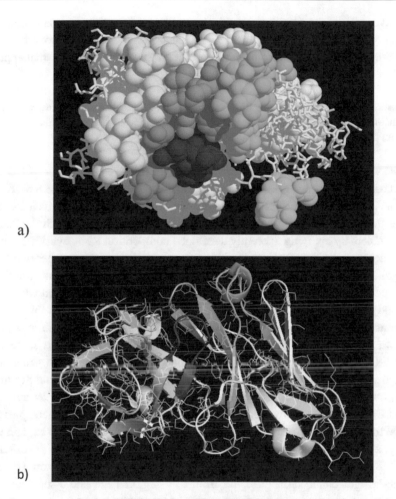

a)

b)

Fig. 2.10 Images created using PyMOL™ from PDB ID: 7FAB (Saul, B. J & Poljak, R.P. Crystal structure of human immunoglobulin fragment Fab New refined at 2.0 Å resolution. 1992, Proteins, 14 (3): 363–37). Note: this is a refined x-ray structure based on the original lower-resolution structure of New published in 1973 with the addition of new x-ray data. In these images the CDRs are as follows: heavy chain, green (H1), yellow (H2), red (H3); light chain, cyan (L1), lemon (L2), pink (L3). CDRL2 (lemon) has a deletion in this myeloma protein (New) and as a result is moved away from its normal contact with the other CDRs. (**a**) shows the CDRs in a space filling representation to illustrate their mutual close packing; (**b**) shows a cartoon side view where the anti-parallel beta strands of the VL and VH domains and the CDRs in the loops at the ends of the anti-parallel beta strands can be seen

Using this rather simplistic genome analysis, they concluded that each light chain could be encoded by 1600 variable region (V-region) genes and 3200 heavy chain V-region genes which would only represent 0.2% of the genomic DNA [38]. Had the complete sequence of the human genome and its number of genes been known at the time, they would have realized their estimate would take up about 13% of the human genome, clearly a non-starter! Elvin Kabat had a different explanation. As indicated above he postulated that variation occurred via "episomal insertion" of short sequences into a smaller number of genes:

> An insertion mechanism involving only short linear sequences provides a substantial simplifi-cation of the problem of providing a seemingly limitless number of complementary sites without the use of very large amounts of DNA [37].

But these were hypotheses and what was needed was proof. In the winter of 1971 after an invitation from Dulbecco in Basle, Susumu Tonegawa moved from the Salk Institute in California to the newly formed Basel Institute of Immunology. He was given the project of studying the basis of antibody diversity. Tonegawa decided to approach the "number of genes" problem by applying a recently developed competition RNA hybridization method he had learned at the Salk that would aid in the quantitation of the number of different V-region genes. His first experimental results from 1973 to 1974 put the number of light chain variable genes at somewhere between 1 and 200 [39]. A wide range and while a lot less than Hood and Talmage's 1600 estimate, he retained some concern that this was still too small to account for the required V-region variability. Using DNA from myeloma cells, Tonegawa repeated the hybridization experiments and arrived at the same range of numbers. Conclusion? If there is a much smaller number of germline genes, diversity must then somehow be generated by a somatic mechanism. In a 1976 Cold Spring Harbor meeting on "The Origins of Lymphocyte Diversity," Tonegawa, with an audience that included Donald Capra, Leroy Hood, Phil Leder, Tasuku Honjo, and others, presented his evidence for a limited repertoire of germline V-genes and argued that to explain antibody diversity, a somatic process must be operating on this small repertoire in the differentiated cell [40]. By 1978 Tonegawa had produced the evidence for the combinatorial assembly of the lambda light chain variable region from a V-region gene and J-segment and their assembly with a constant region to form a complete light chain. In a follow-on publication, he completed the picture by demonstrating that the order of assembly was V-J-C and that because the mature light chain sequence differed in some positions from the germline sequence, mutations observed in the hypervariable regions must have arrived by a process of somatic hypermutation. In the same year, Phil Leder at the NIH (USA) and Lee Hood (Caltech, USA) had come to the same conclusion working on kappa chains, and while Leder [41] had provided more detail on the relationship between the V, J, and C genes on the chromosome and also that recombination only occurred on one of the two light chain alleles, Hood [42] had produced a mechanistic explanation of the recombination process that was as close to the mechanism of light chain assembly as any of his contemporaries.

During the next 2 years, attention switched to the heavy chain variable region. By 1980 Hood had discovered the D segment as a third component in the heavy chain variable region assembly and noted the presence of conserved nonamer and heptamer DNA

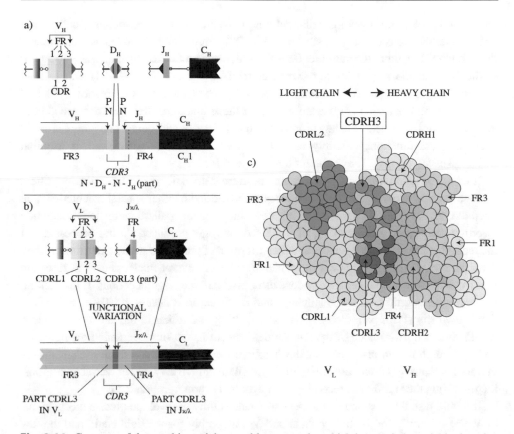

Fig. 2.11 Cartoons of the combinatorial assembly process by which heavy chain variable domains (**a**), and light chain variable domains (**b**), are assembled in the B cell. The spatial relationship between the framework regions and the CDRs in each chain is shown in (**c**). Reproduced from Fig. 1 in [50] with permission from Taylor & Francis Ltd. (www.tandfonline.com)

sequences (later became the recombination signal sequences, RSS) separated by fixed spacer distances of either 12 or 23 nucleotides (leading to the 12/23 rule). This led Hood to propose a model of recombination mediated by these sequence motifs [43]. The same year Tonegawa had explained class switching from IgM to IgG. Exactly 1 year after Hood's publication, Tonegawa [44] provided the experimental evidence for the order and mechanism by which the heavy chain VH-D-JH recombination takes place mediated by the RSS at either end of the various gene segments. This ensured recognition by a specific recombinase enzyme that catalyzed assembly of the VH-D and the DH-J segments, forming the complete heavy chain variable region (see Fig. 2.11). A similar process was shown to operate for recombination of kappa and lambda light chains.

Of course, that was not the end of the story but in some ways just the beginning. The next phase would take advantage of Köhler and Milstein's monoclonal technology (the subject of another chapter in this book) published in 1975. No longer would the biology have to rely solely on myeloma cells, but now individual antibody-producing cells could come under the molecular biologist's knife. By 1982, Frederick Alt and David Baltimore at the Whitehead Institute (Cambridge, USA) had identified a further diversity mechanism in studies of variable region joining events in fetal liver cells. The VH-DH and DH-JH junctions in some instances were not as expected. One of their key observations was that in some antibodies amino acids were present at these junctions that were not encoded in known DH or JH sequences. They proposed that deletions or additions of nucleotides could occur at these junctions, possibly carried out by exonucleases (deletions) or terminal transferases (additions). They called it the N-region [45].

It would be another 6 years before a rather different diversity contributor was identified. Tonegawa, now working on the T-cell receptor, identified a rather special mechanism that operated at the VH-DH junction. This new mechanism, called templated nucleotide insertion, originated from nicking of the DNA hairpins produced during processing of the recombination signal sequences and then filling in of the single strand extensions if they were present. If the nicking was off-center by one or several nucleotides, the resulting filled-in overhangs could generate codons differing from the original codons. If the nicking is exactly central, there are no overhangs, and no filling in is necessary. Tonegawa called this P-diversity (P for palindrome since after nicking and filling in the added nucleotides would form a mirror image of the nucleotides present in the single-stranded overhanging ends). P-addition can also occur at the junctions of the light chain coding elements. A pictorial summary of the assembly of light chain and heavy chain variable regions, incorporating these different recombination events, is shown in Fig. 2.11.

It would take the best part of the next 20 years to fully unravel the precise mechanisms of V-D-J assembly for the heavy chain and V-J assembly for the light chain and the key enzymes involved (e.g., the RAG 1 and RAG2 recombinase system—discovered by Baltimore). As for hypermutation, although in 1983 the process of hypermutation had been shown to target framework positions in the variable regions and not just the CDRs as previously thought, the precise sequence positions in the assembled variable regions for the hypermutation apparatus, where and at what point in the antibody assembly process this occurred, were unknown at the time. At the turn of the millennium, another Japanese scientist, Tasuku Honjo, would uncover the answer curiously while searching for the molecular mechanism of IgM to IgG class switching. In Kyoto, Honjo's laboratory identified a key enzyme, activation-induced deaminase (AID), whose activity appeared to coincide with class switching events and the appearance of V-region hypermutations in the germinal centers. In 1999 based on limited evidence, Honjo went out on a limb and speculated on the role of AID in class switching and hypermutation [46]. One year later, he provided the proof for his theory based on studies in transgenic mice whose AID gene had been disrupted [47].

By the 1990s, the "crossover" notion of Edelman and Gally and the "episome" theory of Kabat had been assigned to the "nice idea but wrong" bins of scientific ideas. Were they alive today they would feel some vindication of those ideas when reading the description of two novel mechanisms of antibody germline gene diversification. The first involves a process similar to gene conversion mediated by the AID (activation-induced deaminase) enzyme, first described by Darlow and Scott [48] in 2006 where sequence segments in an unselected VH germline gene may be exchanged with an already rearranged VDJ by homologous recombination. The second, perhaps somewhat bizarre, mechanism can occur where a rearranged VDJ has its VH germline gene completely replaced by a mechanism of "VH invasion." [49]

The historical backdrop to many of these momentous discoveries on antibody diversity mechanisms, related rather briefly above but with many players not mentioned whose contributions were important to these discoveries, is too long to recite here. The interested reader can find more details of the historical context in ([2], Chapter 12) and further details on the current debates on the human antibody repertoire in Rees [50].

2.10   The Fc Region and Effector Functions

While the structural studies on Fab fragments explained how the hypervariable sequences were juxtaposed in three dimensions and provided an explanation of how antibodies recognized antigens (e.g., IgG, IgM, IgA) or allergens (IgE), the variability, if it existed, and the structure and function(s) of the Fc region and its relationship to the Fab regions were still unknown. In 1976 Robert Huber and colleagues at the Max-Planck Institut für Biochemie in Martinsried, Germany, reported the structure of an isolated Fc fragment at a resolution of 3.5 Å, sufficient to describe the two structural domains on each heavy chain constant region chain (CH_2 and CH_3). The structure showed the Fc domain structural similarity (the "immunoglobulin fold") to the variable or constant domains present in the Fab fragment, their packing relationships to one another, and the location of the carbohydrate, attached to the CH_2 domains [51]. A year later David Davies at NIH, USA, returned to his earlier structural studies at low resolution on the myeloma IgG, Dob, and using new techniques was able to generate a higher-resolution structure showing the first picture of a complete IgG. The shape of an IgG had been suggested as a Y-shape from the electron microscopy experiments of Valentine and Green at NIMR, London, a decade earlier, but Davies' structure was T-shaped. As with many myeloma-derived antibodies, deletions in the sequences were common. Dob had a deleted hinge region restricting the orientations of the two Fab arms with respect to the Fc region. While the orientation of the Fab arms with respect to the Fc region may not have accurately reflected that in a normal IgG, the Fc region in Dob was closely similar to the isolated Fc fragment from Huber's study.

But what did the Fc region do? Early experiments on the hemolysis effects of serum antitoxins by Bordet and Ehrlich that formed the basis of their "alexine" and "complement" theories did not really explain the relationship between the antitoxin and whatever substance was doing the hemolysis. During the first decade of the 1900s, various researchers identified multiple components associated with the complement effect where the lysing activity in the presence of antitoxin would only occur if the separated components were recombined. By the mid-1920s, four different components had been identified, some of which were heat sensitive and showed evidence for enzyme activity. But it was a long, complicated road, and even by 1929, two experts in complement research, Eagle and Brewer at Johns Hopkins Medical School, while having established a direct relationship between the level of hemolysis induced by complement in serum and the amount of antibody bound to cells, admitted: "...the terms alexin and complement still denote unexplained properties of serum rather than a chemical entity." [52] It took another decade and more for the teasing out of the different components, and despite the creative protein chemistry of many research groups during the 1940s—four of the complement components (C1', C2', C3', and C4') had been isolated, partially characterized, and a common nomenclature proposed—the complement genie was still in the bottle at the end of the decade. In an attempt to unravel what was known, Michael Heidelberger, President of the American Association of Immunologists the previous year, gathered all the complement experts at a conference on Shelter Island, New York, in June 1950 to discuss how to take the field forward. Despite some progress, the following 8 years saw little advance, so much so that Rowley and Wardlaw at St Mary's Hospital in London observed in 1958:

> If, in conclusion, one compares the present state of knowledge of serum bactericidal action with that existing around 1900, it is apparent that we yet know little more about the actual mechanism of the reaction [53].

By the early 1960s, a solution to the puzzle would begin to emerge with two key breakthroughs. Hans Joachim Müller-Eberhard working between Uppsala, Sweden, and the Rockefeller in New York identified a new complement component with the size of γ-globulin (11S by ultracentrifugation) which formed a direct interaction with antibody, was heat labile (so probably a protein), and was involved in the initial stage of complement activation [54]. Two years later Irwin Lepow in Cleveland, USA, showed that Müller-Eberhard's 11S component was one (the first) of three proteins in what had been thought of as the single component C'1, naming them C1q, C1r, and C1s [55]. By the beginning of the 1970s, the direct and preferred interaction of C1q with aggregated antibody was shown by Müller-Eberhard, and the basis for its interaction with multiple antibodies was beautifully illustrated in an electron micrograph of the C1q molecule carried out by Robert Stroud at the University of Alabama (see [2], p. 159). The famous "bunch of tulips" structure of C1q, as it became known, was shown by Porter to consist of six copies of three different protein chains. The fibrillar stems of the tulips were formed by the N-terminal regions of the chains and contained a high level of collagen, the single multichain stem supporting the six globular heads. The understanding of the complement system now began to advance rapidly, and by 1980 Porter and Reid in Oxford produced a cartoon of their proposed C1q-antibody assembly complex in which the heads of the C1q tulips bound to the CH_2 domains in the Fc regions of multiple antibodies (Fig. 2.12; see [2], Chapter 9, for more details) [56]. A requirement to avoid activation of the complement system by a single soluble antigen-antibody complex was that only after binding of multiple antibodies to the six heads on the cell surface would the structure of C1q be triggered to bind the remaining

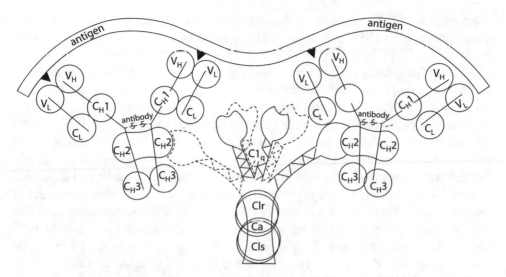

Fig. 2.12 Suggested assembly of early component of complement onto antibody molecules bound to cell surface antigens. The heads of the C1q molecules bind to the CH2 domain of several antibody molecules, and a tetramer of C1r2-Ca-C1s2 binds to the collagen stems of C1q (Ca = calcium). Reproduced from Porter, R.R. The complex proteases of the complement system, Croonian Lecture, Proceedings of the Royal Society B., Volume 210, pp. 477–498, by permission of the Royal Society, London

C1 components, C1r and C1s. These latter components would bind to the now receptive collagen stems of C1q with the help of calcium ions and begin the cascade that would lead to cell membrane disruption.

2.11   The Modern Era

The 1980s was a decade of "enlightenment" in molecular immunology where the antibody began to show its credentials as an immunological "Swiss knife." Rapid advances led to an understanding of exactly how antibodies recognized their antigens and how that recognition could be altered by "engineering," the nature of the molecular species that mediated antibody effector functions and novel methods that opened up a path to the development of therapeutic antibodies.

By 1987 the differential activation of complement by the human IgG isotypes (qualitatively looked at in 1976 by Müller-Eberhard in an ultracentrifuge study), IgM, IgE, and IgA2, had been clarified by Michael Neuberger, Hermann Waldmann, and co-workers who established a ranking for activation in the order IgG1 ≥ IgM > IgG3 (with its two allotypes showing different behavior) and low or almost no activation by IgG2 and IgG4 or IgE [57]. They also resolved the apparent paradox where IgG3 C1q binding was higher than IgG1 but cell lysis was more effective with IgG1. The explanation was that IgG1 was more effective than IgG3 in activation of the subsequent protease stage in the cascade. The higher binding of C1q by IgG3 was thought to be due to the longer hinge of IgG3 allowing better access of the C1q heads [58].

A further function of the Fc region was also unravelled in the 1980s. Early studies in the 1950s on transport of IgG from mother to young in mice, rats, and rabbits had suggested that a specific molecular mechanism may be at work. Richard Rodewald, working at the University of Virginia in Charlottesville, had produced elegant microscopy studies in 1980 showing specific binding of labelled IgG to the intestinal brush borders of neonatal mice whose ability to transport maternal IgG specifically lasted only for around 20 days before the molecular transport system switched off. By 1983 Rees and Simister had shown the receptor was a protein with a size of ~100 kDa and in 1985 isolated the receptor which had a subunit size of 50 kDa, so likely a dimer in the membrane, but also containing a smaller 18 kDa component suggested by the authors to be beta-2 microglobulin [59]. In 1989 Simister and Mostov had cloned and sequenced what was now called the FcRn (for Fc receptor neonatal) and shown the larger subunit to be related to the MHC Class 1 family while the smaller subunit was indeed beta-2 microglobulin. In 1994 a structural study of the receptor confirmed the MHC classification, while in the same year a complex between the receptor and Fc was published showing the binding site for the FcRn as a region between the CH2 and CH3 domains of the Fc [60]. In the same year, the human FcRn was cloned from human placenta. The eventual role of FcRn in controlling the catabolism of IgG would come from a number of different genetic manipulations in mice where either the beta-2 microglobulin or MHC-like subunit would be "knocked out" accompanied by a dramatic reduction in the levels of circulating IgG. This control mechanism, now known to be mediated by FcRn in, for example, endothelial cells, has been and still continues to be a target for manipulation of therapeutic antibody half-life of IgG in vivo.

But FcRn was not the only receptor that bound to the Fc region. In 1986 the work of two giants of the emerging Fc field, Jeffrey Ravetch at Memorial Sloan Kettering Cancer Center and Jay Unkeless at the Rockefeller, began to explain the multitude of functions immunoglobulins appeared to exhibit via their constant regions but with little understanding of how these effector events occurred. Ravetch and Unkeless discovered two different Fc receptors in a study of mouse immune cells, one expressed in macrophage lines and normal peritoneal macrophages and a second that in addition to macrophages was also found on bone marrow (B)- and thymus (T)-derived lymphocytes. The two receptors bound IgG via the Fc region with low affinity and were named Fcγα and Fcγβ [61]. Studies by Allen and Seed in 1989 identified different receptors from human cells that exhibited high affinity for the Fc region and containing an additional domain that differed from either of the two low-affinity receptor domains [62]. By 1994 the Fc receptor story, their structure and function, was beginning to be understood. In his review Ravetch noted:

These studies demonstrated that FcRs have a central role in initiating immunocomplex-triggered inflammation and are likely to be major contributors to the process of lymphocyte regulation of antibody production. [63]

Table 2.2 Properties of human IgG subclasses. Adapted from Vidarsson et al. 2014 [64] under CC BY

General features	IgG1	IgG2	IgG3	IgG4
Mass (KDa)	146	146	170	146
Hinge region length (residues)	156	12	62	12
Inter-heavy chain S-S bonds	2	4	11	2
Relative abundance (%)	60	32	4	4
Serum levels[a] (g/L)	~7	~4	~0.5	~0.6
Half-life in vivo (days)	21	21	7/21[b]	21
Placental transfer efficiency	++++	++	++/++++[b]	+++
Antigen selectivity				
Proteins	++	±	++	++
Polysaccharides	+	+++	±	±
Allergens	+	−	−	++
Fc effector behavior				
C1q binding	++	+	+++	−
FcgR1	+++	−	++++	++
FcgRIIa	+++	++/+[b]	++++	++
FcgIIb/c	+	−	++	+
FcgIIIa	++/+++[b]	+/−[b]	++++	−/++[b]
FcgIIIb	+++	−	++++	−
FcRn (at pH <6.5)	+++	+++	++/+++[b]	+++

[a]Approximate values in adult serum
[b]Depends on allotype

A summary of what we now know about the functional properties of the different IgG subclass constant regions is illustrated in Table 2.2. A more extensive historical review of the constant region functions in IgG but also in IgE, IgM, and IgA can be found in ([2], Chaps. 8–10).

2.12   The Antibody Engineering Revolution

Two key advances, the development of recombinant DNA methods in the 1970s and the discovery of a means by which site-specific mutations could be introduced into genes in the 1980s, opened extraordinary opportunities in the field of molecular immunology and in particular in the fast-developing field of therapeutic antibodies. The recombinant DNA methods began to be exploited for the construction of chimeric antibodies, in which murine constant regions were replaced by human constant region sequences. The earliest thera-peutic antibody application had focused on clinical use of the mouse monoclonal antibody OKT3, an antibody that recognized the CD3 antigen on T cells and was used to treat acute renal allograft rejection in humans, early clinical studies on which had already started in

1981. The potential for anti-mouse responses in humans for such wholly murine antibodies was high, and with that in mind, Sherrie Morrison and others constructed the first example of a chimeric antibody in 1984 [65]. By 1994 the first FDA-approved antibody, ReoPro, entered clinical use based on a mouse-human chimeric IgG1. But the variable regions were still murine with the potential for development of human anti-mouse V-region antibodies. To circumvent this, Greg Winter at the Laboratory of Molecular Biology in Cambridge, UK, produced an immensely creative solution. He surmised that if the CDRs from the mouse antibody were able to maintain their structure on another variable region framework, then they could be transplanted onto a human V-region, the light chain CDRs onto a human VL framework and the heavy chain CDRs onto a human VH framework. After several attempts and protein engineering manipulations that employed the recombinant technology and the site-directed mutagenesis technology, Winter published the results in two reports in 1986 and 1988 showing that the humanization procedure, called "CDR grafting," worked [66, 67]. With refinements it has become one of the major "humanization" methods used for therapeutic antibodies that start their life as murine (murine = mouse or rat) antibodies. For this outstanding development, Winter received the Nobel Prize in 2019. Also in 1987, Roberts and Rees published the first example of in vitro mutagenesis of antibody CDRs, used to improve the affinity of an existing anti-protein antibody by an order of magnitude [68]. Armed with all of these techniques, and the soon to be developed "phage technology" that would provide an alternative to hybridomas, the antibody world was on the cusp of a new therapy revolution that would begin to gather momentum during the 1990s and into the new millennium. Today, therapeutic antibodies represent the fastest growing therapeutic molecules, outpacing their historical small molecule rivals. The remainder of this book will take you though some of these exciting methods and their applications that year on year are improving global health by more effective treatment of both old and new diseases.

Further Reading
1. Fundamental Immunology (7th Edn). William E. Paul (Ed.). Walters Kluwer: Lippincott Williams and Wilkins. 2013—contains online access to fully searchable text and references.
2. The Antibody Molecule: From antitoxins to Therapeutic Antibodies. Anthony R Rees, Oxford University Press, 2014.
3. A History of Immunology (2nd Edn). Arthur M. Silverstein. Academic Press. 2009.
4. Species and Specificity. Pauline M.H. Mazumdar. Cambridge University Press. 2002.

References

1. Jenner E. An inquiry into the causes and effects of the *variolae vaccinae*, a disease discovered in some of the western counties of England, particularly Gloucestershire, and known by the name of the cow pox. London: Sampson Low; 1798.

2. Rees AR. The antibody molecule: from anti-toxins to therapeutic antibodies. Oxford: Oxford University Press; 2014. p. 1–16.
3. Note: the concept of a virus at this time was as a 'toxic agent' that was not easily defined or visible in the microscope. The discovery of viruses (the plant virus, tobacco mosaic virus) didn't arrive until the late 1880s.
4. Pasteur, L. Méthode pour prévenir la rage après morsure. Comptes rendus de l'Académie des science, séance du 26 octobre 1885, CI, p. 765–773 et p. 774.
5. Shibashaburo Kitasato's contribution was not acknowledged by the Nobel Committee, or if it was, they did not consider it of equal status.
6. Ehrlich, P. On immunity with special reference to cell life. Royal Society Croonian Lecture. R Soc Proc. 1900;66:424–448.
7. Bordet J. Agglutination et dissolution des globules rouges par le serum: deuxieme mémoire. Annales de l'Institute Pasteur. 1899;13:273–97.
8. Rees, A. R op cit pp 34, 35.
9. Obermayer F, Pick EP. Über die chemischen Grundlagen der Arteigen schaften der Eiweisskörper. Bildung von Immunoräzipitin durch chemisch veränderte Eiweisskörper. Wiener Klin. Wochenschr. 1906;19:327–33.
10. Landsteiner, K. The specificity of serological reactions. 1962, Revised Edition, Dover Publications Ch. II.
11. Ibid Ch. III.
12. Ibid Ch. VII.
13. Svedberg T. The ultra-centrifuge and the study of high-molecular compounds. Nature. 1937;139 (3529):1061.
14. Pauling L, Niemann C. The structure of proteins. J Amer Chem Soc. 1939;61:1860–7.
15. Tiselius A, Kabat EA. An electrophoretic study of immune sera and purified antibody preparations. J Exp Med. 1939;69:119–31.
16. Pauling L. A theory of the structure and process of formation of antibodies. J Amer Chem Soc. 1940;62:2643–57.
17. Ibid.
18. Burnet FM. The production of antibodies. Melbourne: Macmillan; 1941.
19. Boivin, A. Directed mutation in colon bacilli, by an inducing principle of desoxyribonucleic nature: its meaning for the general biochemistry of heredity. 1947. Cold Spring Harbor Symp., 11, 7–17. Note: Mirsky's comment in the discussion following Boivin's paper.
20. Porter RR. A chemical study of rabbit anti-ovalbumin. Biochem J. 1950;46:473–8.
21. Sanger F. The arrangement of amino acids in proteins. Adv Prot Chem. 1952;7:1–67.
22. Porter RR. The hydrolysis of rabbit γ-globulin and antibodies with crystalline papain. Biochem J. 1959;73:119–26.
23. Edelman G, Poulik MD. Studies on structural units of the γ-globulins. J Exp Med. 1961;113:880.
24. Fleischman JB, et al. The arrangement of the peptide chains in γ-globulin. Biochem J. 1963;88:220–8.
25. Edelman GM, Gally JA. A model for the 7S antibody molecule. Proc Natl Acad Sci. 1964;51:846–53.
26. Edelman GM, Gally JA. Somatic recombination of duplicated genes: an hypothesis on the origin of antibody diversity. Proc Natl Acad Sci. 1967;57:353–8.
27. Milstein C. Linked groups of residues in immunoglobulin κ chains. Nature. 1967;216:330–2.
28. Edelman GM, et al. The covalent structure of an entire γG immunoglobulin molecule. Proc Natl Acad Sci. 1969;63:78–85.
29. Kunkel HG, et al. Notation for human immunoglobulin subclasses. Bull Wld Hth Org. 1966;35:953.

30. Natvig JB, Kunkel HG. Human immunoglobulins: classes, subclasses, genetic variants and Idiotypes. Adv Immunol. 1973;16:1–59.
31. Oudin J. L'allotypie de certains antigènes protéidiques du sérum. Compt Rend Acad Sci France. 1956;242:2606–8.
32. Oudin J. Allotypy of rabbit serum proteins I and II. J Exp Med. 1960;112:107–24; 125–142.
33. Kunkel HG, et al. A relationship between the H chain groups of 7S gamma globulin and the Gm system. Nature. 1964;203:413–4.
34. http://www.imgt.org/IMGTrepertoire/Proteins/allotypes/human/IGH/IGHC/G1m_allotypes.html
35. Ternant D, et al. IgG1 allotypes influence the pharmacokinetics of therapeutic monoclonal antibodies through FcRn binding. J Immunol. 2016;196(2):607–13.
36. Franěk F. Developmental aspects of antibody formation and structure. In: Šterzl J, Řiha I, editors. Proceedings of a symposium held in Slapy, vol. I and II. Prague: Czechoslovak Academy of Sciences; 1969.
37. Wu TT, Kabat EA. Analysis of the sequences of variable regions of Bence Jones proteins and myeloma light chains and their implications for antibody complementarity. J Exp Med. 1970;132:211–49.
38. Hood L, Talmage DW. Mechanism of antibody diversity: germ line basis for variability. Science. 1970;168:325–34.
39. Tonegawa S. Evidence for somatic generation of antibody diversity. Proc Natl Acad Sci. 1974;71:4027–31.
40. Tonegawa S, Hozumi N, Matthyssens G, Schuller R. Somatic changes in the content and context of immunoglobulin genes. Cold Spring Harbor Symp Quant Biol. 1977;41:877–89.
41. Seidman JG, Leder P. The arrangement and rearrangement of antibody genes. Nature. 1978;276:790–5.
42. Loh E, et al. Rearrangement of genetic information may produce immunoglobulin diversity. Nature. 1978;276:785–90.
43. Early P, et al. An immunoglobulin heavy chain variable region gene is generated from three segments of DNA: VH, D and JH. Cell. 1980;19:981–92.
44. Sagano H, et al. Identification and nucleotide sequence of a diversity DNA segment (D) of immunoglobulin heavy chain genes. Nature. 1981;290:562–5.
45. Alt FW, Baltimore D. Joining of immunoglobulin heavy chain gene segments: implications from a chromosome with three D-JH fusions. Proc Natl Acad Sci. 1982;79:4118–22.
46. Muramatsu M, et al. Specific expression of activation-induced cytidine deaminase (AID), a novel member of the RNA-editing deaminase family in germinal center B cells. J Biol Chem. 1999;274:18470–6.
47. Muramatsu M, et al. Class switch recombination and hypermutation require activation-induced cytidine deaminase (AID), a potential RNA editing enzyme. Cell. 2000;102:553–63.
48. Darlow JM, Stott DI. Gene conversion in human rearranged immunoglobulin genes. Immunogenetics. 2006;58:511–22.
49. Meng W, et al. Trials and tribulations with VH replacement. Front Immunol. 2014;5(10):1–12.
50. Rees AR. Understanding the human antibody repertoire. mAbs. 2020. 12:1. https://doi.org/10.1080/19420862.2020.1729683.
51. Deisenhofer J, et al. Crystallographic structural studies of a human Fc fragment. II. A complete model based on a Fourier map at 3.5 a resolution. Hoppe Seylers Z Physiol Chem. 1976;357(10):1421–34.
52. Eagle H, Brewer G. Mechanism of hemolysis by complement I. Complement fixation as an essential preliminary to hemolysis. J Gen Physiol. 1929;12:845–62.
53. Rowley D, Wardlaw AC. Lysis of gram-negative bacteria by serum. J Gen Microbiol. 1958;18:529–83.

54. Müller-Eberhard HJ, Kunkel HG. Isolation of a thermolabile serum protein which precipitates γ-globulin aggregates and participates in immune hemolysis. Proc Exptl Biol Med. 1961;106:291–5.
55. Lepow IH, et al. Chromatographic resolution of the first component of human complement into three activities. J Exp Med. 1963;117:983–1008.
56. Porter RR. The complex proteases of the complement system. Croonian Lecture Proc Roy Soc B. 1980;210:477–98.
57. Brüggemann M, et al. Comparison of the effector functions of human immunoglobulins using a matched set of chimeric antibodies. J Exp Med. 1987;166:1351–61.
58. Binden CI, et al. Human monoclonal IgG isotypes differ in complement activating function at the level of C4 as well as C1q. J Exp Med. 1988;168:127–42.
59. Simister NE, Rees AR. Properties of immunoglobulin G-Fc receptors from neonatal rat intestinal brush borders. Eur J Immunol. 1985;15:733–8.
60. Burmeister W, et al. Crystal structure of the complex of rat neonatal Fc receptor with Fc. Nature. 1994;372:379–83.
61. Ravetch JV, et al. Structural heterogeneity and functional domains of murine immunoglobulin G Fc receptors. Science. 1986;234:718–25.
62. Allen JM, Seed B. Isolation and expression of functional high-affinity Fc receptor complementary DNAs. Science. 1989;243:378–81.
63. Ravetch JV. Fc receptors: Rotor redux. Cell. 1994;78:553–60.
64. Vidarsson G. IgG subclasses and allotypes: from structure to effector functions. Front Immunol. 2014;5(520):1–17.
65. Morrison SL, et al. Chimeric human antibody molecules: mouse antigen-binding domains with human constant region domains. Proc Natl Acad Sci. 1994;81:6851–5.
66. Jones PT, et al. Replacing the complementarity-determining regions of a human antibody with those from a mouse. Nature. 1986;321:522–5.
67. Reichmann L, et al. Reshaping human antibodies for therapy. Nature. 1988;332:323–7.
68. Roberts S, et al. Generation of an antibody with enhanced affinity and specificity for its antigen by protein engineering. Nature. 1987;328:731–4.

Monoclonal Antibodies and Hybridomas

3

Florian Rüker

Contents

F. Rüker (✉)
Institute of Molecular Biotechnology, Department of Biotechnology, University of Natural
Resources and Life Sciences, Vienna (BOKU), Vienna, Austria
e-mail: florian.rueker@boku.ac.at

© Springer Nature Switzerland AG 2021
F. Rüker, G. Wozniak-Knopp (eds.), *Introduction to Antibody Engineering*,
Learning Materials in Biosciences,
https://doi.org/10.1007/978-3-030-54630-4_3

Keywords

Monoclonal antibody · Hybridoma · Cell fusion · Polyethylene glycol · HAT medium · Limited dilution · Humanized antibody · Chimeric antibody · Fully human antibody · Transgenic technology

What You Will Learn in This Chapter
This chapter will start with briefly describing the state of science leading to the ground-breaking work on monoclonal antibodies that was published in 1975 by Köhler and Milstein [1] under the title *Continuous cultures of fused cells secreting antibody of predefined specificity*. This paper is the foundation and one of the pillars of today's modern antibody technology, documented by the fact that 6 monoclonal antibodies are among the top 10 best-selling drugs worldwide as of 2018.[1] We will take a look at the original techniques that were used by Köhler and Milstein and how these techniques evolved over time to the current standard protocols. Methods such as antibody chimerization and humanization, which sparked the development of human monoclonal antibodies for broad therapeutic use, will be discussed as well. Hybridoma technology soon saw itself in competition with in vitro techniques for antibody discovery such as phage display and various other display methods (described in detail in another chapter in this book), but received constant and even increased interest when transgenic mice and rats became widely accessible that had, instead of their own antibody-encoding genes, those of human origin integrated in their genomes.

3.1 The First Monoclonal Antibodies

The 1960s and 1970s saw exciting scientific developments in the field of immunology. One of the burning questions of the time was how the antibody-encoding genes were organized on the genomic level, since it was not yet known exactly how the vertebrate genome could encode a vast number of different antibodies given the limited size of the genome. It was also not perfectly clear yet how the amino acid chains of antibodies were structurally assembled, or how exactly the contact between antibody and antigen was established to name just a few of many more enigmas that still needed to be resolved. The history of this exciting time in science is excellently told in great detail in the book *The Antibody Molecule: From Antitoxins to Therapeutic Antibodies,* by Anthony R. Rees [2].

In the 1970s, the experimental material that was available for research on specific antibodies had so far been rather limited. This is because all antibody-based immune

---

[1]http://www.pmlive.com/top_pharma_list/Top_50_pharmaceutical_products

Fig. 3.1 César Milstein (left) and Georges Köhler (right). Reproduced with kind permission of Celia Milstein/MRC Laboratory of Molecular Biology

responses are of oligo- to polyclonal nature, and antibody-producing cells isolated from the blood will not grow clonally in culture, because of their primary nature. Myelomas, which are spontaneous tumours of the B-cell lineage, were the best tools available for research and also allowed the establishment of myeloma cell lines that could be grown indefinitely in the laboratory. These cell lines were used as a source of antibody protein which consisted either of light or heavy chains or in some cases also of complete IgGs. The target molecules of these myeloma antibodies were however usually not known. Georges Köhler (Fig. 3.1, right), then 27 years old, was a postdoc when he met César Milstein (Fig. 3.1, left) and subsequently started working in his lab at the MRC Laboratory of Molecular Biology (LMB) in Cambridge, UK.

There, Köhler was working with myeloma cell lines, trying to find their antigen, in order to allow further experimentation, for example, by altering their antigen-binding properties by mutagenesis. It is told [2] that Köhler proposed an alternative strategy to Milstein, which consisted of immortalizing antibody-producing B-cells, that would otherwise not grow in culture, by fusing them with immortal myeloma cell lines. In order to select for the fused cells, Köhler had to find a way to select hybrid cells and inhibit unfused myeloma cells from growing. This could be accomplished by the use of HAT medium, which will be explained in detail below. Köhler and Milstein then proceeded to immunize mice with an easily accessible antigen, namely, sheep erythrocytes, isolated spleen cells from the immunized mice, fused them to myeloma cells, selected for hybrid cells and showed that *clones* of continuously growing hybrid cells were secreting *monoclonal* antibodies that specifically recognized sheep erythrocytes. This work was submitted to the journal *Nature* for publication and appeared in its 7 August 1975 issue [1]. The first page of this now classical paper is reproduced in Fig. 3.2. Subsequently in 1984, Köhler and Milstein,

Nature Vol. 256 August 7 1975   495

Continuous cultures of fused cells secreting antibody of predefined specificity

THE manufacture of predefined specific antibodies by means of permanent tissue culture cell lines is of general interest. There are at present a considerable number of permanent cultures of myeloma cells[1,2] and screening procedures have been used to reveal antibody activity in some of them. This, however, is not a satisfactory source of monoclonal antibodies of predefined specificity. We describe here the derivation of a number of tissue culture cell lines which secrete anti-sheep red blood cell (SRBC) antibodies. The cell lines are made by fusion of a mouse myeloma and mouse spleen cells from an immunised donor. To understand the expression and interactions of the Ig chains from the parental lines, fusion experiments between two known mouse myeloma lines were carried out.

Each immunoglobulin chain results from the integrated expression of one of several V and C genes coding respectively for its variable and constant sections. Each cell expresses only one of the two possible alleles (allelic exclusion; reviewed in ref. 3). When two antibody-producing cells are fused, the products of both parental lines are expressed[4,5], and although the light and heavy chains of both parental lines are randomly joined, no evidence of scrambling of V and C sections is observed[4]. These results, obtained in a heterologous system involving cells of rat and mouse origin, have now been confirmed by fusing two myeloma cells of the same mouse strain,

The protein secreted (MOPC 21) is an IgG1 (κ) which has been fully sequenced[7,8]. Equal numbers of cells from each parental line were fused using inactivated Sendai virus[9] and samples contining 2×10^5 cells were grown in selective medium in separate dishes. Four out of ten dishes showed growth in selective medium and these were taken as independent hybrid lines, probably derived from single fusion events. The karyotype of the hybrid cells after 5 months in culture was just under the sum of the two parental lines (Table 1). Figure 1 shows the isoelectric focusing[10] (IEF) pattern of the secreted products of different lines. The hybrid cells (samples c–h in Fig. 1) give a much more complex pattern than either parent (a and b) or a mixture of the parental lines (m). The important feature of the new pattern is the presence of extra bands (Fig. 1, arrows). These new bands, however, do not seem to be the result of differences in primary structure; this is indicated by the IEF pattern of the products after reduction to separate the heavy and light chains (Fig. 1B). The IEF pattern of chains of the hybrid clones (Fig. 1B, g) is equivalent to the sum of the IEF pattern (a and b) of chains of the parental clones with no evidence of extra products. We conclude that, as previously shown with interspecies hybrids[4,5], new Ig molecules are produced as a result of mixed association between heavy and light chains from the two parents. This process is intracellular as a mixed cell population does not give rise to such hybrid molecules (compare m and g, Fig. 1A). The individual cells must therefore be able to express both isotypes. This result shows that in hybrid cells the expression of one isotype and idiotype does not exclude the expression of another: both heavy chain

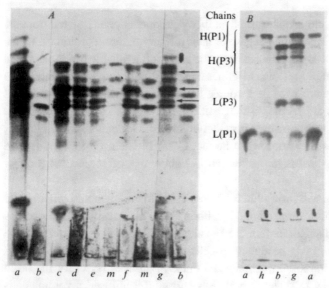

Fig. 1 Autoradiograph of labelled components secreted by the parental and hybrid cell lines analysed by IEF before (A) and after reduction (B). Cells were incubated in the presence of ^{14}C-lysine[11] and the supernatant applied on polyacrylamide slabs. A, pH range 6.0 (bottom) to 8.0 (top) in 4 M urea. B, pH range 5.0 (bottom) to 9.0 (top) in 6 M urea; the supernatant was incubated for 20 min at 37 °C in the presence of 8 M urea, 1.5 M mercaptoethanol and 0.1 M potassium phosphate pH 8.0 before being applied to the right slab. Supernatants from parental cell lines in: a, P1Bul; b, P3-X67Ag8; and m, mixture of equal number of P1Bul and P3-X67Ag8 cells. Supernatants from two independently derived hybrid lines are shown: c–f, four subclones from Hy-3; g and h, two subclones from Hy-B. Fusion was carried out[4,9] using 10^6 cells of each parental line and 4,000 haemagglutination units inactivated Sendai virus (Searle). Cells were divided into ten equal samples and grown separately in selective medium (HAT medium, ref. 6). Medium was changed every 3 d. Successful hybrid lines were obtained in four of the cultures, and all gave similar IEF patterns. Hy-B and Hy-3 were further cloned in soft agar[14]. L, Light; H, heavy.

and provide the background for the derivation and understanding of antibody-secreting hybrid lines in which one of the parental cells is an antibody-producing spleen cell.

Two myeloma cell lines of BALB/c origin were used. P1Bul is resistant to 5-bromo-2′-deoxyuridine[4], does not grow in selective medium (HAT, ref. 6) and secretes a myeloma protein, Adj PC5, which is an IgG2A (κ), (ref. 1). Synthesis is not balanced and free light chains are also secreted. The second cell line, P3-X63Ag8, prepared from P3 cells[2], is resistant to 20 μg ml^{-1} 8-azaguanine and does not grow in HAT medium.

ˈsotypes (γ1 and γ2a) and both V_H and both V_L regions (idiotypes) are expressed. There are no allotypic markers for the C_K region to provide direct proof for the expression of both parental C_K regions. But this is indicated by the phenotypic link between the V and C regions.

Figure 1A shows that clones derived from different hybridisation experiments and from subclones of one line are indistinguishable. This has also been observed in other experiments (data not shown). Variants were, however, found in a survey of 100 subclones. The difference is often associated with changes

---

together with Niels K. Jerne, were awarded the Nobel Prize in Physiology or Medicine "for theories concerning the specificity in control and development of the immune system and the discovery of the principle for production of monoclonal antibodies".

3.2   Sequence, Structure and Function of Antibodies

3.2.1   Overall Structure

Without going into the details of how target-specific antibodies are generated by the immune system in response to challenge with an antigen, which is a topic that is dealt with in great detail in immunology textbooks (e.g. *Janeway's Immunobiology* [3] or *Cellular and Molecular Immunology* [4]), a brief outline of the amino acid sequences of immunoglobulins, the resulting 2D and 3D structures and their functions shall be given here. We will focus on IgG, since this is currently the most important immunoglobulin class used in the field of antibody engineering and antibody therapy.

IgG molecules consist of 2 identical heavy chains, each with a length of approximately 440 amino acids and 2 identical light chains of approximately 220 amino acids in length. The chains are organized in domains of approximately 110 amino acids each. All domains share the classical immunoglobulin fold that consists of a two-layer sandwich of seven (for constant domains) or nine (for variable domains) antiparallel β-strands arranged in two β-sheets with a Greek key topology [5]. Loops connect the β-strands. The heavy and the light chains are closely connected to each other by an interchain disulphide bond as well as by non-covalent interactions between the V_H (variable heavy) and the V_L (variable light) domains and between the C_H1 (constant heavy 1) and C_K (constant kappa) or C_Λ (constant lambda) domains, respectively. The two heavy chains are connected covalently by disulphide bonds (two in the case of human IgG1) in the hinge region as well as by non-covalent interactions between the C_H2 and the C_H3 domains, respectively. Human IgG1 has one site for N-linked glycosylation in the C_H2 domain (see dots in Fig. 3.3b).

A schematic as well as a molecular presentation of an IgG1 is shown in Fig. 3.3. IgG molecules consist of two identical heavy chains, each with a length of approximately 440 amino acids and two identical light chains of approximately 220 amino acids in length.

3.2.2   Variable Domains

Variable domains are responsible for antigen binding. The V_H domain together with the V_L domain is called the Fv fragment of an antibody. Three loops on each of the V_H and the V_L domain together form the antigen-binding site. These loops are called CDRs (complementarity-determining regions), or hypervariable loops, since their amino acid sequences differ between different antibodies. The sequences of amino acids in the CDRs determine the chemical and structural properties of the antigen-binding site of antibodies and thus allow

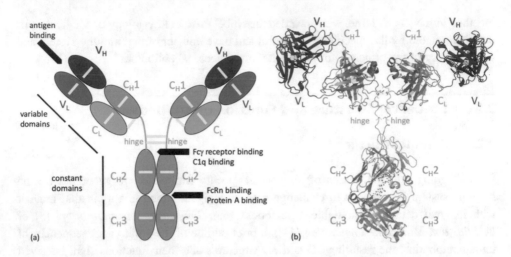

Fig. 3.3 Structure of a human IgG1 molecule. (**a**) Schematic presentation, (**b**) ribbon structure. Heavy chains are marked in red colours; light chains are marked in blue colours. The hinge region is in green; intradomain and interchain disulphides are indicated by yellow lines. Antigen-binding sites, binding sites for Fc receptors and for complement factor C1q, as well as for FcRn, are indicated with black arrows. The ribbon presentation was made with PyMOL Molecular Graphics System, Version 2.0 Schrödinger, LLC. The model structure file is from [6]

them to bind and adapt to a virtually unlimited number of different antigen structures. The specific spot on an antigen molecule where an antibody binds is called "epitope", whereas the corresponding spot on the antibody is the "paratope". Figure 3.4 shows the Fv of a human anti-HIV1 antibody in 3D ((a) and (b)) and the VL (c) and the VH (d) domains in 2D presentations.

3.2.3 Constant Domains

Constant domains of antibodies are, as the name says, constant in sequence, which means that all antibodies of a specific class (e.g. IgG1) have identical sequences which do not change during an immune response. The constant domains can be divided into two regions: the pair of C_H1 and C_K domains is part of the Fab (*f*ragment *a*ntigen-*b*inding) fragment of the antibody and is located between the Fv and the hinge, whereas the homodimer of C_H2 and C_H3 domains of the heavy chains forms the Fc (*f*ragment *c*rystallizable) fragment. A region in the Fc fragment which is on the tip of the C_H2 domains close to the hinge region is where Fc receptors like FcγR1 (=CD64), FcγR2 (=CD32) and FcγR3 (=CD16) bind. These receptors are found on a number of immune cells, including natural killer (NK) cells and macrophages, and are responsible for eliciting effector functions such as ADCC (*a*ntibody-*d*ependent *c*ellular *c*ytotoxicity), ultimately leading to the elimination of the target that is bound by the antibody. Furthermore, C1q, the first component of the

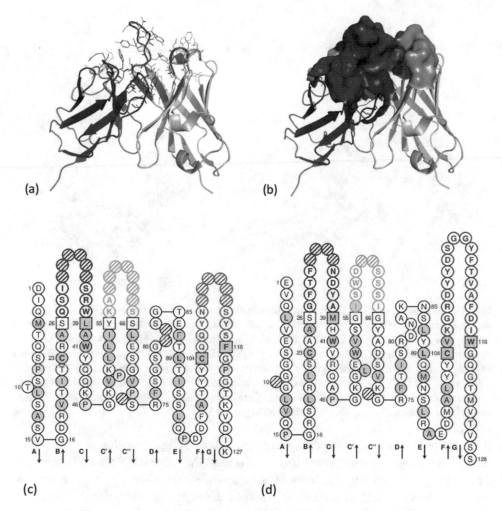

Fig. 3.4 Structure of the antigen-binding Fv region of an antibody. (**a**) Ribbon presentation (**b**) same as (**a**) but with the CDRs shown with their solvent accessible surface. Heavy chain (V_H domain) is shown in light grey; light chain (V_L domain) is in dark grey. CDRs of the heavy chain are coloured in three shades of red, CDRs of the light chain in three shades of blue. The figure was made with PyMOL Molecular Graphics System, Version 2.0 Schrödinger, LLC, using the structure file 1DFB. pdb [7]. (**c**) and (**d**) Collier de Perles presentation of the secondary structure of the V_L (**c**) and the V_H (**d**) domains of 1DFB.pdb. CDR loops are shown on the top sides of the domains in different colours. The Collier de Perles presentations were taken from IMGT®, the International ImMunoGeneTics information system®, at www.imgt.org [8, 9]

complement cascade, has its binding site here and can elicit CDC (*complement-dependent cytotoxicity*) or CDCC (*complement-dependent cellular cytotoxicity*) to kill the antibody-bound target. For further details on these effector functions, the reader is referred to immunology textbooks. Antibodies have a remarkably long in vivo half-life (e.g. up to

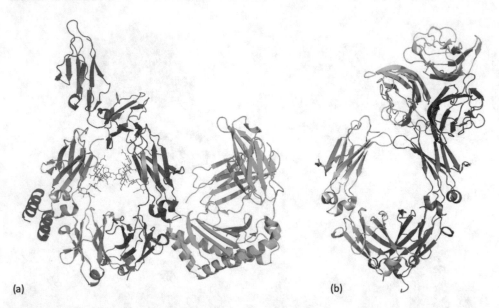

(a)   (b)

Fig. 3.5 Composite structure of an Fc fragment in contact with its main ligands. The two halves of the Fc are in light and dark grey, respectively, and carbohydrate chains are in magenta. (**a**) Crystal structure of Fc bound to FcγR1 (red), FcRn (green) and the Z domain of protein A (blue). For FcRn and protein A, only one of the two binding sites per Fc molecule is shown. Crystal structure files used to compose this part of the Figure are 1OQO.pdb (Fc and protein A [10]), 1IIS.pdb (FcγR1 [11]) and 1FRT.pdb (FcRn [12]). (**b**) Complement C1q subcomponent subunits A, B and C bound to Fc. Cryo-electron microscopy data from 6FCZ.pdb [13] are shown. The figure was made with PyMOL Molecular Graphics System, Version 2.0 Schrödinger, LLC

3 weeks for human IgG1) which is mediated by another receptor binding site. This site is located in the region at the interface between C_H2 and C_H3 domains. Here, FcRn (neonatal Fc receptor) has its binding site. In the same region, we find the binding site for Protein A, which is routinely used in research and also in the biotech industry for the facile and efficient purification of antibodies by affinity chromatography. Figure 3.5 shows a composite structure of an Fc with its different ligands bound to it.

3.3 The Hybridoma Method

3.3.1 The Starting Point: Antibody-Producing B-Cells

As mentioned briefly above, the aim of Köhler's and Milstein's experiment was to establish cell lines that would grow continuously in culture and secrete a single type of antibody molecule, i.e. a monoclonal antibody. Splenocytes are primary, antibody-producing cells that can be isolated from an immunized animal. Other sources of suitable cells with the same property are blood lymphocytes or cells isolated from tonsils, which are more easily accessible, especially when the donor is not an animal but a (convalescent or vaccinated)

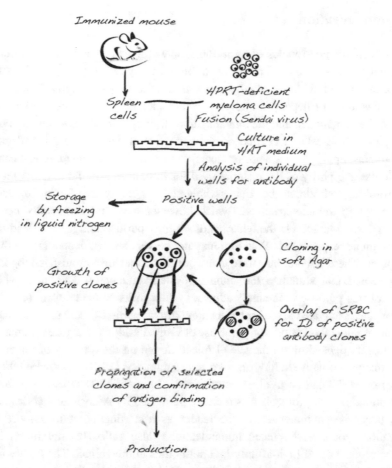

Fig. 3.6 The hybridoma method. Source: data from [1]. Cartoon drawn by Anthony R. Rees [2]. Reproduced with permission of Oxford Publishing Limited through PLSclear, Ref No: 33734

human. These primary cells however do not have the ability to grow in culture for extended periods of time. On the other hand, tumour-derived myeloma cell lines have exactly this ability; however they do not produce meaningful antibodies. The solution was to fuse splenocytes and myeloma cells, thereby producing hybrid cells, so-called hybridomas, that united both desirable properties, antibody production and continuous growth. It is worth noting that the term "hybridoma" was not proposed originally by Köhler and Milstein. César Milstein published a narrative of the "hybridoma revolution" in which he recalls that the term "hybridoma" was proposed by Leonard Herzenberg during a sabbatical in his laboratory in 1976/1977 [14].

A brief graphical summary of the hybridoma method as devised by Köhler and Milstein is shown in Fig. 3.6.

The different steps of the hybridoma method are described in detail below, together with a description of today's corresponding standard laboratory resources and procedures.

3.3.2   Immunization

In order to achieve good yields of hybridomas after the cell fusion, and in order for the resulting hybridomas to be good producers of high-affinity IgG antibodies, it is essential to immunize the animal, which is usually a mouse or a rat, very efficiently. The outcome of the immunization will depend on several factors including the mouse strain, type and strength of the immunogen and the immunization strategy and schedule, among others. As immunogens, whole cells or cellular components, but of course also purified molecules such as proteins, lipids, nucleic acids or others, can be used. As a general rule, it can be said that the higher the purity of the material used for immunization, the more target-specific will the resulting hybridomas be. Using whole cells or cell lysates for immunization can produce a strong immune response, which however must not necessarily be directed against the desired target. On the other hand, a highly purified antigen will lead to a very targeted immune response, which in turn may not be very strong, depending on the nature of the antigen. There is therefore a need to use adjuvants that are administered together with the antigen and that stimulate the immune system efficiently. The most widespread adjuvant for this purpose is Freund's adjuvant either in its complete form (consisting of mineral oil and inactivated *Mycobacterium tuberculosis* particles) or in its incomplete form (consisting of mineral oil only). Since the use of Freund's adjuvant can be traumatic for the animals, mainly depending on the site of injection and on the quantity of material being injected, many alternative adjuvants like aluminium, ALD/MDP (alhydrogel/muramyl dipeptide) or MF59 can be used. A comparison of the antibody repertoire outcomes of different immunization protocols was recently published by Asensio and colleagues [15], and this paper is recommended to the reader as a starting point to explore detailed immunization protocols. Repeated immunizations (also called booster immunizations) induce the production of high-affinity IgG antibodies in the animal. The highly activated lymphoblasts are in turn very efficient in fusing with myeloma cells.

3.3.3   Myeloma Cell Lines Suitable for Fusion

As of today, a large number of myeloma cell lines are available to be used as fusion partners for the generation of hybridomas. All of these cell lines must carry a genetic marker that allows them to be killed following the fusion procedure in case they are not fused to a B-cell. The marker is usually the lack of expression of the enzyme hypoxanthine-guanine phosphoribosyltransferase (HGPRT$^-$), which is found in all normal cells. Lack of HGPRT is based on an inactivating mutation that is easily selected for by cultivating previously mutagenized cells in the presence of a toxic analogue of guanine, such as 6-thioguanine or 8-azaguanine. Only cells with an HGPRT$^-$ genotype will survive under this selective pressure and are thus suitable for use in hybridoma experiments. Since HGPRT is part of a salvage pathway for the synthesis of purines and pyrimidines, HGPRT$^-$ cells depend on de novo synthesis of purines. If de novo purine synthesis is blocked by a drug named aminopterin, HGPRT$^-$ cells will however not survive. Upon fusion of an HGPRT$^-$ myeloma cell with a normal, HGPRT$^+$ splenocyte, an HGPRT$^+$ hybridoma cell is the result, which, when supplied with exogeneous hypoxanthine, can grow very well, even in the presence of aminopterin. Unfused splenocytes do not have the capacity to divide in culture and are therefore eliminated very quickly without any selective measure. This strategy for selection of fused cells and elimination of unfused cells was developed by John W. Littlefield in 1964 [16] and has become famous by the name of the selective medium, HAT medium, where H stands for hypoxanthine, A for aminopterin and T for thymidine.

The myeloma cell line used by Köhler and Milstein in 1975 was the P3-X63Ag8 line that had been established by Kengo Horibata and Alan Harris [17]. This cell line is resistant to 8-azaguanine and does not grow in HAT medium.

A desirable property of a fused hybrid cell is genomic stability over many cell divisions, since loss of chromosomes can change the basic growth properties of the cells or lead to loss of antibody production when Ig gene-carrying chromosomes are affected. Therefore, numerous cell lines of mouse, rat or human origin have been developed that can be chosen in dependence of the source of primary B-cells used for fusion. Another important property is the production of endogenous antibody chains by certain myeloma cell lines, which is undesirable since it will obscure the expression of the desired monoclonal antibody or lead to the production of hybrid antibodies with perturbed functionality. HAT-selectable myeloma cells have therefore been selected also for the absence of endogenous antibody production. Finally, media requirements are an important factor. Typical cell culture media will contain a certain percentage of serum (usually derived from cows or (foetal) calves) as a growth supplement. Since it is generally desirable to avoid such animal-derived components in cell culture media, myeloma cells that can grow in protein-free, fully synthetic culture media are also available. Many companies, but also non-profit organizations such as the European Collection of Authenticated Cell Cultures (ECACC; https://www.phe-culturecollections.org.uk/) and the American Type Culture Collection (ATCC; www.atcc.org), are validated suppliers of myeloma cell lines suitable for the production of hybridomas.

3.3.4 Cell Fusion

In their original work of 1975, Köhler and Milstein applied Sendai virus [18] (also called *hemagglutinating virus of Japan*, or at present *murine respirovirus*) to induce cell fusion between splenocytes and myeloma cells. Although efficient, this procedure was not very practical nor very reliable or reproducible in the lab, and in a methods paper published in 1981, Galfrè and Milstein described the use of polyethylene glycol (PEG) as a fusogenic agent suitable for the production of hybridomas [19], a method that had been described for cell fusion a few years earlier [20]. PEG has dehydrating properties which leads to cell membranes coming in close cell-to-cell contact. Plasma membrane fusion then occurs spontaneously and in an uncontrolled manner, so that not only fusions between one splenocyte and one myeloma cell take place but also fusions of three or more cells in all possible combinations. This topic will be discussed later in this chapter ("when monoclonal antibodies are not monoclonal"). Anyway, in the ideal outcome of one splenocyte having fused with one myeloma cell, a heterokaryon is formed, which is a cell with two cell nuclei. In a further uncontrolled step, the two nuclei must fuse, forming a synkaryon. During the following cell divisions, this double set of chromosomes gets reduced to a mixed set of chromosomes that is stabilized over subsequent generations of the cell clone. Further chromosome loss can never be excluded, and if heavy- or light-chain encoding chromosomes are affected, this will ultimately lead to an unproductive hybridoma clone.

Besides PEG-induced cell fusion, other methods are also available to produce hybridomas. Among these, electrofusion is probably the most frequently applied one. Electrofusion is based on the fact that short high-voltage field pulses can destabilize cell membranes by inducing an unordered status of the lipids, which rearrange within milliseconds after the pulse to form normal lipid bilayer membranes again. When the membranes of two cells are in close contact to each other, this can lead to cell-to-cell fusion. The simplest way to get cells in close contact with one another is centrifugation. A more elaborate way is to apply the principle of dielectrophoresis, in which a high-frequency alternating current leads to the formation of pearl chains of cells, which can then be fused efficiently. This method was first described in the early 1980s (see [21] for a review), and a recent protocol has been described by Greenfield [22].

3.3.5 Identification of Individual Hybridoma Clones

After selection of hybridomas, elimination of unfused myeloma cells takes place in HAT medium (see above). At the same time, stepwise disappearance of non-dividing unfused B-cells occurs, and hybridoma clones start to grow. In order to separate individual clones from each other, so that monoclonal cultures are achieved, it is essential that the cells are plated after fusion in microtiter plates such that only about one fused cell is distributed per well. This process is called "limiting dilution", and numerous papers have been published on how to best perform this procedure, including a recent very detailed methods paper [23].

Once hybridoma clones are established and can be seen in the microscope, their supernatants will contain sufficient amounts of antibodies that allow screening for productivity and specificity. Various well-known methods are available for this purpose, including ELISA, surface plasmon resonance, biolayer interferometry, flow cytometry as well as biological assays. Many papers have been published on such methods, and the reader is referred to these. A few examples are found here [24–31].

3.3.6   Monoclonality

It is generally assumed that monoclonal antibodies produced from hybridomas are mono-specific. Considering the rather uncontrolled conditions during cell fusion, and also because of the genetic setup of myeloma cells used for fusion, this is not necessarily always the case. During cell fusion it is not excluded that two or more antibody-producing B-cells fuse with the same myeloma cell, which is then the carrier of two or more genes encoding heavy and light chains. Furthermore, the myeloma fusion partner may itself produce endogenous antibody chains which could form hybrid antibodies together with the chains encoded by the primary B-cells. Anecdotal evidence has already pointed in this direction, which prompted a large group of scientists from all over the world to explore this issue [32]. 185 random hybridomas were analysed in this large multicentre study out of which 68% were monospecific without additional productive chains. The remaining 32% however contained one or more additional productive heavy or light chains. This should always be kept in mind when working with hybridomas, and great care should be taken when cloning the cDNAs from hybridomas for further recombinant expression.

3.4   From Mouse to Human Monoclonal Antibodies

So far, we have focused on hybridomas generated from splenocytes of immunized mice, giving rise to mouse antibodies. Such antibodies are extremely useful as reagents for application in various in vitro settings, such as in ELISA, Western blot, immunohis-tochemistry and others. However, they are unsuitable as therapeutic agents for the treatment of humans, since they will be recognized by the patient's own immune system as foreign and will give rise to *human anti-mouse antibodies* (a so-called HAMA response). These antidrug antibodies will render the therapeutic antibody ineffective over the continued course of the treatment.

There are several ways in which this problem can be tackled. The most obvious would be to use antibody-producing B-cells from humans for fusion to myeloma cells. Since for ethical reasons it is of course not possible to immunize humans against various interesting antigens, this strategy is of limited use. The only useful source of antigen-specific B-cells is peripheral blood lymphocytes (or cells isolated from organs that are removed surgically for other reasons, such as in routine tonsillectomy or splenectomy). The spectrum of specificities that are addressable like this is naturally very limited. It mainly covers diseases against which active vaccination is available, or (mainly infectious) diseases from which patients have recovered due to neutralizing antibodies. For example, various broadly neutralizing human anti-HIV antibodies have been established in such a way [33].

Much more broadly applicable techniques were developed in the 1980s that made it possible to use mouse antibodies as a starting material and derive more or less human antibodies from them. These two types of antibodies are *chimeric* and *humanized* antibodies and will be described briefly in the following subchapters, followed by methods

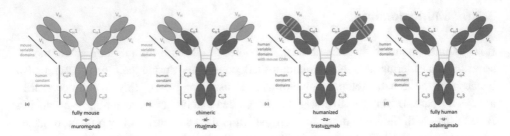

Fig. 3.7 From mouse to human monoclonal antibodies: (**a**) fully mouse antibody, (**b**) mouse-human chimeric antibody, (**c**) humanized antibody, (**d**) fully human antibody; mouse regions are coloured green; human regions are coloured blue. Below each antibody, the source substems as well as the name of an exemplary antibody are given

to produce fully human antigen-specific monoclonal antibodies. Figure 3.7 gives a graphical representation of these differently engineered antibodies.

3.4.1 Chimeric Antibodies

Due to the highly conserved modular structure of antibodies, domains of antibodies as well as global antibody structure share a high degree of similarity between different species. It soon came into the focus of scientists that these similarities should allow to use antibody domains as modular building blocks and to combine mouse and human domains to produce antibody chimera, with the aim of retaining their binding properties on the one hand and making them as human as possible on the other hand. In 1984, almost simultaneously, two reports were published on the engineering of "chimeric" monoclonal antibodies, in which mouse genes encoding the variable domains of an antibody were fused to human genes encoding the constant part of the antibody. Sherrie Morrison and colleagues [34] worked on a mouse antibody directed against phosphocholine and fused its variable domains to either human IgG1 or IgG2 constant regions, while Gabrielle Boulianne and colleagues [35] fused the variable region genes of a mouse anti-trinitrophenyl IgM to the constant regions of human IgM. In both cases, the mouse-derived antigen-binding capacity was fully retained.

Chimeric antibodies quickly made their way to the clinic, and today some of them, e.g. rituximab and cetuximab, are blockbuster antibodies, which means that their annual sales are higher than 1b$.

3.4.2 Nomenclature of Monoclonal Antibodies

We must now briefly discuss the nomenclature of monoclonal antibodies. They follow a naming scheme for assigning generic, or nonproprietary, names to monoclonal antibodies.

The naming scheme is defined in the "World Health Organization's International Nonproprietary Names (INN)" (www.who.int). All antibody names end with the stem -mab, which is placed as a suffix. Just before this stem, a substem is placed which indicates the species on which the immunoglobulin sequence of the mAb is based. Among others, this is -o- for mouse, -xi- for chimeric, -zu- for humanized and -u- for human antibodies. Directly before this, one or two letters indicate the target (molecule, cell, organ) class of the antibody. Examples for this designation would be -l(i)- for immunomodulating (e.g. ada*li*mumab) or t (u) for tumour (e.g. ri*tu*ximab)

3.4.3   Humanized Antibodies

Chimeric antibodies were a huge step towards antibodies that are less immunogenic in humans than mouse antibodies, since out of the 12 domains that make up an IgG, they contain only 4 domains of mouse origin. However, scientists soon realized that this was maybe not good enough for sustainable long-term therapy in humans, in which immunogenicity of the antibody can reduce effectiveness of the therapy. In 1986, 2 years after the publication of the first chimeric antibodies, Greg Winter and his research group at the MRC in Cambridge published a paper that was a major breakthrough in the field of antibody engineering. They took a mouse antibody against a hapten antigen (NP-cap) and introduced the CDR sequences of this antibody into the framework sequences of human V_H and V_L domains, respectively. This procedure was termed "antibody humanization" and is today still one of the cornerstones of modern antibody technology [36]. Numerous humanized antibodies have been approved for clinical use, e.g. trastu*zu*mab or bevaci*zu*mab.

3.4.4   Fully Human Monoclonal Antibodies

The late 1980s and early 1990s brought two further giant advances in antibody engineering, which enabled scientists to generate fully human antibodies without using any human-derived B-cells or other human material, except for genes encoding antibodies and the corresponding mRNA.

Since these methods are very important still today, a detailed chapter for each of them is included in this book, and only a very short description is included here in this chapter.

In 1989, Marianne Brüggemann and her research group [37] published a paper in which they described the introduction of unrearranged human immunoglobulin segments (variable, diversity, joining and constant elements) into the germ line of mice. Immunization of these transgenic mice led to the maturation of B-cells producing fully functional antibodies in which the heavy chain was human, demonstrating that this is a way to generate human antibodies in mice. Hybridoma technology allowed the immortalization of the B-cells from these mice. This technology of introducing human antibody-encoding gene segments in the genome of mice and deriving human antibodies from such transgenic mice subsequently took a very dynamic development and is now widespread in many laboratories and companies. Today, numerous human antibodies from transgenic mice have been approved for therapy in humans; one example for them is nivolumab, an antibody recognizing PD-1, which is used to treat a number of different cancer types.

In another approach, which even made it unnecessary to immunize animals at all, Greg Winter and his group showed in 1990 that it is possible to create large (with 10^9 or more independent clones) libraries of variable domains of antibodies (in the form of so called single-chain Fv fragments, scFvs, which are described in the section below), and to display these libraries on the surface of filamentous phage using *E. coli* [38]. Such libraries can be selected for antigen-specific clones by a process called "panning". See the chapter on Antibody Display Systems for details on these methods. scFvs that are isolated in such a way can then be reformatted into complete antibodies, by combining their variable domain-encoding regions with the corresponding constant regions. When the sequences of the scFvs are human, human antibodies can thus be created with relative ease. Human antibodies derived from phage display have long entered the clinic, and the best-selling drug worldwide in 2018, adalimumab, is an example for such a phage library-derived human antibody.

3.5   Antibody Fragments

When working with antibodies today, chances are great that this will involve the use not only of complete antibody molecules but also that of antibody fragments. The most important ones, fragment antigen-binding (Fab) and single-chain Fv (scFv), will be discussed in this section.

In the 1950s and 1960s, when the structure and functional organization of antibody molecules was not yet entirely clear, scientists used proteolytic enzymes such as papain or pepsin with great success to split the antibody molecule into functional fragments. The history of this ground-breaking science is described in great detail in the book *The Antibody Molecule: From Antitoxins to Therapeutic Antibodies,* by Anthony R. Rees [2].

It was found out that papain has a cleavage site that is located in the hinge region above the disulphide bonds that connect the heavy chains. Three fragments were generated, and their properties were analysed: two of the fragments could bind the antigen in a monovalent manner and were thus termed Fab, which stands for *fragment antigen-binding*. Fab fragments consist of the variable and constant domains of the light chains (V_L and C_L) and of the variable and the first constant domains of the heavy chain (V_H and C_H1). The third fragment that was generated had no antigen-binding properties; however, it was found that it could be crystallized with relative ease, leading to its designation as Fc (*fragment crystallizable*).

On the other hand, the digestion of antibodies with pepsin yielded several fragments: one that was approximately double the size of an Fab and could bind antigen in a bivalent manner (termed F(ab')$_2$) and a number of smaller fragments that derived from cleavage of

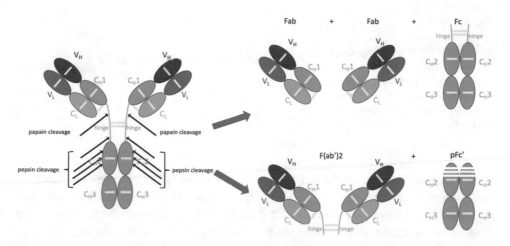

Fig. 3.8 Fragmentation of an IgG with papain (upper panel) or pepsin (lower panel)

the Fc. This was because pepsin has several cleavage sites in antibodies, all of which are located below the disulphide bonds in the hinge.

The results of the digestion of antibodies with papain or pepsin are shown schematically in Fig. 3.8.

Fab fragments retain the antigen binding properties of the antibody from which they were generated. However, due to the absence of the Fc, they lack the ability to elicit effector functions or to activate the complement system. They also have a much shorter in vivo half-life because of their smaller size (especially for the Fab with a molecular weight of approximately 50–60 kD) and because the binding site for FcRn is absent.

Because of the small size, the limited number of disulphide bonds and the absence of N-linked glycosylation, Fab fragments can be readily expressed recombinantly in the full range of expression hosts by co-expression of the two chains, which is described in more detail in another chapter of this book.

Applications of Fab and F(ab')$_2$ fragments are manifold: they are used as reagents in a wide variety of immunological test systems, but also as therapeutic molecules for treatments where their small size is advantageous because it promotes faster and deeper tissue penetration, or where absence of effector functions is desired.

Furthermore, antibody formats that are useful for the selection of antibodies from phage display or yeast display libraries encompass Fab fragments, as well as an even smaller fragment that represents the antigen-binding properties of an antibody, namely, the Fv fragment.

Already in the early 1970s, Givol and coworkers derived the Fv (*fragment variable*) fragment of an antibody by limited proteolytic digestion of an Fab and could show that it comprised the V_H and V_L domains and could bind antigen in a manner comparable to the complete antibody, similar to an Fab [39].

Fig. 3.9 Schematic representation of a single-chain Fv with the arrangement V_H-linker-V_L. The linker which connects the C-terminus of the V_H domain with the N-terminus of the V_L domain is shown as a green line

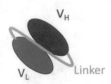

It took until the late 1980s that the usefulness of Fv fragments, then produced by recombinant means, was fully realized and exploited. In 1988, Skerra and Plückthun [40] described the co-expression of VH and VL domains in the periplasmic space of *E. coli* and observed the formation of an antigen-binding Fv fragment, which was however of limited stability due to the absence of an inter-domain disulphide bond, as it would be present in an Fab.

This issue was elegantly solved 1988 by two groups independently of each other, who had the idea of connecting the two variable domains with a peptide linker to form a single-chain polypeptide which could be expressed recombinantly. The linker greatly increased the stability of the V_H–V_L heterodimer, and its length was determined by the distance between the C-terminus of the first domain to the N-terminus of the second domain, in order to allow the two domains to heterodimerize in their native conformations.

Jim Huston and colleagues [41] described a protein which they called single-chain Fv (scFv), consisting of the V_H domain followed by a linker with a length of 15 amino acids, followed by the V_L domain. The amino acid sequence of the linker was chosen by these authors on the basis flexibility and solubility and consisted of three consecutive glycine-serine motifs: $(GGGGS)_3$. This scFv format with its glycine-serine linker has since proven to be very robust and widely applicable and is still today, more than 30 years after its design, the most widely used scFv format. At about the same time, another research group, led by Robert Bird, constructed a similar single-chain polypeptide, with the design V_L-linker-V_H, employing a slightly longer and more complex linker (Fig. 3.9) [42].

scFv molecules have since gained great importance in diverse fields where the specific recognition of antigens by antibodies is exploited. Because of their simple structure consisting of a single polypeptide, scFvs can usually be easily expressed in *E. coli* and other expression systems, and they are thus the antibody format of choice for the phage display of antibodies, similar to Fab fragments. Please refer to the chapter on surface display technologies in this book for further details.

scFvs and more elaborate variations thereof have also entered the arena of therapeutic applications. Especially when it comes to bispecific antigen-binding fragments, this most simple basic format of an antibody has proven to be an extremely valuable molecule, as manifested in the chapter on bispecific antibodies in this book.

Questions

1. What is the structure of an antibody?
2. What are CDRs?
3. What is the meaning of "paratope" and "epitope"?
4. What is a myeloma cell line?
5. What is the HAT selection system?
6. What is "limiting dilution"?
7. What is a chimeric antibody?
8. What is a humanized antibody?
9. How can fully human monoclonal antibodies be generated?
10. What are Fab and F(ab')$_2$ fragments?
11. What are scFv fragments?

Answers to the Questions

1. An antibody of the IgG class is composed of four chains: two identical light and two identical heavy chains. The variable domains (V_L and V_H) determine the antigen specificity of the antibody. The constant domains are responsible for effector functions and the long in vivo half-life of the antibody.
2. CDRs, or complementarity-determining regions, are loop regions located at the tips of the variable domains of an antibody. The V_H and the V_L domains each contribute three CDRs to the paratope of an antibody and thus determine its specificity.
3. The paratope is the region of an antibody (located in the CDRs) that actually contacts the antigen. The region of the antigen which is directly contacted by the paratope is the epitope.
4. A myeloma cell line is derived from a myeloma tumour and has been adopted to growth in vitro.
5. The HAT selection system is based on a mutation in cell lines (such as myeloma cell lines) that inactivates the gene coding for the enzyme HGPRT (hypoxanthine-guanine phosphoribosyltransferase). HGPRT is part of the salvage pathway for the production of purines and pyrimidines. HGPRT$^-$ cells therefore depend on de novo synthesis of purines. When this de novo pathway is blocked by the drug aminopterin, HGPRT$^-$ cells will die. When an HGPRT$^-$ cell line is fused to an HGPRT$^+$ cell, an HGPRT$^+$ hybrid cell is generated that is able to grow in HAT (*h*ypoxanthine-*a*minopterin-*t*hymidine) medium.
6. After fusion, newly generated hybridoma cells are plated in a very dilute way such that on average only one cell will start growing in each well of the culture plates, thus producing only one particular, monoclonal, antibody.

7. A chimeric antibody is an antibody that is composed of variable domains from a mouse antibody and constant domains of a human antibody. The mouse variable domains determine the specificity of the antibody, while the constant domains are responsible for the overall structure of the antibody and its biological functions, such as the elicitation of effector functions and long in vivo half-life.

8. A humanized antibody is an antibody that is almost entirely composed of human sequences. Only the CDRs stem from a mouse antibody.

9. There are two main ways of generating fully human monoclonal antibodies: one is by using transgenic mice that carry human immunoglobulin-encoding genes in their genome, and the other is to use phage display libraries (or other display systems) of human antibody fragments.

10. An Fab fragment can be produced by digestion of an antibody molecule with the protease papain, but it can also be expressed recombinantly. It is composed of the variable and constant domains of the light chains (V_L and C_L) and of the variable and the first constant domains of the heavy chain (V_H and C_H1). Fab fragments have the same antigen-binding characteristics as the antibody from which they are derived, with the difference that they are monovalent. In contrast $F(ab')_2$ fragments represent two Fab molecules which are connected covalently through the disulphide bonds of the hinge region.

11. scFvs or single-chain variable fragments are the smallest antigen-binding fragment that can be derived from an antibody. They are expressed recombinantly, mostly in the conformation V_H-linker-V_L.

Take Home Message

The production of monoclonal antibodies was first described by Köhler and Milstein in 1975. They fused splenocytes from mice (that had previously been immunized with sheep erythrocytes) with myeloma cells. The resulting clones of hybridomas were found to produce monoclonal antibodies that specifically bound to sheep erythrocytes.

Myeloma cells that are used as fusion partners for the production of hybridomas need to carry a genetic marker that allows to kill unfused myeloma cells after fusion. Typically, this marker is $HGPRT^-$, which indicates the absence of a functional HGPRT gene. In HAT medium, such cells cannot survive, while $HGPRT^+$ hybridoma cells will grow continuously.

Antibodies are highly conserved among species in their primary, secondary and tertiary structure. This allows to use them in a modular way and to engineer chimeric antibodies. A typical chimeric antibody is composed of variable domains from a mouse antibody and human constant domains.

(continued)

After fusion, hybridoma cells are plated in a very diluted way, which is called "limiting dilution". This is to make sure that in each well of the culture plate, only one clone of hybridoma cells is growing, producing a monoclonal antibody.

Antibodies are secreted into the culture supernatant, which can be analysed by a vast array of analytical methods in order to test for productivity, specificity and functionality of the monoclonal antibodies.

Initially, only mouse antibodies were generated using the hybridoma technology. Because of the need for human antibodies for therapy, various strategies were developed to produce more human or even 100% human antibodies. These included the engineering of chimeric antibodies, humanized antibodies as well as fully human antibodies.

Fragments of antibodies have gained great attention and importance, especially the Fab and the $F(ab')_2$ fragment, as well as the smallest antigen-binding fragment of the an antibody, the so-called scFv.

Further Reading and Internet Resources

1. Abul K. Abbas AHL and SP. Cellular and Molecular Immunology, Ninth Edition. Cellular and Molecular Immunology. 2018, ISBN-10: 9780323479783 and ISBN-13: 978-0323479783.
2. IMGT®, the International ImMunoGeneTics information system® at www.imgt.org
3. Murphy K. Janeway's Immunobiology. Janeway's Immunobiology. 2016. ISBN-10: 0815345054 and ISBN-13: 978-0815345053.
4. Online Macromolecular Museum (OMM): http://earth.callutheran.edu/Academic_Programs/Departments/BioDev/omm/gallery.htm
5. Rees AR. The antibody molecule: from antitoxins to therapeutic antibodies. 2015. ISBN-10: 0199646570 and ISBN-13: 978–0199646579.
6. The Antibody Society: https://www.antibodysociety.org/about/

References

1. Köhler G, Milstein C. Continuous cultures of fused cells secreting antibody of predefined specificity. Nature. 1975;256(5517):495–7.
2. Rees AR. The antibody molecule: from antitoxins to therapeutic antibodies. Oxford: Oxford University Press; 2015. p. 364.
3. Murphy K. Janeway's immunobiology. 2016.
4. Abul K, Abbas AHL. Cellular and molecular immunology. 9th ed; 2018. p. 564.
5. Bork P, Holm L, Sander C. The immunoglobulin fold: structural classification, sequence patterns and common core. J Mol Biol. 1994;242:309–20.
6. Padlan EA. Anatomy of the antibody molecule. Mol Immunol. 1994;31:169–217.

7. He XM, Rüker F, Casale E, Carter DC. Structure of a human monoclonal antibody Fab fragment against gp41 of human immunodeficiency virus type 1. Proc Natl Acad Sci U S A. 1992;89 (15):7154–8.

8. Ehrenmann F, Kaas Q, Lefranc MP. IMGT/3dstructure-DB and IMGT/domaingapalign: a database and a tool for immunoglobulins or antibodies, T cell receptors, MHC, IgSF and MHcSF. Nucleic Acids Res. 2009;38(1)

9. Ehrenmann F, Giudicelli V, Duroux P, Lefranc MP. IMGT/collier de perles: IMGT standardized representation of domains (IG, TR, and IgSF variable and constant domains, MH and MhSF groove domains). Cold Spring Harb Protoc. 2011;6(6):726–36.

10. Raju TS, Scallon BJ. Glycosylation in the Fc domain of IgG increases resistance to proteolytic cleavage by papain. Biochem Biophys Res Commun. 2006;341(3):797–803.

11. Radaev S, Motyka S, Fridman WH, Sautes-Fridman C, Sun PD. The structure of a human type III Fcγ receptor in complex with Fc. J Biol Chem. 2001;276(19):16469–77.

12. Burmeister WP, Huber AH, Bjorkman PJ. Crystal structure of the complex of rat neonatal Fc receptor with Fc. Nature. 1994;372(6504):379–83.

13. Ugurlar D, Howes SC, De Kreuk BJ, Koning RI, De Jong RN, Beurskens FJ, et al. Structures of C1-IgG1 provide insights into how danger pattern recognition activates complement. Science. 2018;359(6377):794–7.

14. Milstein C. The hybridoma revolution: an offshoot of basic research. BioEssays. 1999;21:966–73.

15. Asensio MA, Lim YW, Wayham N, Stadtmiller K, Edgar RC, Leong J, et al. Antibody repertoire analysis of mouse immunization protocols using microfluidics and molecular genomics. MAbs. 2019;11(5):870–83.

16. Littlefield JW. Selection of hybrids from matings of fibroblasts in vitro and their presumed recombinants. Science. 1964;145(3633):709.

17. Horibata K, Harris AW. Mouse myelomas and lymphomas in culture. Exp Cell Res. 1970;60 (1):61–77.

18. Harris H, Watkins JF. Hybrid cells derived from mouse and man: artificial heterokaryons of mammalian cells from different species. Nature. 1965;205(4972):640–6.

19. Galfrè G, Milstein C. Preparation of monoclonal antibodies: strategies and procedures. Methods Enzymol. 1981;73(C):3–46.

20. Pontecorvo G. Production of mammalian somatic cell hybrids by means of polyethylene glycol treatment. Somatic Cell Genet. 1975;1(4):397–400.

21. Zimmermann U, Vienken J. Electric field-induced cell-to-cell fusion. J Membr Biol. 1982;67:165–82.

22. Greenfield EA. Electro cell fusion for hybridoma production. Cold Spring Harb Protoc. 2019;2019(10):684–8.

23. Greenfield EA. Single-cell cloning of hybridoma cells by limiting dilution. Cold Spring Harb Protoc. 2019;2019(11):726–8.

24. Akagi S, Nakajima C, Tanaka Y, Kurihara Y. Flow cytometry-based method for rapid and high-throughput screening of hybridoma cells secreting monoclonal antibody. J Biosci Bioeng. 2018;125(4):464–9.

25. Lu M, Chan BM, Schow PW, Chang WS, King CT. High-throughput screening of hybridoma supernatants using multiplexed fluorescent cell barcoding on live cells. J Immunol Methods. 2017;451:20–7.

26. Lad L, Clancy S, Kovalenko M, Liu C, Hui T, Smith V, et al. High-throughput kinetic screening of hybridomas to identify high-affinity antibodies using bio-layer interferometry. J Biomol Screen. 2015;20(4):498–507.

27. Long TTA, Kalantarov G, Chudner A, Trakht I. A dual-chamber system for screening cytotoxic effects of hybridoma-produced antibodies. Monoclon Antib Immunodiagn Immunother. 2013;32 (4):246–54.
28. Debs BE, Utharala R, Balyasnikova IV, Griffiths AD, Merten CA. Functional single-cell hybridoma screening using droplet-based microfluidics. Proc Natl Acad Sci U S A. 2012;109 (29):11570–5.
29. Chen P, Mesci A, Carlyle JR. CELLISA: reporter cell-based immunization and screening of hybridomas specific for cell surface antigens. Methods Mol Biol. 2011;748:209–25.
30. Cervino C, Weber E, Knopp D, Niessner R. Comparison of hybridoma screening methods for the efficient detection of high-affinity hapten-specific monoclonal antibodies. J Immunol Methods. 2008;329(1–2):184–93.
31. Listek M, Hönow A, Gossen M, Hanack K. A novel selection strategy for antibody producing hybridoma cells based on a new transgenic fusion cell line. Sci Rep. 2020;10(1):1–12.
32. Bradbury ARM, Trinklein ND, Thie H, Wilkinson IC, Tandon AK, Anderson S, et al. When monoclonal antibodies are not monospecific: hybridomas frequently express additional functional variable regions. MAbs. 2018;10(4):539–46.
33. Binley JM, Wrin T, Korber B, Zwick MB, Wang M, Chappey C, et al. Comprehensive cross-clade neutralization analysis of a panel of anti-human immunodeficiency virus type 1 monoclonal antibodies. J Virol. 2004;78(23):13232–52.
34. Morrison SL, Johnson MJ, Herzenberg LA, Oi VT. Chimeric human antibody molecules: mouse antigen-binding domains with human constant region domains. Proc Natl Acad Sci U S A. 1984;81(21):6851–5.
35. Boulianne GL, Hozumi N, Shulman MJ. Production of functional chimaeric mouse/human antibody. Nature. 1984;312(5995):643–6.
36. Jones PT, Dear PH, Foote J, Neuberger MS, Winter G. Replacing the complementarity-determining regions in a human antibody with those from a mouse. Nature. 1986;321(6069):522–5.
37. Bruggemann M, Caskey HM, Teale C, Waldmann H, Williams GT, Surani MA, et al. A repertoire of monoclonal antibodies with human heavy chains from transgenic mice. Proc Natl Acad Sci U S A. 1989;86(17):6709–13.
38. McCafferty J, Griffiths AD, Winter G, Chiswell DJ. Phage antibodies: filamentous phage displaying antibody variable domains. Nature. 1990;348(6301):552–4.
39. Inbar D, Hochman J, Givol D. Localization of antibody-combining sites within the variable portions of heavy and light chains. Proc Natl Acad Sci U S A. 1972;69(9):2659–62.
40. Skerra A, Plückthun A. Assembly of a functional immunoglobulin Fv fragment in *Escherichia coli*. Science. 1988;240(4855):1038–41.
41. Huston JS, Levinson D, Mudgett-Hunter M, Tai MS, Novotny J, Margolies MN, et al. Protein engineering of antibody binding sites: recovery of specific activity in an anti-digoxin single-chain Fv analogue produced in *Escherichia coli*. Proc Natl Acad Sci U S A. 1988;85(16):5879–83.
42. Bird RE, Hardman KD, Jacobson JW, Johnson S, Kaufman BM, Lee SM, et al. Single-chain antigen-binding proteins. Science. 1988;242(4877):423–6.

Antibody Display Systems

4

Janina Klemm, Lukas Pekar, Simon Krah, and Stefan Zielonka

Contents

Keywords

Antibody hit discovery · B-cell cloning · cDNA display · *E. coli* display · Genotype
phenotype coupling · Mammalian display · Phage display · Ribosome display · Yeast
surface display

Authors "Janina Klemm" and "Lukas Pekar" contributed equally.

J. Klemm · L. Pekar · S. Krah · S. Zielonka (✉)
Protein Engineering and Antibody Technologies, Merck Healthcare KGaA, Darmstadt, Germany
e-mail: stefan.zielonka@merckgroup.com

F. Rüker, G. Wozniak-Knopp (eds.), *Introduction to Antibody Engineering*,
Learning Materials in Biosciences,
https://doi.org/10.1007/978-3-030-54630-4_4

What You Will Learn in This Chapter
Antibodies are promising tools for biomedical applications. With respect to function-ality, scientists nowadays have a plethora of options to tailor-make therapeutic antibodies with prescribed properties. Besides hybridoma technology, which is still state of the art for antibody hit discovery but that also has intrinsic limitations regarding throughput, several other promising methodologies emerged. All of these have in common that they couple the phenotype, i.e., the binding properties of an antibody to its genotype, the genetic information that encodes for that particular moiety. In this chapter, the reader will learn about the most commonly applied technologies for antibody hit discovery.

4.1   Introduction

Antibodies are multifunctional components of the immune system. They facilitate numer-ous cellular and humoral reactions to a plethora of different antigens. Protein-based therapeutics, such as monoclonal antibodies (mAbs), have proven to be successful treat-ment options for patients in various indications, e.g., in cancer. Since 1986 over 80 mAb therapeutics have been granted marketing approval either in the USA or the EU [1]. In this respect, one of the most successful moieties referred to as adalimumab (Humira) was approved for arthritis therapy. Adalimumab is a human IgG1 with specificity for human tumor necrosis factor α (TNFα) and was developed by Abbott and Cambridge Antibody Technology [2]. While human antibody repertoires yield a large source for human mAbs, the search for appropriate antibodies with prescribed properties during preclinical phase remains laborious and challenging.

   In general, two approaches are predominant for antibody discovery: animal immuniza-tion and surface display methodologies. Since display systems can overcome the limitations of animal immunizations due to a possible immune tolerance, different systems, based on the in vitro selection from naïve, immunized or synthetic repertoires [3], were developed over the last 30 years. Those in vitro selection platforms mimic the in vivo process in B-cell lymphocytes: the generation of genotypic diversity, the coupling of genotype to phenotype, application of selective pressure, and the amplification of desired candidates. The first description of an in vitro display technology for the selection of antibodies, namely, phage display [4], has been published in the 1990s. Since then, many antibodies have been selected or optimized with the use of display methods. To this date, several display systems were developed, where the polypeptide library is presented at the outside of the cell wall or membrane, allowing for interaction of the library candidates with the antigen. With respect to prokaryotic display systems as well as display on lower eukaryotes, typically antibody derivatives are displayed as fragments such as single chain Fv (scFv) or fragment antigen binding (Fab) since full length display of immuno-globulin G (IgG) is rather difficult.

In vitro display systems can be divided into cellular and acellular approaches. Cellular approaches, for example, yeast display, require the polypeptide libraries to be cloned and expressed from cells, whereas in acellular approaches, for example, cDNA display, expression and display is accomplished without the need for transformation or transfection. Cell-based systems rely on the multi-copy presentation of recombinant peptide libraries linked to their coding sequence on the cell surface in functional form. Repetitions of the screening and selection process with libraries of mutants enable the isolation of variants that are improved in terms of specificity and affinity. Affinities of identified antibodies in the picomolar range [5] are frequently obtained using display technologies. In contrast to the commonly applied hybridoma-technique for the generation of mAbs [6], these in vitro methods, based on microbial systems, have the possibility to automate the selection and screening process, affording the benefit of identifying a multitude of different antibody leads against a given therapeutic target. High-throughput screening allows for the selection of clones presenting an antibody variant with the desired affinity, specificity, and stability. Besides selecting antibodies against protein epitopes, display technologies also allow screening for binders recognizing chemical modifications, small molecules, toxins, and pathogens [7]. In this chapter the reader will learn about the function, applications, advantages, and limitations of the following display systems: phage, yeast, E. coli, mammalian, cDNA, and ribosomal display, as well as B-cell cloning.

4.2   Phage Display

Antibody phage display was the first developed and still is the most widely applied in vitro selection technology and has enabled the generation of the first fully human therapeutic antibody [2]. This platform technology has proven to be a robust and versatile methodology for the discovery of human antibodies and allows scientist to access and the rich and diverse human antibody repertoire in vitro. Phage display is based on the work of George P. Smith on filamentous E. coli phage M13, where Smith fused peptides to phage envelope proteins. The fusion of gene fragments of oligopeptides to the minor coat protein III gene led to genotype-phenotype coupling and consequently enabled affinity purification of the peptide and its corresponding gene (Fig. 4.1) [8]. This groundbreaking work, together with advanced applications of phage display for antibody engineering as developed by Sir Gregory P. Winter, was awarded with the Nobel Prize in Chemistry in 2018 [9]. Nowadays different phage display systems are available, most of them based on antibody::pIII fusion proteins. In this respect, integration of the antibody::pIII encoding gene into the phage genome [10] was not as successful as the commonly used phagemid system [11]. The latter has a useful and very flexible mode of action: the separation of the antibody::pIII fusion from phage protein expression, as well as phage replication. Consequently, antibody expression is uncoupled from phage propagation by keeping the genes encoding the

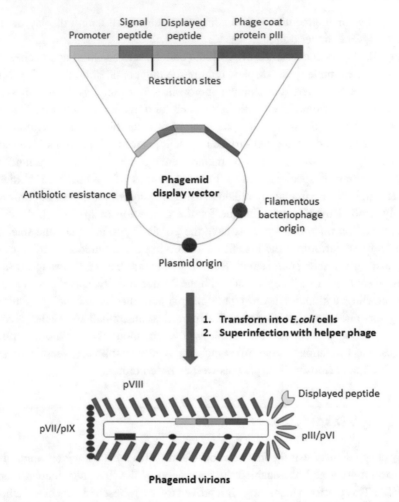

Fig. 4.1 Phage display using phagemid vectors. Typically, phagemids contain a plasmid origin of replication, a filamentous bacteriophage origin, as well as an antibiotic resistance gene for plasmid selection in *E. coli*. The library candidate is expressed as fusion protein with a phage coat protein such as pIII. A promoter enables construct expression and a signal peptide is added N-terminally. The phagemid vector is incorporated into infective phage by superinfection of phagemid bearing *E. coli* cells with helper phage. Modified from [14]

antibody::pIII fusion proteins on a separate plasmid (called "phagemid"). This phagemid contains a phage morphogenetic signal for packing the vector into the assembled phage particles enabling coupling of the genotype (the antibody::pIII fusion) to its phenotype (antibody expression on the phage particle). However, for the production of antibody phage particles helper phage is needed [4, 11].

Inspired by the work on peptide phage display of G.P. Smith antibody phage display was developed independently in Heidelberg by Breitling and Dübel [11], in Cambridge by

McCafferty et al. [10], and in California by Barbas et al. [4] Due to limitations of the *E. coli* folding machinery, only antibody fragments like scFv, Fab, the variable heavy domain of camels (VHH), or humans (dAb) are commonly used formats for antibody phage display [12]. The production of full-length IgG in *E. coli* remains challenging [13].

Essential sources for antibody discovery by phage display are antibody libraries, which can be divided into immune and universal libraries. Immune libraries are built from B-cells of immunized donors via vaccination or from patients suffering from disease. Furthermore, immune libraries can be built from samples of immunized animals, especially transgenic animals harboring the human antibody variable gene repertoire enable the selection of human antibodies without the need for a human donor. In medical research, immune libraries are used to generate an antibody against one target antigen, e.g., of an infectious pathogen [15]. Those libraries have the advantage that their V-genes contain hypermutations and are usually affinity maturated. The term universal or "single-pot" libraries include naïve, semi-synthetic, and synthetic libraries. In theory, they are designed to isolate antibody fragments binding to virtually every possible antigen [16]. Naïve libraries are generated from rearranged V genes of B-cells of non-immunized donors. Semi-synthetic libraries are constructed from unrearranged V genes of germline cells or from one antibody framework in which one or several complementary determining regions (CDR), most frequently CDR-H3, are randomized [17]. Fully synthetic libraries are typically made of human frameworks with randomized CDR cassettes [18]

The first step of phage display is antibody library construction, meaning cloning of the desired antibody repertoire into the phagemid in *E. coli*. Subsequently, this *E. coli* library is infected with helper phage, providing all the necessary viral components that enable single-stranded DNA replication and packaging of the phagemid DNA into phage particles. Integration of antibody genes in phage particles leads to bacteriophages displaying the encoded antibody molecule on their outside. Consequently, the phage particle can be considered as kind of a nanoparticle displaying a distinct antibody fragment on its outside while carrying its corresponding genetic information on the inside (Fig. 4.2). The next important step to obtain a specific antibody with phage display is the "panning" process, referring to the gold washer's tool. Therefore, a mix of billions of different bacteriophages, all displaying another antibody on their surface (in theory), is tested against one immobilized antigen molecule. The antigen can be immobilized on nitrocellulose [19], magnetic beads [20], column matrices [11], plastic surfaces like polystyrene tubes [21] or 96 well microtiter plates [4]. During this incubation step, physical (e.g., temperature), chemical (e.g., pH), and biological (e.g., competitor) parameters can be controlled. These settings enable the selection of antibodies, which can bind the antigen under these defined conditions. Due to washing steps with flexible applied stringencies, only those bacteriophages, displaying an antibody with required specificities and affinities, still bind the antigen while the others are removed by washing. After this step of enrichment, the bound and therewith desired bacteriophages can be eluted (e.g., by trypsination or pH shift) and are used to infect new *E. coli* cells enabling propagation of phagemids harboring enriched library candidates. Afterwards, those phagemid-containing *E. coli* cells are again

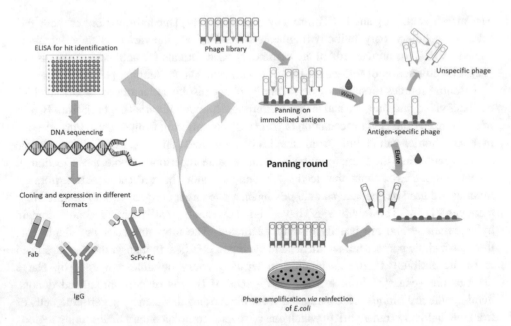

Fig. 4.2 Schematic representation of antibody phage display selection, screening, and reformatting/production. Phage library is selected for target binding, unspecific phages are removed by applying washing steps, and antigen-specific phage are eluted and amplified via reinfection of *E. coli* cells. Usually, after 2–6 panning rounds monoclonal binders are identified by ELISA, DNA-sequencing is performed and after subcloning different antibody formats can be produced. Modified from [23]

infected with helper phage to produce new antibody phage particles (enriched for variants specific to the antigen), which can be used for further panning rounds. Normally two or three panning rounds, rarely up to six, are necessary to select antigen-binding antibody fragments. For the identification of individual binders, an antigen-binding ELISA analysis is commonly used. At this point it is possible to readout the genetic information through sequencing. Further biochemically characterizations can be performed. The panning process can also be performed in a high-throughput manner with automated screenings [22].

Initial identification of antibody candidates is often followed by an optimization step to improve affinity, stability, production yield, or other properties. Affinity maturation is an in vitro technique to evolve high-affinity ligands from polypeptide libraries. The term originates from immunology, where during a natural antibody response to an antigen, the overall affinity of circulating antibodies to the specific antigen improves. In the course of natural affinity maturation, the antibody genes of responding B-cell lymphocytes get hypermutated. Translated to antibody phage display, this means that due to gradually increasing selection stringency (e.g., decreasing antigen concentrations) only high-affinity antibodies "survive." This principle was first demonstrated by a graduate student named Jinan Yu from the lab of George P. Smith who used an affinity maturation technique to overcome the leanness of random peptides in her starting library. She introduced random mutations into the DNA inserts during the amplification step of each round of selection. At the same time, she gradually increased stringency, similar to the decline of antigen concentrations during natural affinity maturation [24]. The successful outcome of an affinity maturation campaign is a so-called "dark horse" peptide: a peptide with higher affinity to the antigen compared to all the other clones, which have evolved from the initial peptide ancestor [25]. Another commonly used strategy is the so called light-chain-shuffling, which combines the heavy chains of a panned subset of antibodies with the whole light-chain repertoire of a naïve or synthetic library [26]. This newly generated library is then re-screened under modified conditions to obtain antibodies with improved properties.

Today the typical task of antibody research facilities is to generate antibodies to many different antigens in a high-throughput manner requiring minimal effort. To compete here, automation and miniaturization are essential. With the use of deep-screening, nowadays, one can typically analyze more than 20,000 individual clones, often resulting in more than 100 different binders. A huge improvement of efficiency was the development of hyperphage, a helper phage significantly enhancing antibody display efficiency [27]. This very elegant approach developed by Stefan Dübel and co-workers has been nicely described elsewhere [27].

At present, multiple phage display-derived mAbs are in clinical or preclinical stages of development, with several antibodies that have been granted marketing access either in the USA or the EU. The most successful phage display-derived mAbs are described in the following paragraph.

Adalimumab (marketed as Humira) was the first phage display-derived monoclonal antibody approved for therapy. It is used as a TNF-inhibiting anti-inflammatory drug and is approved by the FDA for treatment of rheumatoid arthritis, juvenile idiopathic arthritis, psoriatic arthritis, ankylosing spondylitis, Crohn's disease, ulcerative colitis, and plaque psoriasis [28–30]. In 2011 the phage display-derived antibody Belimumab, which inhibits B-cell-activating factor, was approved by the FDA. It is used for treatment of autoimmune diseases, mainly systemic lupus erythematosus [31]. Ranibizumab, FDA approved in 2006, is an antibody against VEGF-A and neutralizes its activity. The antibody was affinity matured by phage display and is used for treatment of wet age-related macular edema [32]. Another human mAb, generated with the use of phage display, was FDA approved in the year 2012: Raxibacumab, an antibody binding to the protective antigen (PA83) of *Bacillus anthracis*. Nowadays Raxibacumab is used for treatment of inhalational anthrax raised from *B. anthracis* in combination with antibacterial drugs [33]. In addition to the four herein described phage display-derived mAbs, two more (Necitumumab and Ramucirumab) entities were approved by the US authorities as of May 2016, and a plethora of antibodies is currently undergoing clinical evaluation, clearly demonstrating the versatility of phage display for antibody selection in terms of disease therapy [23].

4.3 Yeast Surface Display

The in vitro selection of human antibody fragment libraries displayed on yeast surfaces represents a well-established technology capable of generating target-specific mAbs. The most commonly used species in this strategy is *Saccharomyces cerevisiae*. This eukaryotic organism normally uses its cell wall surface agglutinins as adhesion proteins for instance during mating events [34]. Yeast surface display (YSD) is based on the genetic fusion of an antibody fragment to such cell wall proteins, e.g., to Aga2p, as developed by Boder and Wittrup in 1997 [35, 36]. In this well-established system employing mating-related proteins Aga1p and Aga2p, the Aga2p subunit, to which the protein of interest is fused, is covalently linked to the yeast cell wall via two disulfide bonds to Aga1p [37]. The subunit Aga1p contains the glycosylphosphatidylinositol (GPI) anchor and consequently is responsible for cell wall anchoring, whereas Aga2p acts as interaction module with Aga1p [38] (Fig. 4.3).

The first step to construct yeast antibody display libraries relies in the cloning of the immunoglobulin gene pool into the yeast display plasmid [39–42] as fusion to the gene encoding for Aga2p. Using this platform technology libraries in the range of 10^7 to 10^9, unique variants can readily be obtained [37, 43]. In the most established system developed by Boder and Wittrup, the expression of both Aga1p (integrated into the yeast genome) and the mAb-Aga2p fusion construct (expressed from the display plasmid) is under the control of the galactose-inducible GAL1 promoter. The presence of galactose as carbon source (instead of glucose) enables the yeast cells to display the antibody fragment fused to Aga2p subunit on their surface [43, 44]. ScFv is the most commonly used antibody format in yeast display; however Fabs [45], whole IgG1s, and camelid antibodies [46] can be displayed as well [44].

It's been known quite well that YSD enables the obtain antigen-specific antibodies [47]. Isolation of antibody clones from yeast libraries can be performed using magnetic-assisted cell sorting (MACS) and fluorescence-activated cell sorting (FACS) [43]. MACS steps are typically used to remove nonbinding mAb fragments to reduce the number of cells for subsequent FACS selection, since throughput is the limiting factor for the latter. These MACS separations are comparable with the phage panning process and can reduce the nonreactive background by approximately 100 times [44]. After MACS-based depletion, the antigen-enriched population of yeast cells is screened by FACS in order to isolate antigen-specific library candidates (Fig. 4.4). The antigen concentration during FACS screening is often kept at tenfold excess above the desired dissociation constant (K_D) of individual clones to allow the majority of the displayed mAb fragments to bind the target at equilibrium. Implementation of epitope tags, for instance, a hemagglutinin (HA) tag or a c-myc epitope at the termini of the respective library candidate allow for immunofluorescent detection of full-length proteins that are displayed on the surface of the yeast cell. This is advantageous since it allows for getting rid of frame-shifted library candidates or candidates containing stop codons. Both malfunctioned versions of library candidates typically have a tendency for aggregation as well as unspecific binding which would result

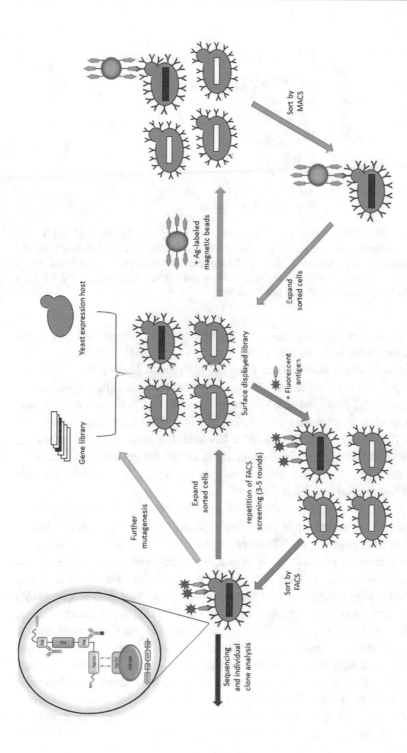

Fig. 4.3 Schematic representation of yeast display and yeast display library selection. In yeast display, the protein of interest (POI) is expressed as fusion protein with Aga2p, which upon secretion forms a covalent linkage with Aga1p that is situated on the cell surface. Tags can be included for the detection of full-length display and the POI can be fused N- or C-terminally to Aga2p. GOI gene of interest. After library construction, the yeast display library can be either selected by MACS using antigen-coated magnetic beads or by FACS using fluorescently-labeled antigen. After a particular selection round, enriched cells are expanded and subsequently utilized for further rounds of selection. Typically, after 3–5 FACS-based selections, individual plasmids are analyzed by sequencing. Moreover, antigen-enriched cells can be further modified by mutagenesis. Modified from [44, 48]

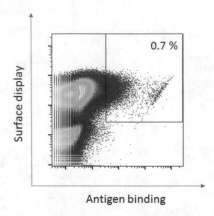

Fig. 4.4 Exemplary blot showing first round of selection of a single domain antibody library constructed from immunized camelids against a given target. The Y-axis indicates full-length surface display by antibody-mediated fluorescence labeling of a tag that has been included in plasmid design and the X-axis represents antigen binding (using a fluorescently labeled antigen). By applying a sorting gate, the double-positive population can be sorted

in false-positive events. The screened yeast cells are labeled for the simultaneous detection of antibody fragment expression, i.e., display on the cell surface (using, e.g., the c-myc tag) and antigen binding (using a fluorescence label on the target). During the first FACS selection round, typically the complete double-positive population is collected in order to enrich the complete antigen-specific diversity. In subsequent FACS rounds, the selection gate can then be adjusted to increase sorting stringency [43]. Typically, after three to five rounds of these FACS selections, the plasmid DNA of the enriched antigen binding population is collected and sent out for sequencing to obtain the genetic information for the respective paratopes. These can further be subcloned into appropriate plasmids enabling antibody expression in mammalian cells.

When comparing yeast and phage antibody library technologies, both advantages and disadvantages exist. A big advantage of yeast display is the precise control over selection parameters during FACS screening [47]. This ability to prescribe binding criteria during the FACS process represents an advantage over phage display platforms, where variant discrimination is dependent on washing steps rather than real-time kinetic observations [44]. Moreover, due to the utilization of a eukaryotic expression host with proper quality control machineries, proteins are believed to be more properly folded compared to phage display, where a prokaryotic expression host is employed [44]. However, compared to phage display, there are also some obvious drawbacks. Total library sizes using yeast libraries tend to be significantly lower (several orders of magnitude) than typical phage display library sizes. The reason therefore relies in lower transformation efficiencies observed in yeast cells. While yeast libraries with up to 10^{10} total clones can be constructed with a reasonable effort, phage libraries with more than 10^{11} variants can readily be obtained [43, 49]. Another potential disadvantage of YSD is the high number of antibody copies displayed per cell (ranges between 10^4 and 10^5 copies per cell), leading to a selection based on antibody avidity, rather than affinity (at least in theory. However, this is not true by all means, as shown in the next paragraph). Modern selection strategies make use of both yeast and phage display in combinatorial approaches. The two-system strategy takes advantage of the strengths of both technologies: the large library size and affinity-based selections of phage display and the well-controlled isolation of variants from yeast display [50].

A conventional application of YSD is affinity maturation of antibody fragments. Wittrup and co-workers were able to demonstrate the effectiveness of YSD in protein affinity maturation with their studies on an anti-fluorescein scFv in 1997 [36]. Flow cytometry allows for the detection as well as quantification of the fluorescence signal from each yeast cell [36]. This feature allows for precise affinity measurements and rapid enrichment of high-affinity populations within mutant libraries [36, 51]. These mutant libraries are often screened with a limiting concentration of soluble antigen to select mutants having higher affinities. In this way Boder et al. developed an antibody with affinity among the highest reported so far (K_D = 48 fM) [51]. Chao et al. gave a detailed methodical protocol for affinity engineering by YSD [43]. They described the use of random mutagenesis through error-prone PCR, enabling the control of mutation frequencies by varying the number of PCR cycles [43]. Other methodologies of generating diversity, such as DNA shuffling [52], are also compatible with YSD. Besides affinity maturation of antibodies, the feasibility of this approach has been extended to T-cell receptors [53, 54]. Moreover, YSD can also be used for epitope mapping of antibodies [55].

A more recent application of YSD are cell-based selections. Shusta et al. showed that yeast cells displaying high-affinity single-chain T-cell receptors (scTCRs) can build a cell-cell complex with an antigen presenting cell (APC) [56]. These results give clear evidence that the YSD can be used to screen yeast polypeptide libraries against cell surface ligands. Indeed, several reports show that such cell-based selection can be effectively performed. In this respect Wang et al. identified 34 unique antibodies binding and in some cases internalizing into rat brain endothelial cells [57]. Cell-based selections enable the identification of antibodies against complex membrane proteins, avoiding the requirement for sophisticated and difficult membrane protein expression and purification.

As of January 2020, one yeast display-derived mAb was approved by the Chinese authorities. This antibody, referred to as Sintilimab, is a fully human therapeutic entity blocking the interaction of PD-1 to PD-L1 and PD-L2 [58].

4.4   Mammalian Display

Phage display and yeast display enable the isolation of antigen-specific antibodies with desired functionalities. However, essential intrinsic properties that qualify an antibody to become a potential therapeutic such as high-level expression in mammalian cells, aggregation behavior, or accurate glycosylation (that might impact immunogenicity) can only insufficiently be controlled using microbial display technologies. To address these characteristics more adequately, mammalian-based selection systems have been developed. In this section, we aim at giving the reader an overview of some of the key principles of this selection strategy.

Mammalian cell display systems can be broadly categorized into methods employing transient protein candidate expression and those taking advantage of stable cell lines. Transient expression is a very effective way for short-term expression of proteins. Yet, it results in loss of overall DNA within the course of a few days and consequently to inefficient mid-term and long-term expression. Typically, such transient expression systems are utilized for a single round of selection followed by plasmid rescue. In contrast to this, stable integration would result in constant mid-term (and long-term) expression of library candidates. Nevertheless, stable integration is a relatively rare event and such systems are commonly used the construction of rather small libraries. Selection from large libraries using stably integrated gene expression used to be extremely inefficient [59]. With the event of modern technologies of genetic engineering and gene editing, however, huge progress has been in this respect during the last years.

Higuchi and colleagues were one of the first groups developing methods for mammalian antibody display (Fig. 4.5). They constructed a novel screening system, which allows to obtain variable region (V) genes of antigen-specific antibodies using mammalian cells as an expression host. To this end, they displayed an antibody library on the surface of COS cells by utilization of a plasmid allowing for expression of membrane-bound antibodies. In more detail, the first step was library generation, which was prepared from peripheral blood lymphocytes of a human donor vaccinated against hepatitis B surface antigen (HBsAg). mRNA extraction and cDNA synthesis were performed, followed by gene amplification of antibody variable regions via PCR and cloning into a display vector. Afterwards COS7 cells were transiently transfected. Membrane-anchoring was achieved by *in frame* genetic fusion of the antibody variant to the transmembrane domain of human thrombomodulin. The generated combinatorial library had a size of approximately 5×10^6 independent clones. In order to select for antigen-specific antibodies, the COS library was stained with biotin-labeled antigen and sorted by flow cytometry. Afterwards, plasmids were extracted from antigen-enriched cells followed by re-amplification in *E. coli* and repetition of the whole process for three rounds. Overall, this led to a 450-fold enrichment of V genes coding for HBsAg-specific antibodies. Finally, variable region genes were isolated and inserted into an expression allowing for soluble antibody expression, resulting ultimately into the isolation of four unique clones specific for HBsAg [60].

In recent years, several different technologies for mammalian display emerged. In 2009 Ho and Pastan showed that human embryonic kidney cells (HEK293T) can be used for surface display of scFvs and affinity maturation of those [61]. Moreover, Zhou et al. described the presentation of full-length antibodies on the surface of CHO cells by applying the Flp-In technology (Invitrogen). In this approach, recombinase-mediated chromosomal (stable) DNA integration was used to generate library diversity, and antibodies with significantly improved neutralizing activity and affinities could be isolated [62].

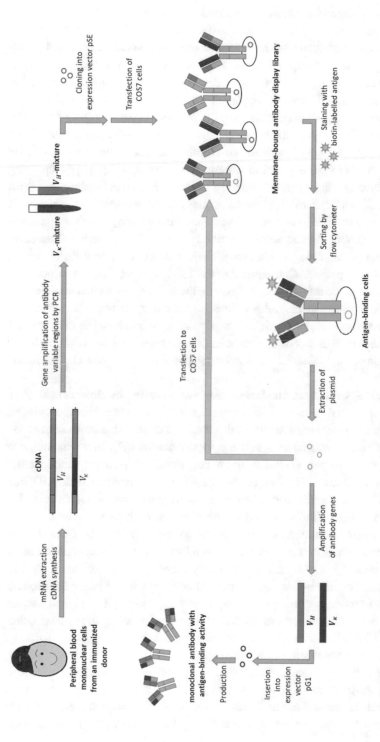

Fig. 4.5 Scheme of the preparation of a cell display library and the selection of antigen-specific antibody V genes by FACS. Messenger RNA was extracted from peripheral blood mononuclear cells of an immunized human donor. Antibody V genes were obtained from mRNA by RT-PCR and inserted into expression vector pSE. A cell display library was prepared by transfecting COS7 cells with the resultant plasmids. The library was stained with biotin-conjugated antigen followed by PE-conjugated streptavidin, and then positively stained cells were sorted by flow cytometry. Plasmids containing antigen-specific antibody V genes were recovered from the collected cells, and this enrichment process was repeated several times. Antigen-specific antibody V genes were isolated from the finally obtained plasmids and inserted into expression vector pG1 to produce monoclonal antibodies with antigen-binding activity. Modified from [60]

Retrocyte display, a technique where the natural expression apparatus of B-cells is exploited for retroviral B lymphocyte display, was developed in 2014 [63]. Here, libraries in the range of 10^9 clones were generated by step-wise transduction of immortalized pre-B-cells with retroviral particles carrying light chain and heavy chain diversities, respectively. Besides remarkable library sizes for mammalian display, this technology is supposed to have the advantage of enhanced antibody display efficacy. This is due to the utilization of naturally antibody producing B-lineage cells with the presence of all necessary co-factors that ensure proper protein folding, antibody heavy-light chain assembly, glycosylation, and disulfide bridge formation. Besides, Beerli and co-workers described a mammalian display system relying on a Sindbis virus expression system [64]. Another technology employing viral components for mammalian display is referred to as vaccinia virus display. This virus infects most mammalian cell lines, leading to expression of recombinant genes with mammalian quality control and post-translational modification [65]. In this approach, mammalian cells were co-transduced with an antibody heavy chain library and an additional light chain library. Interestingly, the heavy chain was produced as a fusion with a vaccinia virus membrane protein. Consequently, the full-length antibody was not only expressed on the surface of the infected cell but also on the surface of the virus particle shed from the cell. Similar to phage display, this allows for "panning" of huge diversities and enrichment of antigen-specific clones. Subsequently, infected mammalian cells can be sorted by FACS enabling precise control over selection. Besides viral transduction in the context of mammalian display, libraries can also be generated using non-viral transposition technology [66].

Another fascinating system of mammalian display was described by Bowers et al. [67]. Essentially, this method makes use of the enzyme named activation-induced cytidine deaminase (AID) that is responsible for the initiation of the natural affinity maturation process in B-cells, which is referred to as somatic hypermutation [67]. In a first step, low affinity antibodies were selected against a given target and assessed for functionality. Subsequently, genes for antigen-positive antibodies were co-transfected with AID into HEK293 cells allowing for simultaneous display and affinity maturation. Ultimately, this enabled the isolation of high-affinity antibodies. Another approach focusing on a natural diversification mechanism of antibody repertoires was presented in 2005 by Ohta and co-workers [68]. Here, the authors utilized a chicken B-cell line for antibody diversification. In chickens, antibody diversity is generated in a process called gene conversion, a kind of homologous recombination. By treating the chicken B-cell line with a histone deacetylase inhibitor, the gene conversion frequency increased massively, resulting in a diversity at the immunoglobulin locus sufficient for the selection of antigen-specific antibodies. Other approaches employ CRISPR/Cas9 or TALE nucleases as tools for library construction as well as for homology-directed mutagenesis [69, 70].

In addition to the selection and affinity maturation of antigen-specific antibodies, mammalian display systems can also be utilized for the selection of functional, e.g., agonistic or antagonistic antibodies. In 2012 Lerner et al. published a new lentiviral screening method that allowed for the selection for erythropoietin agonistic antibodies.

The authors first enriched antibodies specific to EpoR via phage display, followed by lentiviral library preparation. Afterwards, reporter cells were transduced with the pre-enriched lentiviral library and the phenotype, i.e., function of the antibodies was monitored [71]. The authors further demonstrated the applicability of their studies by appropriate reporter cell engineering and by optimizing the autocrine selection system itself [72].

4.5   *E. coli* Display

Besides the often cumbersome application of eukaryotic systems, the use of prokaryotic *Escherichia coli* cells for the cell surface display of heterologous peptides offers a facile and rapid alternative. The availability of numerous genetic tools, mutant strains, and the high transformation efficiencies of *E. coli* makes it an attractive host for screening large peptide libraries. Microbial cell-surface display has many potential applications, e.g., live vaccine development, peptide library screening, antibody production in animals, bioconversion using whole cell biocatalysts, biosensor development, and bioadsorption [73]. To this end, the protein of interest to be presented, referred to as passenger protein, can be fused by N-terminal, C-terminal, or sandwich fusion to the surface anchor, the so-called carrier protein. The utilized fusion arrangement, as well as the characteristics of the carrier protein, passenger protein, and the host cell strain, affects the efficiency of surface display regarding immobilization yield, stability, specific activity, and post-translational modification of the fusion protein [73].

Unfortunately, there is no microbial equivalent of the glycosylphosphatidylinositol (GPI) anchor found in eukaryotic cells for facile display of heterologous proteins in prokaryotic cells. Thus, the carrier protein must be adjusted for each passenger protein as well as for each individual application to enable adequate transport through the inner membrane. Moreover, the carrier protein should be insusceptible for protease degradation, anchor the passenger protein reliable to the membrane, and utilize the fusion or insertion of foreign peptide sequences and therefore their proper orientation and surface presentation [73, 74].

Based on those requirements and the complex cell envelope of gram negative bacteria such as *E. coli*, most carrier proteins developed to date are based on the outer membrane proteins (Fig. 4.6), supporting the accessibility of the passenger protein [74]. These proteins comprise a plethora of different molecules, e.g., bacterial fimbriae, S-layer proteins, ice nucleation protein (INP) [75–78], protein peptidoglycan-associated lipoprotein (PAL) [79], members of the immunoglobulin A (IgA) protease family [80], *Shigella* outer membrane autotransporter protein (IcsA/VirG), and many others more (Fig. 4.6).

Combining the appropriate fusion profile with a suitable carrier protein is therefore essential for microbial surface display.

Jose et al. demonstrated, for example, the surface presentation of *Vibrio cholerae* toxin B subunit (CtxB) as a heterologous passenger via N-terminal fusion to the β domain of IgA protease which mediates the translocation due autotransporter function [80]. Another

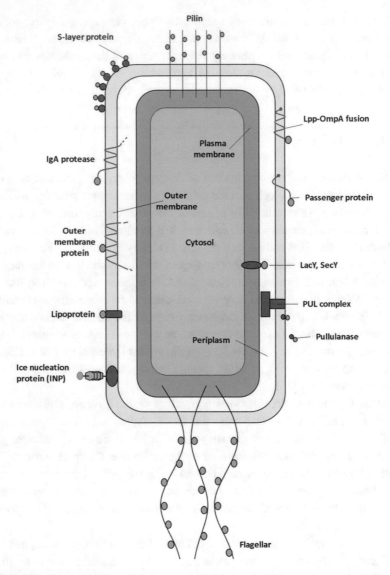

Fig. 4.6 Different surface display systems developed in Gram-negative bacteria; Modified from [73]

example for the N-terminal fusion arrangement was the display of alkaline phosphatase (PhoA) and maltose-binding protein (MalE) by the group of Sasakawa [81] via *Shigella* VirG protein, which is important for the deposition of filamentous actin.

In contrast to that, the Lpp-OmpA-hybrid carrier system is an example for a mechanism constructed with the C-terminal fusion method. It takes advantage of the signal sequence and the first N-terminal residues of the mature *E. coli* lipoprotein and residues of the E. *coli* outer membrane protein A (OmpA). Lpp is responsible for the adequate translocalization of the fusion protein in the periplasm, whereas OmpA is necessary for crossing the outer membrane [82–85].

Despite N-terminal and C-terminal fusion, sandwich arrangement is a suitable strategy for the surface display of peptides on Gram-negative bacteria using whole structure proteins. The lack of anchoring regions does not allow for the truncation of these outer membrane proteins (OMPs), subunit proteins of extracellular extensions, and S-layer proteins (Fig. 4.6). For example, OmpC forms a transmembrane β-barrel structure spanning the outer membrane and allowing its external loops to be used as fusion sites, as shown in 1999 by Zhaohui Xu and Sang Yup Lee [74]. More recently, Bessette et al. developed a system based on a surface-exposed loop of E. coli OmpA allowing rapid isolation of high-affinity protein-binding peptides. Using sequential MACS and FACS analysis, proteins with high binding affinities (K_D in the low nanomolar range) for a variety of distinct antigens (e.g., streptavidin and HIV-1 GP120) could be isolated straightforward [86].

One of the major drawbacks of the microbial display system was the difficulty in presenting complex structures such as antibodies.

In 2008 Georgiou and colleagues addressed this issue by bacterial periplasmic display of so-called E-clonal mAbs [87]. Therefore, antibody heavy and light chains were secreted into the periplasm of E. coli, where they were properly folded and assembled into aglycosylated IgGs. The antibody immobilization on the inner membrane facing the periplasm was mediated by a chrimeric NlpA-ZZ fusion protein. Permeabilization of the outer membrane resulted in spheroplast clones that could be screened via FACS for specific antigen binding.

Another approach of microbial antibody display was published in 2015 by Wang and colleagues. They were able to present the neutralizing single-chain antibody VRC01 against HIV-1 infection on E. coli cells and showed that the engineered bacterial cells could capture HIV-1 particles and inhibit HIV-1 infection in cell culture [88]. Furthermore, in 2016 the group of Fernández isolated human fibrinogen nanobodies (VHHs) using E. coli display by fusing an immune library of VHH domains to the β-domain of intimin [89, 90].

As mentioned before, bacterial display systems are interesting for the development of recombinant bacterial vaccines. They consist of an avirulent bacterial or viral carrier expressing foreign passenger antigens. Bacteria, e.g., E. coli, are attractive candidates as carriers, because they are often strongly immunogenic and capable of colonizing host tissues. Another advantage of bacterial vaccines is their ability to express various foreign antigens in different forms (e.g., soluble, membrane bound, secreted). Moreover, surface-exposed antigens may facilitate the recognition by the immune system with bacterial lipopolysaccharides (LPS) acting as immune-adjuvants due to their immunogenicity [91]. This approach was already used in 1987 by Charbit et al. to display foreign epitopes of hepatitis B virus surface antigen on E. coli. Intravenous injections into rabbits and mice resulted in high titer specific antibody responses indicating a decent immune response [92]. In addition, Agterberg and colleagues immunized guinea pigs with partially purified outer membrane pore protein E (PhoE) carrying inserted epitopes from foot-and-mouth disease virus, which yielded substantial titers of neutralizing antibodies, protecting the guinea pigs completely against challenge with the virus [93].

4.6   Ribosome Display

A representative of the next generation of display technologies is ribosome display, which may overcome some limitations of cell-based display strategies by using a cell-free expression system.

One limitation of conventional cellular systems is the host's transformation efficiency, which can limit the overall number of library variants.

During ribosome display, genotype phenotype coupling is mediated by the formation of ternary antibody-ribosome-mRNA (ARM) complexes in cell extracts, without the requirement of transformation steps. The absence of a stop-codon on the mRNA leads to a stalling of the ribosome at the end of translation. The release of the polypeptide chain cannot be triggered by release factors that normally take the place of the tRNA at that position.

Thereby the individual protein of interest (e.g., an antibody fragment) is connected to its corresponding mRNA, which enables the simultaneous isolation of the functional protein and its encoding mRNA (genotype-phenotype coupling). ARM complexes are stabilized by high concentrations of magnesium ions and low temperature for several days [94]. Selection steps are performed through affinity capture with a specific ligand, e.g., bound on magnetic beads. Afterwards the mRNA is recovered as DNA by reverse transcription (RT)-PCR. Successive screening rounds allow the isolation of high-affinity antibodies. Both prokaryotic and eukaryotic ribosome display systems have been developed (Fig. 4.7) [95, 96].

The biggest advantage of ribosome display over phage display and cell surface display systems is the possibility to generate and screen larger libraries. As uncloned PCR fragments are utilized as source of diversity, libraries with 10^{12-14} members can easily be produced [97]. Restrictions are the number of functional ribosomes as well as the presence of different mRNA molecules in the reaction.

In addition, uncloned PCR fragments can be easily used to introduce new diversity during selection [98, 99]. For example, non-proofreading polymerases [100], error-prone PCR, DNase I shuffling [101, 102] and amplification of templates by PCR in the presence of dNTP-analogs such as 8-oxo-guanosine or dPTP [103] are convenient methods for diversification of the repertoire. This capability of continuously diversifying DNA followed by selection against an antigen provides an optimal system for in vitro antibody maturation and directed evolution of binding proteins [104, 105].

At last, ribosome display is also suited to display proteins that are difficult or impossible to express by cell-based systems, since cell-free systems tolerate also toxic, proteolytically sensitive, or unstable proteins.

In the following, the different types of ribosome display systems (*E. coli* S30 system, eukaryotic systems, and wheat germ system) will be described. In the first publication of

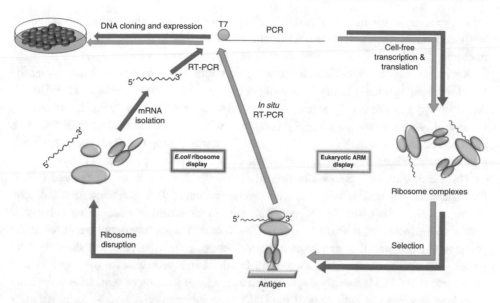

Fig. 4.7 Ribosome display cycles. *E. coli* ribosome display cycle (left side in blue) and eukaryotic ARM display cycle (right side in green). RT: reverse transcription, T7: T7 promoter, ARM: antibody-ribosome-mRNA. Modified from [95]

ribosome display, the *E. coli* S30 system was utilized for peptide selection (Fig. 4.7). The so-called polysome display generates polysome complexes (stalled Ribosome/mRNA complexes) with chloramphenicol [106]. The isolation of polysome complexes displaying interacting peptide epitopes was performed by capture with an immobilized antibody on microtiter wells. The dissociation of the complexes was accomplished by EDTA to release the bound mRNA. Afterwards DNA recovery was performed by RT-PCR. To display single-chain antibody fragments, the system was later modified by Hanes and colleagues. They generated ARM complexes through deletion of the 3′ terminal stop codon from DNA [107]. In 2004, Sawata et al. further refined the *E. coli* S30 system by introducing a protein-mRNA interaction to generate even more stable ribosome complexes, leading to an improved selection efficiency [108].

In the meantime, a eukaryotic ribosome display system was developed for selection of single-chain antibody fragments. The coupled rabbit reticulocyte lysate system was initially termed "ARM display" (Fig. 4.7) [109]. This system also utilized the deletion of the stop codon from the DNA to generate eukaryotic ribosome complexes. A novel recovery procedure was developed, in which RT-PCR was performed directly on the ribosome complexes without the dissociation step used in former systems. For this, in situ RT-PCR primers were designed, which hybridize slightly upstream of the mRNA 3′ end, thereby preventing binding to the ribosome-covered region [109]. In situ RT-PCR facilitates automation of the entire ribosome display process.

The procedure was later modified by inclusion of oxidized/reduced glutathione and Qβ RNA-dependent RNA polymerase for improvement of protein folding and introduction of mutations [110].

Kanamori and colleagues developed a highly controllable ribosome display using the PURE (Protein synthesis Using Recombinant Elements) system in the year 2014. Herein, ARM complexes are highly stable, and the selected mRNA can easily be recovered, because of very low nuclease activity combined with reconstitution of components such as ribosome and translation factors [111]. In 2002, a wheat germ cell-free protein synthesis system has also been developed for ribosome display of folded proteins [112].

There are some key factors affecting ribosome display, which are mentioned below. Constructs for ribosome display require a promoter such as T7 and a translation initiation sequence (Shine-Dalgarno for *E. coli* S30 system or Kozak sequence for eukaryotic systems). Furthermore, a linker of at least 23–30 amino acids in length is required at the proteins C-terminus. This generates a spacer between the ribosome and the displayed protein and abolishes steric hindrance [96]. In addition, a sequence for the hybridization of primers in RT-PCR recovery stage, as well as the incorporation of sequences containing stem-loop structures at both ends of the DNA to stabilize mRNA against degradation, is required [106]. The choice of a cell-free system may affect the formation of ribosome complexes either as monoribosome or polysome complexes, which can carry more than one ribosome translating a single mRNA molecule and can lead to avidity effects. Rabbit reticulocyte lysate can produce mainly monoribosome complexes, whereas *E. coli* S30 usually generates polysomes [95, 106]. Furthermore, cell-free systems can be divided into coupled (simpler and more efficient) and uncoupled systems. In a coupled system, transcription of the DNA template is immediately followed by translation. An uncoupled system translates mRNA, which is generated either by in vitro transcription or isolated from native sources. It allows the optimization of the transcription and translation step separately.

Ribosome display systems have been successfully applied for in vitro antibody selection, evolution, and humanization. Repeated rounds of mutation and selection led to the isolation of antibody variants with improved affinity, specificity, and stability [94, 96, 98, 113]. In 2006, Groves and colleagues could show that ribosome display is suited for affinity maturation of phage display-derived antibody populations. In addition, they compared the output of phage and ribosome display which were able to demonstrate that higher affinity antibodies were isolated using ribosome display [114].

4.7 cDNA Display

After the development of ribosome display [107] (described above) and mRNA display [115], in which the protein of interest is linked to its coding mRNA via a puromycin-linker (Pu-linker), the cDNA display technology was developed utilizing a similar strategy (Fig. 4.8). cDNA display is a powerful in vitro display technology in which complementary

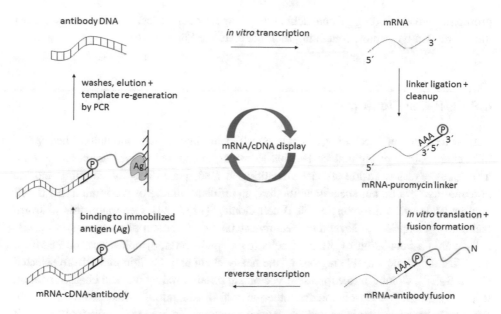

Fig. 4.8 Schematic for mRNA/cDNA display of antibodies. Antibody-encoding DNA is in vitro transcribed to mRNA and ligated to a 3′ puromycin (P) containing DNA linker. Following in vitro translation and fusion formation, reverse transcription is performed to synthesize cDNA. mRNA-cDNA-antibody complexes are exposed to the antigen (Ag) of interest, which is immobilized on a solid-phase affinity matrix. Following washes and elution, PCR is performed to retrieve antibody-encoding amplicons. N/C: N- and C-termini of displayed antibody. Modified from [118]

DNA is fused to its coding peptide or protein via a covalent puromycin linkage [116]. The technology is used to explore functional peptides and proteins from huge libraries by in vitro selection. In contrast to mRNA display, where a puromycin molecule is covalently linked to the mRNA via a DNA spacer, in cDNA display, the DNA is directly fused to the C-terminus of the peptide or protein via puromycin. This makes cDNA display more stable compared to other in vitro selection methods. Besides others, cDNA display was used for selection of peptide-based affinity reagents against cell surface proteins. For example, peptides binding to G protein-coupled receptors (GPCRs) could be selected, using cells with high GPCR expression [117].

cDNA display was further refined during the last years, especially the Pu-linker technology. Mochizuki, Ueno, and colleagues could improve the preparation efficiency of the cDNA display complex by developing additional Pu-linkers [119, 120] and modifying the preparation protocol [121]. These modifications make cDNA display more practical. The novel Pu-linker, developed by the same group., contains 3-cyanovinylcarbazole nucleoside (cnvK) and is called the cnvK-Pu-linker. cnvK is an ultrafast photo-cross-linker using UV irradiation to connect hybridized oligonucleotides, which enables ultrafast cross-linking without the use of an enzymatic ligation reaction [122]. Furthermore, cDNA display can be used as screening method for functional disulfide-rich peptides by solid-phase synthesis, promoting protein folding. Yamaguchi and colleagues could select novel peptides against the interleukin-6 receptor, containing multiple disulfide bonds [116].

4.8   B-Cell Cloning

Similar to hybridoma technology, single B-cell technologies retain the natural heavy and light chain pairing of immunized B-cells. However, in contrast to hybridoma that has the disadvantage of low fusion efficacies resulting in inadequate diversity sampling and the potential loss of rare but specific antibodies, an efficient mining of the immunized B-cell population is enabled by single cell B-cell cloning [123, 124]. Generally, this platform technology comprises of several consecutive steps, i.e., detection of (antigen-specific) B-cells within a population of different cells (e.g., splenocytes, lymphocytes, or PBMCs), amplification of the variable regions of the heavy chain and the light chain from selected single B-cells, cloning into expression vectors, antibody production, and characterization (Fig. 4.9). Single B-cell isolation can either be performed randomly (i.e., without selecting for antigen specificity) or in an antigen-selective manner. Techniques for the isolation of B cells in randomly comprise but are not limited to cell picking by micromanipulation [125], laser capture microdissection, [126] and FACS [127, 128]. Alternatively, single B-cell isolation in an antigen-selective manner has been described with the use of antigen-coated magnetic beads [129], fluorochrome-labeled antigens via multi-parameter FACS [130], the hemolytic plaque assay, [131] and a fluorescent foci method [132]. One of the most commonly applied techniques for the isolation of antigen-specific B-cells is fluorescence-activated cell sorting. Here, a multi-parameter labeling strategy is typically employed: On the one hand, fluorescently labeled antibodies against surface markers of B-cells (such as CD19) are utilized to gate for the specific B-cell population within the tissue sample. On the other hand, a fluorescently labeled antigen allows for the isolation of antigen-positive B cells. Single B-cell technologies have been elegantly reviewed by Rashidian and Lloyd [133].

Tiller et al. developed a highly efficient strategy to generate monoclonal antibodies from human B-cells at different stages of development based on surface marker expression [128]. Antibodies targeting different surface markers of B-cells and labeled with different fluorescence dyes allowed to discriminate between subpopulations of B-cells at different stages of development. A nested RT-PCR approach on sorted B-cells allowed for the amplification of heavy chain as well as light chain variable regions, followed by cloning into appropriate expression plasmids, antibody production, and characterization.

Starkie and co-workers described an efficient strategy for the isolation of antigen-specific antibodies from immunized mice [123]. Here, B-cells from the spleen were first enriched by MACS using anti-mouse CD45R microbeads. Afterwards, a multi-parameter staining strategy involving fluorescently labeled antibodies targeting different mouse isotypes such as IgG, IgM, and IgD as well as antibodies targeting different surface markers

Fig. 4.9 Schematic representation of the B-cell cloning technology from immunized mice. Single B-cells isolated from mouse tissues are sorted by FACS. After reverse transcription, VH and VL genes are amplified from sorted single B-cell cDNA in independent PCRs. Subsequently, variable regions are cloned into expression vectors comprising constant antibody regions. For production of monoclonal antibodies, expression vectors can be con-transfected into mammalian cells such as HEK 293. Modified from [134]

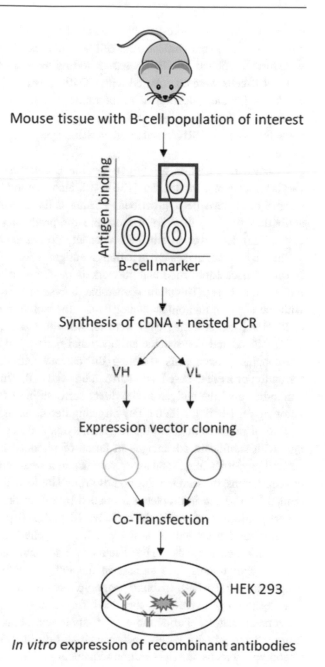

of desired (B-cell) and undesired (e.g., T-cell) populations enabled the isolation of antigen-specific class-switched IgG-positive memory B-cells.

In 2017, Carbonetti et al. described an efficient mAb discovery platform that combines single B-cell selection as well as B-cell cultivation [135]. Initially, after immunization of mice, spleens were harvested, homogenized to single cell suspension, and MACS was employed for the enrichment of B-cell populations. A multi-parameter labeling strategy was further applied for FACS-based sorting of antigen-positive cells. Subsequently, isolated B-cells were co-cultured with CD40L-positive feeder cells as well as supportive cytokines for the induction of proliferation and mAb production. Antibody-containing supernatants were then assessed with respect to reactivity. Afterwards, positive cells were harvested for RNA isolation, variable gene amplification, cloning, and antibody expression.

Besides the selection of antigen-specific B-cells using different methodologies with varying levels of complexities [136–140], also methods for the selection of mAbs from plasma cells have been described. Plasma cells are differentiated B-cells that secrete antibodies (in fact, they are responsible for producing immunoglobulins in the blood circulation). However, plasma cells normally do not express the antibody moiety on their surface, a fact that makes the isolation of antigen-specific plasma cells rather complicated. In this respect, Lightwood and co-workers described a method referred to as fluorescent foci method [141]. This fluorescence-based system was performed on microscope slides with the use of a micromanipulator device for isolation of antigen-specific IgG secreting cells. Here, IgG-secreting cells were mixed with a solid phase antigen as well as with a FITC-labeled anti-Fcγ-specific antibody and plated as a monolayer on glass slides. Antigen-specific plasma cells secreting the desired antibody were visualized through the formation of a FITC-based halo surrounding the cell. This is because the secreted specific antibodies bind the antigen in the direct surrounding of the antigen-specific plasma cell, allowing for labeling with the Fcγ-targeting detection moiety.

Recent progress in microfluidic systems enabled new strategies in single-cell technology with significant advantages in terms of phenotypic screenings for antibodies with desired properties (e.g., total assay volumes in a small range allowing to obtain detectable concentrations of secreted antibodies from single cells). Nowadays, reactions can be compartmentalized in droplets surrounded by oil or air. The droplets typically shaped in a very uniform size (typically between 10 and 200 μm in diameter) contain all assay reagents needed as well as a single cell to be characterized. In this way thousands of individual compartments can be generated per second. The generated droplets can be sorted according to fluorescence signals in an automated fashion via fluorescence-activated droplet sorting (FADS), enabling phenotypic screening of hundreds of samples per second. The technology overcomes the need for any slow mechanical equipment such as micromanipulators. Furthermore, the approach should also be applicable to non-immortalized primary B-cells, because no cell proliferation is required. In addition to the phenotypic screens, next-generation sequencing (NGS)-based approaches can be exploited to gather information about antibody binding. Those approaches are called "Ig-seq" [142]. DeKosky and colleagues developed an entirely emulsion-based technology, where single B-cells were encapsulated in water-in-oil droplets. The technology allowed the sequencing of repertoires of more than 2×10^6 B-cells in a single experiment [143]. In the near to mid-term future, microfluidics might ultimately allow to screen B-cells from humans to generate highly specific therapeutics. Screenings will become more efficient by the use of highly advanced devices for single cell manipulation, which can be rapidly automated and remotely controlled. In this respect, several services and devices from different companies are available [124, 144, 145].

Take Home Message
- Antibody display systems enable the identification of antibodies with prescribed properties from huge repertoires.
- The key feature that facilitates the selection of a favored library candidate relies in coupling the phenotype (i.e., the antibody protein) to its genotype (i.e., the DNA encoding for the entity).
- The most frequently used technology is phage display that enabled the generation or isolation of multiple marketed antibody derivatives.
- Yeast surface display emerged as another platform technology that has proven to be successful for antibody hit discovery and engineering.
- In recent years several different mammalian display systems were developed that make use of higher eukaryotic quality control machineries.
- B-cell cloning can be considered as the most natural form of genotype phenotype coupling utilizing the native protein production and secretion apparatus of antibody producing cells.

References

1. Kaplon H, Reichert JM. Antibodies to watch in 2019. MAbs. 2019;11:219–38.
2. Rau R. Adalimumab (a fully human anti-tumour necrosis factor α monoclonal antibody) in the treatment of active rheumatoid arthritis: the initial results of five trials. Ann Rheum Dis. 2002;61:70–3.
3. Hoogenboom HR. Selecting and screening recombinant antibody libraries. Nat Biotechnol. 2005;23:1105–16.
4. Barbas CF, Kang AS, Lerner RA, Benkovic SJ. Assembly of combinatorial antibody libraries on phage surfaces: the gene III site. Proc Natl Acad Sci U S A. 1991;88:7978–82.
5. Yang WP, et al. CDR walking mutagenesis for the affinity maturation of a potent human anti-HIV-1 antibody into the picomolar range. J Mol Biol. 1995;254:392–403.
6. Köhler G, Milstein C. Continuous cultures of fused cells secreting antibody of predefined specificity. Nature. 1975;256:495.
7. Bradbury ARM, Sidhu S, Mccafferty J, Alamos L. Beyond natural antibodies: the power of in vitro display technologies. Nat Biotechnol. 2011;29:245–54.
8. Smith GP. Filamentous fusion phage: novel expression vectors that display cloned antigens on the virion surface. Science. 1985;228:1315–7.
9. Arnold FH and Smith GP. Nobel prize in chemistry: 2018 press release. 50005 (2018).
10. McCafferty J, Griffiths AD, Winter G, Chiswell DJ. Phage antibodies: filamentous phage displaying antibody variable domains. Nature. 1990;348:552–4.

11. Breitling F, Dübel S, Seehaus T, Klewinghaus I, Little M. A surface expression vector for antibody screening. Gene. 1991;104:147–53.
12. Hust M, Dübel S. Phage display vectors for the in vitro generation of human antibody fragments. In: Burns R, editor. Immunochemical protocols: Humana Press; 2005. p. 71–96. https://doi.org/ 10.1385/1-59259-873-0:071.
13. Mazor Y, Van Blarcom T, Mabry R, Iverson BL, Georgiou G. Isolation of engineered, full-length antibodies from libraries expressed in *Escherichia coli*. Nat Biotechnol. 2007;25:563–5.
14. Naqid I. Investigation of antibody-based immune recognition of infections with *Salmonella enterica* serovars Typhimurium and Enteritidis. (2017).
15. Clackson T, Hoogenboom HR, Griffiths AD, Winter G. Making antibody fragments using phage display libraries. Lett to Nat. 1991;352
16. Hust M, Dübel S. Mating antibody phage display with proteomics. Trends Biotechnol. 2004;22:8–14.
17. Pini A, et al. Design and use of a phage display library. J Biol Chem. 1998;273:21769–76.
18. Knappik A, et al. Fully synthetic human combinatorial antibody libraries (HuCAL) based on modular consensus frameworks and CDRs randomized with trinucleotides. J Mol Biol. 2000;296:57–86.
19. Hawlisch H, et al. Site-specific anti-c3a receptor single-chain antibodies selected by differential panning on cellulose sheets. Anal Biochem. 2001;293:142–5.
20. Moghaddam A, et al. Identification of scFv antibody fragments that specifically recognise the heroin metabolite 6-monoacetylmorphine but not morphine. J Immunol Methods. 2003;280:139–55.
21. Hust M, Maiss E, Jacobsen HJ, Reinard T. The production of a genus-specific recombinant antibody (scFv) using a recombinant potyvirus protease. J Virol Methods. 2002;106:225–33.
22. Buckler DR, Park A, Viswanathan M, Hoet RM, Ladner RC. Screening isolates from antibody phage-display libraries. Drug Discov Today. 2008;13:318–24.
23. Frenzel A, Schirrmann T, Hust M. Phage display-derived human antibodies in clinical development and therapy. MAbs. 2016;8:1177–94.
24. Yu J, Smith GP. Affinity maturation of phage-displayed peptide ligands. In: Methods in enzymology, vol. 267: Academic Press; 1996. p. 3–27.
25. Smith GP. Phage display: simple evolution in a Petri dish (Nobel lecture). Angew Chemie Int Ed. 2019;58:14428–37.
26. Kang AS, Jones TM, Burton DR. Antibody redesign by chain shuffling from random combinatorial immunoglobulin libraries. Proc Natl Acad Sci U S A. 1991;88:11120–3.
27. Rondot S, Koch J, Breitling F, Dübel S. A helper phage to improve single-chain antibody presentation in phage display. Nat Biotechnol. 2001;19:75–8.
28. Kaymakcalan Z, et al. Comparisons of affinities, avidities, and complement activation of adalimumab, infliximab, and etanercept in binding to soluble and membrane tumor necrosis factor. Clin Immunol. 2009;131:308–16.
29. Schirrmann T, Meyer T, Schütte M, Frenzel A, Hust M. Phage display for the generation of antibodies for proteome research, diagnostics and therapy. Molecules. 2011;16:412–26.
30. Nelson AL, Dhimolea E, Reichert JM. Development trends for human monoclonal antibody therapeutics. Nat Rev Drug Discov. 2010;9:767–74.
31. Baker KP, et al. Generation and characterization of LymphoStat-B, a human monoclonal antibody that antagonizes the bioactivities of B lymphocyte stimulator. Arthritis Rheum. 2003;48:3253–65.
32. Steinbrook R. Treatment of macular degeneration. October 1409–1412 (2006).
33. Mazumdar S. Raxibacumab. MAbs. 2009;1:531–8.
34. Lipkel PN, Kurjan J. Agglutination. Jpn J Med Mycol. 2016;57:125.

35. Cherf GM, Cochran JR. Applications of yeast surface display for protein engineering. Methods Mol Biol. 2015:155–75.
36. Boder ET, Wittrup KD. Yeast surface display for screening combinatorial polypeptide libraries. Nat Biotechnol. 1997;15:553–7.
37. Gera N, Hussain M, Rao BM. Protein selection using yeast surface display. Methods. 2013;60:15–26.
38. Huang G, Zhang M, Erdman SE. Posttranslational modifications required for cell surface localization and function of the fungal adhesin Aga1p. Eukaryot Cell. 2003;2:1099–114.
39. Weaver-Feldhaus JM, et al. Yeast mating for combinatorial Fab library generation and surface display. FEBS Lett. 2004;564:24–34.
40. Krah S, et al. Generation of human bispecific common light chain antibodies by combining animal immunization and yeast display. Protein Eng Des Sel. 2017;30:291–301.
41. Engler C, Kandzia R, Marillonnet S. A one pot, one step, precision cloning method with high throughput capability. PLoS One. 2008;3:e3647.
42. Rosowski S, et al. A novel one-step approach for the construction of yeast surface display Fab antibody libraries. Microb Cell Factories. 2018;17:3.
43. Chao G, et al. Isolating and engineering human antibodies using yeast surface display. Nat Protoc. 2006;1:755–68.
44. Boder ET, Raeeszadeh-Sarmazdeh M, Price JV. Engineering antibodies by yeast display. Arch Biochem Biophys. 2012;526:99–106.
45. Schröter C, et al. Selection of antibodies with tailored properties by application of high-throughput multiparameter fluorescence-activated cell sorting of yeast-displayed immune libraries. Mol Biotechnol. 2018;60:727–35.
46. Roth L, et al. Isolation of antigen specific VHH single-domain antibodies by combining animal immunization with yeast surface display. Methods Mol Biol. 2019:173–89.
47. Sheehan J, Marasco WA. Phage display and yeast display. Microbiol Spectr. 2015;3:1–12.
48. Doerner A, Rhiel L, Zielonka S, Kolmar H. Therapeutic antibody engineering by high efficiency cell screening. FEBS Lett. 2014;588:278–87.
49. Benatuil L, Perez JM, Belk J, Hsieh CM. An improved yeast transformation method for the generation of very large human antibody libraries. Protein Eng Des Sel. 2010;23:155–9.
50. Ferrara F, et al. Using phage and yeast display to select hundreds of monoclonal antibodies: application to antigen 85, a tuberculosis biomarker. PLoS One. 2012;7:e49535.
51. Boder ET, Midelfort KS, Wittrup KD. Directed evolution of antibody fragments with monovalent femtomolar antigen-binding affinity. Proc Natl Acad Sci U S A. 2000;97:10701–5.
52. Swers JS. Shuffled antibody libraries created by in vivo homologous recombination and yeast surface display. Nucleic Acids Res. 2004;32:36.
53. Holler PD, et al. In vitro evolution of a T cell receptor with high affinity for peptide/MHC. Proc Natl Acad Sci U S A. 2000;97:5387–92.
54. Weber KS, Donermeyer DL, Allen PM, Kranz DM. Class II-restricted T cell receptor engineered in vitro for higher affinity retains peptide specificity and function. Proc Natl Acad Sci U S A. 2005;102:19033–8.
55. Chao G, Cochran JR, Dane Wittrup K. Fine epitope mapping of anti-epidermal growth factor receptor antibodies through random mutagenesis and yeast surface display. J Mol Biol. 2004;342:539–50.
56. Shusta EV, Holler PD, Kieke MC, Kranz DM, Wittrup KD. Directed evolution of a stable scaffold for T-cell receptor engineering. Nat Biotechnol. 2000;18:754–9.
57. Wang X, Cho Y, Shusta E. Mining a yeast library for brain endothelial cell-binding antibodies. Nat Methods. 2007;135:612. https://doi.org/10.1038/jid.2014.371.

58. Zhang S, et al. Preclinical characterization of Sintilimab, a fully human anti-PD-1 therapeutic monoclonal antibody for cancer. Antib Ther. 2018;1:65–73.
59. Bowers PM, et al. Mammalian cell display for the discovery and optimization of antibody therapeutics. Methods. 2014;65:44–56.
60. Higuchi K, et al. Cell display library for gene cloning of variable regions of human antibodies to hepatitis B surface antigen. J Immunol Methods. 1997;202:193–204.
61. Ho M, Pastan I. Mammalian cell display for antibody engineering. In: Dimitrov AS, editor. Therapeutic antibodies: methods and protocols: Humana Press; 2009. p. 337–52. https://doi.org/10.1007/978-1-59745-554-1_18.
62. Zhou C, Jacobsen FW, Cai L, Chen Q, Shen WD. Development of a novel mammalian cell surface antibody display platform. MAbs. 2010;2:508–18.
63. Breous-Nystrom E, et al. Retrocyte Display® technology: generation and screening of a high diversity cellular antibody library. Methods. 2014;65:57–67.
64. Beerli RR, et al. Isolation of human monoclonal antibodies by mammalian cell display. Proc Natl Acad Sci U S A. 2008;105:14336–41.
65. Smith ES, Zauderer M. Antibody library display on a mammalian virus vector: combining the advantages of both phage and yeast display into one technology. Curr Drug Discov Technol. 2014;11:48.
66. Waldmeier L, et al. Transpo-mAb display: transposition-mediated B cell display and functional screening of full-length IgG antibody libraries. MAbs. 2016;8:726–40.
67. Bowers PM, et al. Coupling mammalian cell surface display with somatic hypermutation for the discovery and maturation of human antibodies. Proc Natl Acad Sci U S A. 2011;108:20455–60.
68. Seo H, et al. Rapid generation of specific antibodies by enhanced homologous recombination. Nat Biotechnol. 2005;23:731–5.
69. Mason DM, et al. High-throughput antibody engineering in mammalian cells by CRISPR/Cas9-mediated homology-directed mutagenesis. Nucleic Acids Res. 2018;46:7436–49.
70. Parthiban K, et al. A comprehensive search of functional sequence space using large mammalian display libraries created by gene editing. MAbs. 2019;11:884–98.
71. Zhang H, Wilson IA, Lerner RA. Selection of antibodies that regulate phenotype from intracellular combinatorial antibody libraries. Proc Natl Acad Sci U S A. 2012;109:15728–33.
72. Xie J, Zhang H, Yea K, Lerner RA. Autocrine signaling based selection of combinatorial antibodies that transdifferentiate human stem cells. Proc Natl Acad Sci U S A. 2013;110:8099–104.
73. Lee SY, Choi JH, Xu Z. Microbial cell-surface display. Trends Biotechnol. 2003;21:45–52.
74. Xu Z, Lee SY. Display of polyhistidine peptides on the Escherichia coli cell surface by using outer membrane protein C as an anchoring motif. Appl Environ Microbiol. 1999;65:5142–7.
75. Chang H, Sheu S, Lo S. Expression of foreign antigens on the surface of Escherichia coli by fusion to the outer membrane protein traT. J Biomed Sci. 1999:64–70.
76. Chang HH, Lo SJ. Modification with a phosphorylation tag of PKA in the TraT-based display vector of Escherichia coli. J Biotechnol. 2000;78:115–22.
77. Lee JS, Shin KS, Pan JG, Kim CJ. Surface-displayed viral antigens on Salmonella carrier vaccine. Nat Biotechnol. 2000;18:645–8.
78. Jung H, Lebeault J, Pan J. Surface display of Zymomonas mobilis levansucrase by using the ice-nucleation protein of Pseudomonas syringae. Nat Biotechnol. 1998;66:1124. https://doi.org/10.1271/nogeikagaku1924.66.1124.
79. Dhillon JK, Drew PD, Porter AJR. Bacterial surface display of an anti-pollutant antibody fragment. Lett Appl Microbiol. 1999;28:350–4.

80. Jose J, Krämer J, Klauser T, Pohlner J, Meyer TF. Absence of periplasmic DsbA oxidoreductase facilitates export of cysteine-containing passenger proteins to the *Escherichia coli* cell surface via the Iga(β) autotransporter pathway. Gene. 1996;178:107–10.

81. Suzuki T, Lett MC, Sasakawa C. Extracellular transport of VirG protein in Shigella. J Biol Chem. 1995;270:30874–80.

82. Chen W, Bae W, Mehra R, Mulchandani A. Enhanced bioaccumulation of heavy metals by bacterial cells with surface-displayed synthetic phytochelatins. ACS Symp Ser. 2002;806:411–8.

83. Francisco JA, Earhart CF, Georgiou G. Transport and anchoring of β-lactamase to the external surface of *Escherichia coli*. Proc Natl Acad Sci U S A. 1992;89:2713–7.

84. Francisco JA, Campbell R, Iverson BL, Georgiou G. Production and fluorescence-activated cell sorting of *Escherichia coli* expressing a functional antibody fragment on the external surface. Proc Natl Acad Sci U S A. 1993;90:10444–8.

85. Stathopoulos C, Georgiou G, Earhart CF. Characterization of *Escherichia coli* expressing an Lpp'OmpA(46-159)-PhoA fusion protein localized in the outer membrane. Appl Microbiol Biotechnol. 1996;45:112.

86. Bessette PH, Rice JJ, Daugherty PS. Rapid isolation of high-affinity protein binding peptides using bacterial display. Protein Eng Des Sel. 2004;17:731–9.

87. Mazor Y, Van Blarcom T, Iverson BL, Georgiou G. E-clonal antibodies: selection of full-length igg antibodies using bacterial periplasmic display. Nat Protoc. 2008;3:1766–77.

88. Wang L-X, et al. *Escherichia coli* surface display of single-chain antibody VRC01 against HIV-1 infection. Virology. 2015;475:179. https://doi.org/10.1016/j.cortex.2009.08.003.Predictive.

89. Salema V, López-Guajardo A, Gutierrez C, Mencía M, Fernández LÁ. Characterization of nanobodies binding human fibrinogen selected by *E. coli* display. J Biotechnol. 2016;234:58–65.

90. Salema V, Fernández LÁ. *Escherichia coli* surface display for the selection of nanobodies. Microb Biotechnol. 2017;10:1468–84.

91. Georgiou G, et al. Display of heterologous proteins on the surface of microorganisms: from the screening of combinatorial libraries to live recombinant vaccines. Nat Biotechnol. 1997;15:29–34.

92. Charbit A, et al. Presentation of two epitopes of the preS2 region of hepatitis B virus on live recombinant bacteria. J Immunol. 1987;139:1658–64.

93. Agterberg M, Adriaanse H, Barteling S, van Maanen K, Tommassen J. Protection of Guinea-pigs against foot-and-mouth disease virus by immunization with a PhoE-FMDV hybrid protein. Vaccine. 1990;8:438–40.

94. Schaffitzel C, Hanes J, Jermutus L, Plückthun A. Ribosome display: an in vitro method for selection and evolution of antibodies from libraries. J Immunol Methods. 1999;231:119–35.

95. He M, Khan F. Ribosome display: next-generation display technologies for production of antibodies in vitro. Expert Rev Proteomics. 2005;2:421–30.

96. He M, Taussig MJ. Ribosome display: cell-free protein display technology. Br Funct Genomic Proteomic. 2002;1:204.

97. Lamla T, Erdmann VA. Searching sequence space for high-affinity binding peptides using ribosome display. J Mol Biol. 2003;329:381–8.

98. Plückthun A, Schaffitzel C, Hanes J, Jermutus L. In vitro selection and evolution of proteins. In: Evolutionary protein design, vol. 55: Academic Press; 2001. p. 367–403.

99. Jermutus L, Honegger A, Schwesinger F, Hanes J, Plückthun A. Tailoring in vitro evolution for protein affinity or stability. Proc Natl Acad Sci U S A. 2001;3:29. https://doi.org/10.2116/analsci.3.29.

100. Hanes J, Jermutus L, Weber-Bornhauser S, Bosshard HR, Plückthun A. Ribosome display efficiently selects and evolves high-affinity antibodies in vitro from immune libraries. Proc Natl Acad Sci U S A. 1998;95:14130–5.
101. Cadwell RC, Joyce GF. Randomization of genes by PCR mutagenesis. Genome Res. 1992;2:28–33.
102. Stemmer W. Rapid evolution of a protein in vitro by DNA shuffling. Nature. 1994;370:389.
103. Zaccolo M, Williams DM, Brown DM, Gherardi E. An approach to random mutagenesis of DNA using mixtures of triphosphate derivatives of nucleoside analogues. J Mol Biol. 1996;255:589–603.
104. Zahnd C, Amstutz P, Plückthun A. Ribosome display: selecting and evolving proteins in vitro that specifically bind to a target. Nat Methods. 2007;4:269–79.
105. Yan X, Xu Z. Ribosome-display technology: applications for directed evolution of functional proteins. Drug Discov Today. 2006;11:911–6.
106. Mattheakis LC, Bhatt RR, Dower WJ. An in vitro polysome display system for identifying ligands from very large peptide libraries. Proc Natl Acad Sci U S A. 1994;91:9022–6.
107. Hanes J, Plückthun A. In vitro selection and evolution of functional proteins by using ribosome display. Proc Natl Acad Sci U S A. 1997;94:4937–42.
108. Sawata SY, Suyama E, Taira K. A system based on specific protein-RNA interactions for analysis of target protein-protein interactions in vitro: successful selection of membrane-bound Bak-Bcl-xL proteins in vitro. Protein Eng Des Sel. 2004;17:501–8.
109. He M, Taussig MJ. Antibody-ribosome-mRNA (ARM) complexes as efficient selection particles for in vitro display and evolution of antibody combining sites. Nucleic Acids Res. 1997;25:5132–4.
110. Forster AC, Cornish VW, Blacklow SC. Pure translation display. Anal Biochem. 2004;333:358–64.
111. Kanamori T, Fujino Y, Ueda T. PURE ribosome display and its application in antibody technology. Biochim Biophys Acta - Proteins Proteomics. 2014;1844:1925–32.
112. Takahashi F, et al. Ribosome display for selection of active dihydrofolate reductase mutants using immobilized methotrexate on agarose beads. FEBS Lett. 2002;514:106–10.
113. Lipovsek D, Plückthun A. In-vitro protein evolution by ribosome display and mRNA display. J Immunol Methods. 2004;290:51–67.
114. Groves MA, Nickson AA. Affinity maturation of phage display antibody populations using ribosome display. Methods Mol Biol. 2012;805:163–90.
115. Nemoto N, Miyamoto-Sato E, Husimi Y, Yanagawa H. In vitro virus: bonding of mRNA bearing puromycin at the 3′-terminal end to the C-terminal end of its encoded protein on the ribosome in vitro. FEBS Lett. 1997;414:405–8.
116. Yamaguchi J, et al. cDNA display: a novel screening method for functional disulfide-rich peptides by solid-phase synthesis and stabilization of mRNA-protein fusions. Nucleic Acids Res. 2009;37:e108.
117. Ueno S, et al. In vitro selection of a peptide antagonist of growth hormone secretagogue receptor using cDNA display. Proc Natl Acad Sci U S A. 2012;109:11121–6.
118. Doshi R, et al. In vitro nanobody discovery for integral membrane protein targets. Sci Rep. 2014;4
119. Mochizuki Y, et al. One-pot preparation of mRNA/cDNA display by a novel and versatile Puromycin-linker DNA. ACS Comb Sci. 2011;13:478–85.
120. Ueno S, Kimura S, Ichiki T, Nemoto N. Improvement of a puromycin-linker to extend the selection target varieties in cDNA display method. J Biotechnol. 2012;162:299–302.
121. Mochizuki Y, Kumachi S, Nishigaki K, Nemoto N. Increasing the library size in cDNA display by optimizing purification procedures. Biol Proced Online. 2013;15:1–5.

122. Mochizuki Y, Suzuki T, Fujimoto K, Nemoto N. A versatile puromycin-linker using cnvK for high-throughput in vitro selection by cDNA display. J Biotechnol. 2015;212:174–80.
123. Starkie DEO, Compson JE, Rapecki S, Lightwood DJ. Generation of recombinant monoclonal antibodies from immunised mice and rabbits via flow cytometry and sorting of antigen-specific IgG+ memory B cells. PLoS One. 2016;11:1–26.
124. Fitzgerald V, Leonard P. Single cell screening approaches for antibody discovery. Methods. 2017;116:34–42.
125. Küppers R, Zhao M, Hansmann ML, Rajewsky K. Tracing B cell development in human germinal centres by molecular analysis of single cells picked from histological sections. EMBO J. 1993;12:4955–67.
126. Obiakor H, et al. A comparison of hydraulic and laser capture microdissection methods for collection of single B cells, PCR, and sequencing of antibody VDJ. Anal Biochem. 2002;306:55–62.
127. Smith K, et al. Rapid generation of fully human monoclonal antibodies specific to a vaccinating antigen. Nat Protocol. 2009;4:372–84.
128. Tiller T, et al. Efficient generation of monoclonal antibodies from single human B cells by single cell RT-PCR and expression vector cloning. J Immunol Methods. 2008;329:112–24.
129. Lagerkvist A, Furebring C, Borrebaeck C. Single, antigen-specific B cells used to generate Fab fragments using CD40-mediated amplification or direct PCR cloning. BioTechniques. 1995:862–9.
130. Battye FL, Light A, Tarlinton DM. Single cell sorting and cloning. J Immunol Methods. 2000;243:25–32.
131. Leslie KB, Babcook JS, Olsen OA, Salmon RA, Schrader JW. The single lymphocyte antibody method (SLAM): a novel strategy for generating monoclonal antibodies form single, isolated lymphocytes producing antibodies of defined specificities. FASEB J. 1996;10:7843–8.
132. Tickle S, et al. High-throughput screening for high affinity antibodies. J Lab Autom. 2009;14:303–7.
133. Rashidian J, Lloyd J. Single B cell cloning and production of rabbit monoclonal antibodies. Methods Mol Biol. 2020;2070:423–41.
134. Tiller T, Busse CE, Wardemann H. Cloning and expression of murine Ig genes from single B cells. J Immunol Methods. 2009;350:183–93.
135. Carbonetti S, et al. A method for the isolation and characterization of functional murine monoclonal antibodies by single B cell cloning. J Immunol Methods. 2017;176:139. https://doi.org/10.1016/j.physbeh.2017.03.040.
136. Di Niro R, et al. Rapid generation of rotavirus-specific human monoclonal antibodies from small-intestinal mucosa. J Immunol. 2010;185:5377–83.
137. Ouisse LH, et al. Antigen-specific single B cell sorting and expression-cloning from immunoglobulin humanized rats: a rapid and versatile method for the generation of high affinity and discriminative human monoclonal antibodies. BMC Biotechnol. 2017;17:3.
138. Wrammert J, et al. Rapid cloning of high-affinity human monoclonal antibodies against influenza virus. Nature. 2008;453:667–71.
139. Scheid JF, et al. Broad diversity of neutralizing antibodies isolated from memory B cells in HIV-infected individuals. Nature. 2009;458:636–40.
140. Mietzner B, et al. Autoreactive IgG memory antibodies in patients with systemic lupus erythematosus arise from nonreactive and polyreactive precursors. Proc Natl Acad Sci. 2008;105:1–6.
141. Clargo AM, et al. The rapid generation of recombinant functional monoclonal antibodies from individual, antigen-specific bone marrow-derived plasma cells isolated using a novel fluorescence-based method. MAbs. 2014;6:143–59.

142. Seah YFS, Hu H, Merten CA. Microfluidic single-cell technology in immunology and antibody screening. Mol Asp Med. 2018;59:47–61.
143. DeKosky BJ, et al. In-depth determination and analysis of the human paired heavy- and light-chain antibody repertoire. Nat Med. 2015;21:86–91.
144. Eyer K, et al. Single-cell deep phenotyping of IgG-secreting cells for high-resolution immune monitoring. Nat Biotechnol. 2017;35:977–82.
145. Winters A, et al. Rapid single B cell antibody discovery using nanopens and structured light. MAbs. 2019;11:1025–35.

Transgenic Animals for the Generation of Human Antibodies

5

Biao Ma and Michael Osborn

Contents

B. Ma (✉)
Apollomics Inc, Xiaoshan, Hangzhou, China

M. Osborn
Teniobio, Newark, CA, USA

LifeArc, SBC Open Innovation Campus, Stevenage, UK

© Springer Nature Switzerland AG 2021
F. Rüker, G. Wozniak-Knopp (eds.), *Introduction to Antibody Engineering*,
Learning Materials in Biosciences,
https://doi.org/10.1007/978-3-030-54630-4_5

Keywords

Transgenic animals · Monoclonal antibodies · Chimeric antibodies · Transchromosomic animals · Humanisation · Antibody repertoires · Gene knockout

What You Will Learn in This Chapter

Currently, 28% of all monoclonal antibody (mAb) therapies and 74% of fully human mAb therapies approved by US Food and Drug Administration (FDA) are derived from transgenic animal platforms. In these platforms, the host antibody heavy chain VDJ and light chain VJ repertoire are substituted by their counterparts which are encoded exclusively by human transgenes. These platforms take advantage of the host's naturally antigen-driven antibody selection and maturation process associated with the primary and secondary immune response. Through immunisations and subsequent antibody discovery using common technologies including hybridoma and single B-cell screening, a large panel of mAbs exhibiting high affinity and specificity against the target antigen along with ideal biophysical characteristics and developability can be obtained. They can be streamlined into clinical development. This chapter gives a comprehensive overview of the transgenic strategies, methods of creating knockout of the endogenous immunoglobulin (Ig) production and generation of transgenic constructs containing human Ig heavy and light chain gene locus. Brief introductions are also given to a list of representative transgenic platforms which have originated in multiple species and can produce human antibodies in various formats. The strategies for antibody discovery in these platforms are also discussed.

5.1   The Concept of Using Transgenic Animals to Generate Human Therapeutic Antibodies

The major drawback of using rodent-derived monoclonal antibodies (mAbs) for human therapies is their immunogenicity. Upon administration of the fully xenogeneic mAbs, patients can develop anti-drug antibodies (ADAs). The ADAs can alter the pharmacokinetics of therapeutic mAbs, leading to their accelerated clearance. A sub-set of ADAs, characterised as neutralisation antibodies, can lead to the loss of efficacy of therapeutic mAbs [1, 2]. Further, life-threatening allergic reactions have been reported in some patients treated with murine-derived mAbs [1–3]. To mitigate these problems, chimeric antibodies which contain the variable domains of a xenogeneic mAb appended onto human Ig constant regions in a functional configuration have been engineered to reduce the immunogenicity and maintain or enhance the efficacy of the original mAb [4, 5]. Up to date, nine chimeric therapeutic mAbs have been approved for human therapies by FDA, accounting for 11% of all approved mAbs in the USA (https://www.antibodysociety.org/resources/approved-antibodies/). Although the immunogenicity is reduced, the use of therapeutic chimeric antibodies can still elicit human anti-chimeric antibody response in patients [2]. To further increase the human content of therapeutic mAbs, sophisticated humanisation processes have been developed since the mid-1980s. Successful methods include grafting the complementarity-determining regions (CDRs) of the variable domains of a xenogeneic mAb into the framework regions (FRs) of a selected human antibody, resulting in transferring the specificity and affinity to the recipient antibody [6, 7]. In recent years, with the increased number of protein structures published for antibodies from various origins, computer modelling and prediction programs have been developed to aid the decision-making during the antibody humanisation process [8]. To date, 40 humanised mAbs have been approved by FDA, accounting for half of antibody drugs on the US market. Despite this success, there are several drawbacks inherent in the humanisation process. The strategy for humanisation needs to be designed case by case, and there is still no generally applicable technology guaranteeing a positive outcome. Frequently, during the process, the humanised mAb candidates exhibit altered affinity, specificity and biophysical characteristics such as solubility, and can be prone to aggregation and decreased expression in production systems comparing to their original counterparts [9–11]. Significant amounts of time and resources are often needed to re-engineer and re-validate candidates to yield a humanised product that retains biological function as well as clinical efficacy and meets the Chemistry, Manufacturing and Control (CMC) requirements. It is usually impractical to put multiple xenogeneic mAbs into the humanisation process in parallel, which generates a bottleneck for drug development.

Another widely used technology for the discovery of fully human mAbs relies on screening large naïve phage display libraries, usually containing over 10^{10} individual phage clones displaying Ig variable regions derived from human B lymphocytes [12, 13]. In this technology, the in vitro selection process carried out in test tubes replaces the target antigen-driven primary and secondary immune response in vivo. With carefully designed panning strategies, a diverse panel of candidate mAbs can be obtained. Further rounds of mutagenesis combined with screening are often required to yield mAbs with high affinity to a given target and desired biological functions. The major drawback of this technology is that the natural pairing of Ig heavy and light chain in clonal B cells are not preserved in the phage clones because during the process of library construction, the Ig heavy chain and light chain variable domains are randomly assembled together. It has been reported that a significant number of naïve library-derived mAbs exhibit developability problems including off-target binding, reduced solubility, being prone to aggregation and unacceptable expression yields [14]. As described for the humanisation process, considerable effort is also required for re-engineering and re-validating naïve library-derived mAb candidates to generate a clinically and commercially successful product. To date, eight fully human mAbs approved by FDA have been discovered through screening naïve phage libraries, accounting for 10% of all marketed mAb drugs and just 26% of fully human mAb drugs in the USA.

It is therefore highly desirable to reconstitute the human humoral immune response in model organisms such as mouse and rat through a transgenic approach. In this system, the human genomic regions encompassing the genes encoding antibody repertoire and the relevant expression regulatory regions are stably integrated into the genome of the host animal. The resulting transgenic animals can be immunised with a target antigen similar as in the wild-type animal. The naturally target-driven antibody selection and maturation process associated with the primary and secondary immune response can generate significant numbers of antibodies encoded exclusively by the human transgenes. These fully human antibodies with high affinity and specificity to the target can be discovered using well-established technologies such as hybridoma technology. Consequently, a diverse panel of candidate mAbs can be streamlined into the pre-clinical development with minimal further engineering work. It has been suggested the numerous checkpoints during natural B-cell development may indirectly select for antibody variable domains with desirable qualities for product development such as solubility [14, 15]. So far, 23 fully human mAbs approved by FDA have been discovered using transgenic platform based in mouse, accounting for 28% all marketed mAb drugs and 74% of fully human mAb drugs in the USA.

Thirty years have passed since the first attempt to introduce human Ig genes into an animal using transgenic technology. In 1989, Brüggemann and colleagues reported the creation of a transgenic mouse line carrying a human/mouse hybrid IgH minilocus [16]. The transgene in this study included one functional human and one functional mouse V_H segment, one mouse D segment, two human-like D segments derived from mutated mouse D, one human D segment and all six human J_H segments followed by a chimeric human/mouse C_μ gene. This line of mouse rearranged the transgene locus in the lymphoid tissue, and serum expression of human IgM could be readily detected. To extend from this work, in 1991, the same team reported another transgenic mouse line carrying 100 kilobase pairs (kbps) of the human IgH locus mostly in the germline configuration [17]. The transgene in this instance was all human, including two functional V_H, seven D and all six J_H followed by the complete C_μ gene. Rearrangement involving both V_H segments was observed, and extensive junction diversity in the V_H-(D)-J_H joining was also demonstrated. In 1992, Taylor and colleagues from then GenPharm International Inc. reported a new transgenic mouse line carrying both human IgH and Igκ minilocus [18, 19]. The fully human light chain transgene consisted of one functional V_κ, all five J_κ and C_κ segment followed by a downstream regulatory sequence. One fully human heavy chain construct used in this study consisted of one functional V_H, ten D, all six J_H, μ switch region, entire C_μ gene, γ1 switch region, entire $C_{\gamma 1}$ gene and a downstream regulatory sequence. Serum analysis showed the expression of both human IgM and IgG1 in the transgenic mice. Extensive rearrangement involving all segments in the transgene and substantial diversity in both V_κ-J_κ and V_H-(D)-J_H joining were demonstrated. Importantly, the study showed that the functional gene segment in the IgH transgene could undergo antibody class switch from IgM to IgG1. These early exploratory studies demonstrated there were only minor differences between the mouse and human humoral immune system. With respect to gene rearrangement and antibody class switch, the trans-acting factors in mouse can exert their functions on cis-elements in the human transgene loci. In all the studies described above, the transgenic constructs were microinjected into the pronuclei of fertilised mouse oocytes, resulting in the transgene loci being integrated into mouse genome in a random fashion with one or more copies. Hence, the reported data also suggest that the trans-acting factors in mouse for rearrangement and class switch can function irrelevant of chromosomal locations and varied gene copy numbers. These studies validated the concept and paved the way for developing transgenic animals for the discovery of fully human antibodies.

5.2   Strategies in Developing Transgenic Platforms for Human Antibody Discovery

To recapitulate the immensely diverse human antibody repertoire and ensure the high expression of human antibodies in transgenic animals, it is anticipated to be necessary to incorporate as many functional gene segments in the human Ig heavy and light chain genomic loci as well as the associated cis-regulatory elements into the transgenic constructs. It is also critical to inactivate the endogenous Ig heavy and light chain loci in the host to minimise the competition between endogenous antibodies and human antibodies derived from transgenes. Different aspects in transgenic strategies and designing the human transgenic constructs as well as Ig-knockout host are discussed below.

5.2.1   Transgenic Strategies

Owing to the collective efforts of the scientific community in human genome research, large human genomic regions, ranging from 10 to greater than 1000 kbps, have been cloned into cosmid, P1-derived artificial chromosome (PAC), bacterial artificial chromosome (BAC) and yeast artificial chromosome (YAC) vectors and can be acquired from commercial companies or academic laboratories. Specific clones that contain human Ig heavy and light chain gene loci can be identified using conventional molecular biology techniques including end sequencing followed by alignment to reference genome sequence and amplifying specific regions by PCR. PACs and BACs can host up to 300 kbps of mammalian DNA and be stably propagated in *Escherichia coli*. Precise modification of PACs and BACs can be achieved in *E. coli* using homologous recombination systems derived from λ phage such as the well-established Red/ET Recombination system [20–22]. The cloning capacity of YACs is open-ended and ranges from less than 100 to greater than 1000 kbps, providing an extensive coverage of human Ig gene loci. YACs can be shortened, extended and modified in *Saccharomyces cerevisiae* by utilising the robust homologous recombination system in this organism together with a range of selectable markers [23]. In addition, human artificial chromosomes (HACs) ranging from 10 to 20 megabase pairs (Mbps) were generated by truncating human chromosome 2, 14 and 22 containing Igκ-, IgH- and Igλ-gene loci, respectively, and maintained in mouse embryonic stem cells (ES cells) and homologous recombination-proficient chicken DT40 cells [24, 25]. When designing and building transgenic constructs, the target site of site-specific recombinase such as LoxP (for Cre recombinase), FRT (for flippase) and attP (for phiC31 integrase) and that for PiggyBac transposase have been frequently incorporated into the constructs [20–25]. These sites can facilitate the deletion, insertion, translocation and inversion at specific sites in the DNA of cells that carry the transgenic constructs.

One of the most commonly used methods for introducing transgenic constructs into the host genome is microinjecting purified linear DNA directly into the pronuclei of fertilised oocytes of the host [26, 27]. The transgenes can integrate into the host genome in a random fashion and usually in one or more copies. Subsequently, the microinjected oocytes are implanted into pseudo-pregnant foster mothers to produce the transgene-carrying founders at varying frequencies. The founders are typically identified by either a Southern blot or PCR assay using transgene-specific probes or primer pair, respectively, on genomic DNA extracted from ears or tail, and positives used to establish independent transgenic lines through breeding. The profile and strength in transgene expression may vary among different lines due to the nature of different transgene integration sites as well as gene copy numbers. Those transgenic lines with the most desired pattern of transgene expression can be chosen for further breeding to homozygosity. Even though it has low efficiency, this method is simple, relatively speedy and proven to be successful in many mammals including mice, rats and rabbits. Due to the physical nature of this method and the difficulties in purifying and handling large DNA fragments, the size of individual linear DNA for microinjection is usually restricted to under 300 kbps.

The well-established mouse ES cells provide an unparalleled system for generating transgenic strains. Gene targeting by the deletion of large murine genomic region and subsequent replacement with human counterparts in situ can be achieved in the ES cells [22, 28]. Using this strategy, mouse Ig heavy and light chain gene loci can be precisely humanised in situ, which renders the human Ig transgenes under the authentic temporal and spatial regulation of mouse trans-acting factors. To generate transgenic mouse ES cells, constructs as plasmids or BACs comprising designated transgenes, appropriate mammalian cell selectable markers and specially positioned LoxP or FRT sites flanked by murine genomic sequences to allow homologous integration are transfected or electroporated into the cultured cells. Upon selection, stable ES cell clones with the desired transgene integration can be identified via assays such as Southern blot, PCR including real-time quantitative PCR and fluorescence in situ hybridisation (FISH) [28]. Through the transient expression of corresponding site-specific recombinase, the desired chromosome modification such as insertion, deletion or inversion can be achieved and verified using similar assays. Subsequently, the verified ES cells are injected into recipient blastocysts which can be transplanted into surrogate mothers to yield chimeric mouse. Transgenic lines with germline transmission of the intended genetic features can then be established through several rounds of breeding from the initial chimeras. ES cell-based gene targeting and genetic engineering as described here for mouse are currently not established in rat, rabbit, cattle and other non-human mammals. Furthermore, spheroplasts derived from yeast cells that carry a modified YAC with appropriate mammalian cell selectable markers can be fused to mouse ES cells to create yeast-ES cell hybrid [23]. These ES cells with entire human transgenes harboured in the YAC integrated into the genome can be selected and verified using Southern blot, PCR and FISH analysis. Through this technology, transgenic mouse lines carrying megabase-sized human transgenes have been established [23, 29]. In this case, the integration site is random, and there is usually just one copy of the YAC integrated into the mouse genome due to its large size. Additionally, mouse ES cells harbouring a HAC can be directly used for generating transchromosomic (Tc) lines. In this instance, the HAC behaves as an autonomously replicating entity, independent from the endogenous chromosomes. The stability during mitosis and frequency of germline transmission vary significantly among different HACs when developed into Tc mouse lines.

For large mammals such as cattle, transgenic constructs can be introduced into cultured foetal fibroblast cells via electroporation. Additionally, through microcell-mediated chromosome transfer, HACs which have been modified and maintained in DT40 cells can be introduced into foetal fibroblast cells to establish Tc cells. Subsequently, selected fibroblast cells which have been verified for carrying the transgenes can be used for whole animal cloning to yield transgenic offspring [30, 31]. In chickens, isolated primordial germ cells can be cultured and genetically modified using transgenic constructs together with traditional transfection and selection techniques. These modified cells still possess the ability to develop into sperms and eggs when injected into the recipient chicken embryos. Transgenic chickens with germline transmission of the transgenes can be established through this strategy [32, 33].

5.2.2   Animals with Knockout of Endogenous Ig Production

To guarantee the optimal expression of human Ig gene-encoded antibodies in transgenic animals and to facilitate the downstream antibody discovery with minimal interference by both endogenous Ig heavy and light chains, it is essential to inactivate the endogenous Ig production in the host. This is rather crucial when the transgenic constructs carrying human Ig gene loci are integrated randomly into the host genome or maintained on a separate chromosome. The knockout strategies for host Ig gene loci are closely associated with the transgenic strategies described above. Using mouse ES cell-based gene targeting technology, a variety of constitutive IgH knockout mice which lack B-cell development and antibody production have been generated. These include the μMT^- ($Ighm^{tm1Cgn}$/J) mice in which the first exon encoding the transmembrane domain of IgM was disrupted [34], J_HT ($Igh\text{-}J^{tm1Cgn}$/J) mice in which the genomic region encompassing all J_H gene segments was deleted [35] and homozygous C-gene deletion mice in which a 200 kbps genomic region accommodating all C_H genes, namely, C_μ-C_δ-$C_{\gamma3}$-$C_{\gamma1}$-$C_{\gamma2b}$-$C_{\gamma2a}$-C_ε-C_α, was deleted by targeted integration of two LoxP sites followed by Cre-loxP-mediated in vivo deletion [36].

The ES cell-based gene targeting was also applied to knockout mouse IgK and created lines with features including C_κ disruption [37] and targeted deletion of all J_κ and C_κ gene segments [38]. The mouse Igλ gene locus, spanning 200 kbps, contains two sets of V_λ, J_λ and C_λ gene segments separated by an intergenic region of approximately 100 kbps. A large deletion that removes all functional C_λ gene segments was accomplished in several stages using ES cell-based gene targeting combined with Cre-loxP-mediated in vivo deletion [39].

In large mammals such as cattle and pig, targeted gene disruption and deletion of large genomic region can be achieved initially in cultured primary fibroblast cells. Through animal cloning, the offspring with intended genotype can be derived from the engineered fibroblast cells. Using this strategy, both bovine IgM genes, b*IGHM* and b*IGHML1*, were disrupted sequentially to yield double knockout cattle in which the endogenous B-cell development was blocked. Tc cattle was established in the double knockout background using HACs comprising the entire unrearranged human IgH, Igκ and Igλ germline loci [40]. However, because in this genetic background the endogenous bovine Ig light chain gene loci were still intact, significant levels of chimeric antibodies comprising human IgH derived from transgene and the endogenous bovine Igκ or Igλ could be found. To improve fully human Ig production in Tc cattle, the entire genomic region encompassing bovine J_λ and C_λ gene segments was deleted using the fibroblast approach to yield Tc cattle in triple knockout background (b*IGHM*$^{-/-}$, b*IGHML1*$^{-/-}$ and b*IGL*$^{-/-}$) [41]. Furthermore, by using a similar strategy, knockout pigs with the deletion of all J_H gene segments or the C_κ gene were separately established in 2011 [42, 43]. Targeted deletions in chicken Ig heavy and light chain gene loci were also achieved using the primordial germ cells approach described above [44].

Neither ES cell-based gene targeting nor animal cloning is well established in rat and rabbit. In both species, gene disruption and deletion can be achieved using specially designed zinc finger nucleases (ZFNs). In particular, the messenger RNAs encoding a pair of engineered ZFNs that introduce double-strand break in a specific DNA region can be directly injected into fertilised oocytes. In these cells, repairment of the DNA break by the error-prone non-homologous end joining can usually introduce nucleotide insertion or deletion ranging from 1 bp to several kbps into the break point. Animals born which carry the ensuing gene disruption or deletion can be identified through PCR, sequencing and Southern blot analysis [45, 46]. Using this method, rat lines carrying deletions of the entire J_H gene segments, C_κ gene and all functional J_λ-C_λ gene segments were generated to establish the triple knockout strain which completely lacks the production of endogenous Ig heavy and light chains [47, 48]. Rabbit lines with disruption in exon 1 and 2 of IgM gene locus were also established [46].

5.2.3   Human IgH Gene Loci in Transgenic Constructs

To date, there are documented 56 functional V_H, 23 functional D and 6 functional J_H gene segments in humans (http://www.imgt.org/genedb/) [49]. In the genome, the functional V_H segments are distributed among many highly related pseudogenes with a total number of identified V_H segments reaching 159. The V_H gene locus, spanning nearly 1 Mbps, is highly polymorphic among different people. This is reflected not only in the abundance of alleles for each V_H gene but also in the variation of V_H gene counts among individuals. Duplication and deletion of genomic regions containing V_H segments are frequently observed. For instance, a study on the haplotypes of 9 human donors showed there were between 35 and 46 functional V_H segments in 18 unique haplotypes [50]. When developing transgenic animals, great efforts have been made to incorporate nearly all known functional human V_H gene segments into the host's genome. However, since the majority of transgenic constructs containing human IgH locus have been generated from cloned genomic regions from one or more donors, they inevitably represent the genetic makeup of those individuals. This makes it very difficult for any transgenic construct to capture the complete functional V_H gene repertoire of the human species. In addition, there are some issues in antibody drug discovery that argue against using the complete repertoire of human functional V_H genes. Firstly, it is known that some V_H genes are more likely to encode antibody fragments prone to misfolding or aggregation [51]. One example is VH7–4 which contains an unpaired cysteine residue in its framework 3 whose reactivity may promote aggregation [52]. Secondly, if an antibody drug were to be given to individuals who don't possess the corresponding V_H gene used in that drug, there can be serious concerns for immunogenicity due to the likely absence of tolerance in these patients. It has been suggested when designing the transgenic construct, it may be preferential to incorporate the functional V_H gene repertoire that are most common to the bulk of the human population [53]. The functional D and J_H gene segments used in humans are covered in a relatively compact genomic region. Thus, the majority of transgenic constructs have this entire region incorporated.

The first-generation transgenic constructs were designed to produce fully human antibodies in animals, including the genomic loci encoding human IgM, IgD and one of the IgG constant regions [53]. Although, upon immunisation, transgenic animals carrying this type of construct elicited pronounced immune responses, and fully human therapeutic mAbs were subsequently discovered, there have been some concerns about the suboptimal B-cell development and relatively low antibody serum titres in these animals [48, 54, 55]. This drawback is underpinned by several possible sources of inefficiencies. Firstly, when functioning as a B-cell receptor (BCR), the interspecies interaction between human IgM constant region and the host's signalling proteins such as Igα and Igβ may be inefficient [56]. This could compromise the level of signalling which are normally required for B-cell maturation, survival and proliferation in the host. Secondly, the secreted fully human immunoglobulins from B cells in transgenic animals may interact inefficiently with various host Fc receptors, further compromising the magnitude of humoral immune response against target antigens [57]. Thirdly, even though the trans-acting molecular machinery in the host can exert its function on regulatory elements in the human IgH loci such as the recombination signal sequences and switch regions which govern gene rearrangement, antibody class switch and appropriate expression in a temporal and cell lineage-specific manner, such interactions may not be optimal [55]. Lastly, due to the size limit imposed by cloning and transgenic strategies, some of the first-generation transgenic constructs may lack the extended locus control region such as the complete $3'$ enhancer located down-stream of all constant region genes which might affect antibody expression and class switch [48, 58]. To overcome these inefficiencies, the second-generation transgenic constructs have been created which are chimeric, comprising the human IgH locus encoding antibody variable regions followed by the host locus encoding constant regions which provides a functional replacement to the endogenous locus especially in the knockout background [48, 59]. In particular, the host locus may contain the intronic enhancer region at the $3'$ of J_H and $5'$ of C_μ, the entire C_μ gene, part of or all the rest of constant region genes including one or more C_γ genes, all intact switch regions corresponding to the constant region genes and the extensive locus control region. Furthermore, in the in situ humanisation approach through ES cell manipulation in mouse, the gene replacement has been limited to the mouse genomic region encompassing V_H, D and J_H gene segments while leaving the constant region genes completely intact [22, 54, 60]. As a result, the second-generation transgenic animals can efficiently produce chimeric Ig with only the variable regions encoded exclusively by the human gene repertoire. Following the discovery process, such chimeric antibodies can be easily converted to a desired fully human format through antibody engineering.

5.2.4 Human Igκ and Igλ Gene Loci in Transgenic Constructs

Sequencing of the human Igκ gene locus has identified 41 functional V_κ genes which are found in 2 clusters arising through segmental duplication [61, 62]. The proximal cluster is located ~24 kbps upstream of 5 J_κ genes and a single C_κ gene and spans ~444 kbps. There are 23 functional V_κ genes in this cluster, with all but the 2 most $3'$ V_κ genes sharing the same orientation as the J_κ and C_κ genes. The distal cluster is located almost ~1 Mbps upstream of the J_κ and C_κ genes and contains 18 functional V_κ genes. The orientation of all these V_κ genes is opposite to that of J_κ and C_κ genes. Expression analysis has shown that the distal cluster contributes little to the diversity of the human Igκ repertoire, and in some people, this cluster appears to be missing [63]. Hence, many transgenic human Igκ constructs only contain part of or the entire proximal cluster together with human J_κ and C_κ genes [22, 48, 53]. Additionally, it is crucial to incorporate the regulatory regions at the $3'$ end of the C_κ gene, especially the kappa deleting element, as these regions ensure the proper expression of human Igκ in animals [64]. Further, in situ humanisation of mouse V_κ genes by replacement with the corresponding human genes has been achieved [22, 54, 60].

The human Igλ gene locus contains 33 functional $V_λ$ genes and 5 downstream tandem cassettes, each made up of a $J_λ$ gene followed by a $C_λ$ gene. The majority of transgenic constructs expressing human Igλ in animals contain part of or all human $V_λ$ genes, all $J_λ$-$C_λ$ cassettes and the 3′ enhancer region [48, 60, 65]. In mammals, the proportion of serum antibodies containing κ or λ light chains varies considerably between different species. In mice and rat, the (% κ:λ) ratio can be more than 95:5; in humans, the ratio is approximately 67:33; and in cattle, the ratio can reach 1:99 (https://www.sigmaaldrich.com/technical-documents/articles/biology/antibody-basics.html) [41, 66, 67]. As mice make few antibodies with a λ light chain, the Igλ gene locus in this species is considered to be relatively silent. Therefore, since human Igκ-transgenes can achieve significantly high expression, it is regarded as preferable but not essential to include human Igλ-transgenes in transgenic mice platforms.

5.3 Examples of Currently Available Transgenic Platforms for Human Antibody Discovery

Table 5.1 contains a list of transgenic platforms in different species currently in use for human antibody discovery. Several platforms with published data in peer-reviewed journals are described in detail below.

5.3.1 HuMAb-Mouse® and XenoMouse®

Developed in the 1990s, HuMAb-Mouse® and XenoMouse® are both history-making first-generation transgenic platforms for generating fully human antibodies. HuMAb-Mouse® was originally created by Genpharm Inc. which later became Medarex and is now part of Bristol-Myers Squibb. This platform was also branded as UltiMAb® transgenic mouse technology by Medarex and licensed to Genmab and ImClone Systems (now part of Eli Lilly and Company). This platform employed fully human IgH- and Igκ-miniloci transgenes with the following distinctive features: deletion of all murine J_H gene segments as well as the entire $J_κ$-$C_κ$ genomic region; a single 80 kbps IgH transgene locus containing 4 functional human V_H, 16 D, all 6 J_H, $C_μ$ and $C_{γ1}$ gene segments with associated regulatory regions such as intronic enhancer, switch regions and locus control region; and a reconstituted Igκ-transgene locus (~500 kbps) through co-integration of 3 constructs into the mouse genome including 26 functional human $V_κ$, all $J_κ$ and $C_κ$ followed by a downstream regulatory region (Fig. 5.1a) [70, 71]. The diversity of the V_H and $V_κ$ gene repertoire of this platform was later expanded to include additional variable gene segments [53]. Even with its initially limited variable gene diversity, HuMAb-Mouse® is the most prolific transgenic platform so far, yielding 11 FDA-approved fully human mAb drugs, accounting for nearly half of those derived from transgenic animals [72].

Table 5.1 A list of currently available transgenic platforms for human antibody discovery

Platform name	Transgenic strategy	Features disclosed in scientific publications or online	Reference
Platforms producing conventional fully human or chimeric antibodies:			
5′ feature mouse (MRC Laboratory of Molecular Biology, UK)	ES cells fused to YAC-containing spheroplasts	1. Background: $\mu MT^{-/-}$ and $MoIg\kappa^{-/-}$ (disruption of $C\kappa$) 2. IgH construct contains 5 functional human V_{Hs}, complete D_s and J_{Hs} and human C_μ-C_δ, but no C_γ gene 3. $Ig\kappa$ and $Ig\lambda$ constructs contain part of human loci	[68, 69]
HuMAb-Mouse®	Pronuclei microinjection	See Sect. 5.3.1	[70, 71]
XenoMouse®	ES cells fused to YAC-containing spheroplasts	See Sect. 5.3.1	[73–75]
Tc Mouse™	ES cells carrying HACs	See Sect. 5.3.2	[25, 76]
Tc Bovine	Engineered HACs transferred into fibroblast cells followed by animal cloning	See Sect. 5.3.2	[31, 40, 41, 55, 77]
THP rabbit (Therapeutic Human Polyclonals, now Roche)	Pronuclei microinjection	For the generation of polyclonal antibodies encoded by human Ig gene loci on transgenic constructs	[82]
VelociMouse®	ES cells, in situ humanisation	See Sect. 5.3.4	[22, 54]
H2L2 Transgenic Mice (Harbour Antibodies)	Pronuclei microinjection	1. Endogenous Ig expression inactivated 2. IgH construct contains 18 human V_{Hs}, 19 human D_s, 6 human J_{Hs}, rat C_μ, $C\gamma_{2c}$, $C\gamma_1$, $C\gamma_{2b}$, $C\gamma_\alpha$ and mouse locus control region 3. $Ig\kappa$ construct contains 11 human $V\kappa_s$, 5 human $J\kappa_s$, rat $C\kappa$ and mouse enhancer regions 4. No $Ig\lambda$ transgene	https://harbourantibodies.com/science-technology/h2l2/#.Xek7kej7SUk
OmniRat® and OmniMouse®	Pronuclei microinjection	See Sect. 5.3.3	[48, 59]

(continued)

Table 5.1 (continued)

Platform name	Transgenic strategy	Features disclosed in scientific publications or online	Reference
Kymouse™	ES cells, in situ humanisation	See Sect. 5.3.4	[60]
AlivaMab Mouse (Ablexis)	(information not available)	1. Autonomously functioning synthetic Ig transgenes were designed and synthesised 2. Curated repertoires of human V_H, V_κ and V_λ gene segments were selected for developability and lowered risk of immunogenicity 3. To better maintain the native conformation of human F(ab) fragment, the transgenes yield human/mouse chimeric antibodies with the heavy chain variable domain-CH1-upper hinge region and the entire light chain being fully human	https://www.ablexis.com/technology/
Trianni Mouse®	ES cells, in situ humanisation	See Sect. 5.3.4	https://trianni.com/technology/transgenicmouse/
OmniChicken®	Genetic modification in cultured primordial germ cells, then injection into recipient embryos	See Sect. 5.3.3	[44, 83]
Tc goat (caprine)	HACs transferred into fibroblast cells followed by animal cloning	See Sect. 5.3.2	[81]
Platforms producing chimeric antibodies with a fixed human Ig light chain:			
OmniFlic®	Pronuclei microinjection	See Sect. 5.3.5.1	[85]
OmniClic®	As in OmniChicken®	See Sect. 5.3.5.1	https://www.omniab.com/
MeMo® mouse (Merus)	(information not available)	1. Knockout of endogenous murine Ig loci 2. Transgenes contain the human IgH gene locus and a rearranged fragment encoding a single Ig light chain	https://merus.nl/technology/

(continued)

Table 5.1 (continued)

Platform name	Transgenic strategy	Features disclosed in scientific publications or online	Reference
Platforms producing chimeric heavy chain only antibodies (HCOAs):			
HCAb transgenic mice	Pronuclei microinjection	See Sect. 5.3.5.2	[100, 101]
Humabody® VH platform	ES cells fused to YAC-containing spheroplasts and pronuclei microinjection	See Sect. 5.3.5.2	[102]
UniRat®	Pronuclei microinjection	See Sect. 5.3.5.2	[103]

XenoMouse® was originally created by Cell Genesys Inc. which later became Abgenix and now part of Amgen. The technology of fusing mouse ES cells with spheroplasts carrying modified YACs which contain megabase-sized human Ig transgene locus was employed to generate this platform. XenoMouse® carries the following features: double knockout background through deleting all mouse J_H gene segments and C_κ region; stable integration into the genome of a 970 kbps fully human IgH locus encompassing in germline configuration 66 consecutive V_H gene segments of which 34 are functional; all 30 D, 6 J_H, C_μ and C_δ followed by 1 C_γ ($C_{\gamma1}$, $C_{\gamma2}$ or $C_{\gamma4}$) in conjunction with all associated regulatory regions; and stable integration of a 800 kbps human Igκ locus encompassing the entire proximal cluster of V_κ gene segments, all J_κ and C_κ followed by the downstream regulatory region (Fig. 5.1b) [73–75]. Three different XenoMouse® strains were generated, each producing fully human antibodies with only one of the three isotypes, IgG1κ, IgG2κ and IgG4κ, allowing the generation of antigen-specific mAbs with the desired effector functions for specific disease indications. Later, the entire human Igλ locus carried on a YAC was also introduced into existing XenoMouse® strains to generate three additional strains, each producing both human IgGκ and IgGλ antibodies with the corresponding isotype [53]. To date the XenoMouse® platform has yielded 7 FDA-approved fully human mAb drugs, accounting for nearly one third of those derived from transgenic animals [72].

5.3.2 Tc Mouse™, Tc Bovine and Tc Goat (Caprine)

The Tc Mouse™ which stands for 'TransChromo mouse' was engineered by Kyowa Hakko Kirin Co., Ltd. from the mid-1990s to early 2000s. This platform was generated by using mouse ES cells that harbour HACs derived from the truncated human chromosomes 2, 14 or 22 containing the entire Igκ-, IgH- and Igλ-gene locus, respectively, to establish the Tc lines which were then bred with a double knockout mouse line with the disruption of endogenous IgM and Igκ gene locus [25]. The lack of stability of the HACs was a major issue of this platform, which limited its use for antibody discovery. It appears

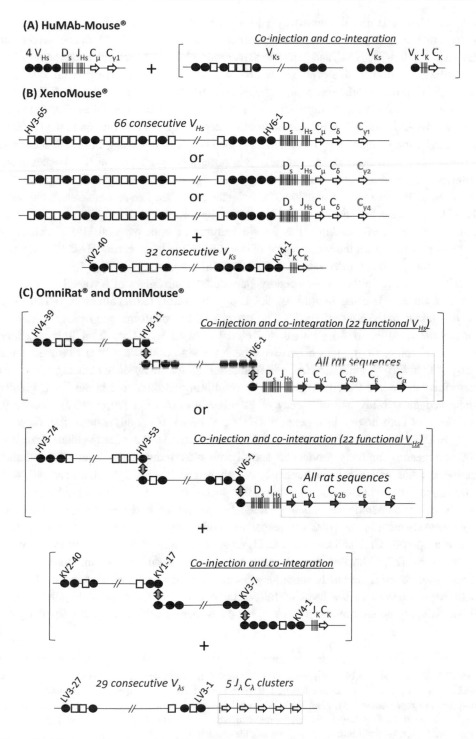

Fig. 5.1 Diagram showing the composition of transgenes in (**a**) HuMAb-Mouse®, (**b**) XenoMouse®, (**c**) OmniRat® and OmniMouse®. Functional V_H, V_K and V_λ genes are represented

that the HAC carrying the human IgH locus is relatively stable mitotically and can be transmitted through germline, whereas significant mitotic and meiotic loss is observed for Igκ- and Igλ-bearing HAC, and germline transmission have not been achieved for both. An immediate solution to this problem was to crossbreed the Tc Mouse™ with HuMAb-Mouse® to produce offspring which retains both the stable IgH-bearing HAC from the former and the Igκ locus stably integrated into the genome from the latter. This strategy yielded the KM Mouse™ [76]. Another more generally applicable solution to the problem is through further engineering of the HACs in the DT40 cell line using Cre-LoxP-mediated chromosome translocation. In particular, via homologous integration, LoxP sites were inserted into the designated sequence on IgH- and Igκ- or Igλ-bearing HAC, respectively, in DT40 cells. Through the function of transiently expressed Cre recombinase, a chimeric HAC that contains both the human IgH locus and Igκ or Igλ locus could then be generated. Such chimeric HAC combines the genetic features of both parental HACs and, most importantly, possesses the same degree of stability as the IgH-bearing HAC [25]. So far, the Tc Mouse™ has yielded one FDA-approved fully human mAb drug.

Tc Bovine was initially developed by Hematech, a subsidiary of Kyowa Hakko Kirin Co. Ltd. and later became Sanford Applied Biosciences LLC (SAB) after its acquisition by North Dakota-based Sanford Health. The key aim of this platform is to rapidly produce a substantial quantity of fully human polyclonal antibodies (pAbs) from immunised or vaccinated cattle. Continuous efforts have been made to optimise this platform since the early 2000s. Firstly, chimeric HAC via fusing human Igλ or Igκ locus with IgH locus as described above and the latest construct by combining all three loci in one HAC greatly improved its stability and efficiency of germline transmission [30, 31, 77]. Secondly, knockout of both bovine IgM genes, b*IGHM* and b*IGHML1*, and, additionally, deletion of J_λ and C_λ gene segments to disrupt the dominantly expressed bovine Igλ locus abolished the endogenous antibody production and significantly reduced the level of circulating chimeric pAbs which contain human heavy chain and bovine light chain [31, 40, 41]. Thirdly, to circumvent the human-bovine interspecies incompatibility between human IgM and bovine transmembrane signalling proteins - Igα and Igβ in the pre-BCR complex, the IgM gene segment on the HAC was partially bovinised by replacing the human sequences encoding part of CH2 and the entire CH3, CH4, TM1 as well as TM2 with their bovine counterparts [77]. The resulting chimeric IgMs in the membrane-bound form showed notably improved functions in supporting B-cell activation and proliferation, and led to a significant increase in the level of fully human pAbs in the plasma. Finally, further bovinisation to counter human-bovine interspecies incompatibility included replacing the

←──

Fig. 5.1 (continued) by black circles, and pseudogenes are represented by blank squares. The first and last V_H, V_K and V_λ genes are labelled. Regions in continuous sequences but not shown in the diagram are represented by broken lines. All variable region, constant region genes and intergenic regions are derived from human except the rat sequences specified in (**c**). In (**c**), the overlaps between transgenic constructs are indicated as green double arrows

class switch regulatory element preceding human IgG1 gene segment with the counterpart associated with bovine IgG1 gene and the sequence encoding human IgG1 transmembrane as well as cytoplasmic domains with their counterparts from bovine IgG1 gene [55]. These species-specific modifications at the chromosome level greatly improved the fully human pAbs production, especially with a desired dominance of human IgG1, κ sub-class in the resulting Tc cattle. Plasma levels of up to 15 g/L fully human IgGs were reached in these transgenic animals, close to the serum antibody levels in a healthy human. Following immunisations or vaccinations, Tc bovine has demonstrated its remarkable potential in producing a substantial quantity of highly potent pAbs against several deadly human viral pathogens including Ebola virus and Middle East respiratory syndrome (MERS) coronavirus [78–80]. The same HAC optimised via a couple rounds of bovinisation to establish the latest Tc cattle was also used to produce Tc goat (caprine) and proven to function efficiently in the latter species due to the high similarity of genomic sequences in both ungulates [81]. It has been shown that Tc goat could work as an economical platform for producing fully human pAbs and upon immunisation, specific antibodies against Avian influenza virus subtype H7N9 could be obtained from this platform.

5.3.3 OmniRat®, OmniMouse® and OmniChicken®

OmniRat® and OmniMouse® are second-generation transgenic platforms producing chimeric antibodies comprising human variable and rodent Fc regions. Both were developed between 2008 and 2015 by Open Monoclonal Technologies Inc. (OMT) which became part of Ligand Pharmaceuticals Inc. in 2016. Technological advances in three areas underlined the rapid development of OmniRat® [48, 59]. Firstly, a circular YAC (cYAC)/BAC shuttle vector was developed which allows multiple DNA fragments with overlapping ends to be assembled as a self-replicating cYAC in a single step in *S. cerevisiae*. Following extraction, such cYAC can be transformed into *E. coli* as a BAC. The high fidelity associated with the robust homologous recombination system and the stability of highly repetitive DNA in yeast have been especially advantageous when constructing the human-rat chimeric transgene locus. This method ensures the integrity of the highly complex regulatory regions linked to the rat IgH constant gene segments which include switch regions, various intronic enhancers and the 3′ locus control region. In addition, the resulting BACs can be efficiently purified in a large quantity and high purity from *E. coli* to facilitate the transgenic process. Secondly, the gene knockout technology via specifically designed ZFNs was employed to generate a triple knockout rat strain carrying deletions of the entire J_H gene segments, C_κ gene and all functional J_λ-C_λ gene segments in a short timeframe. The endogenous antibody production is completely inactivated in this strain. Thirdly, co-injection of three large DNA fragments with overlapping ends of 6–20 kbps into the pronuclei of fertilised rat oocytes resulted in the co-integration of these fragments into the host genome in the expected configuration with relatively high efficiency. This method allows the reconstitution of a transgene locus up to 500–600 kbps in size via one round of microinjection combined with subsequent founder identification. As shown in Fig. 5.1c, two OmniRat®️ strains have been generated, each harbouring a distinct chimeric IgH transgene locus composed of a series of human V_H with 22 being functional, all human D_H and J_H followed immediately by rat intronic enhancer region at the $3'$ of J_H and $5'$ of C_μ, the entire rat C_μ-$C_{\gamma 1}$-$C_{\gamma 2b}$-C_ε-C_α in conjunction with all associated regulatory regions and the $3'$ locus control region [48, 59, 83]. Both strains also contain the reconstituted human Igκ locus encompassing all functional V_κ gene segments within the proximal cluster, all J_κ and C_κ followed by the downstream regulatory region and a human Igλ locus encompassing 16 functional V_λ genes, all 5 J_λ-C_λ cassettes and the $3'$ enhancer region ([48, 59] and unpublished data). The two OmniRat®️ strains can be bred together to increase the diversity of the V_H gene repertoire. The same transgenic constructs for establishing OmniRat®️ were also used to develop OmniMouse®️, and the latter was generated in the double knockout background by deleting all mouse J_H gene segments and C_κ region via gene targeting in ES cells.

OmniChicken®️ is also a second-generation transgenic platform producing chimeric IgGs comprising human variable and chicken IgY Fc regions. It was developed by Crystal Bioscience Inc. which later became part of Ligand Pharmaceuticals Inc. As human and chicken are much more phylogenetically distant than human and rodent, this platform potentially provides a way to raise mAbs against antigens which are highly conserved between mammals. Further, cross-reactive mAbs against a given human antigen and its mouse homolog can be obtained from this platform, which is highly desirable for pre-clinical drug development. The chicken Ig heavy chain gene locus contains a single functional V_H, a small number of highly related D_H and a single J_H segment, and its only light chain locus contains a single functional V_L and J_L segment. Rearrangement in both loci produces limited diversity in the somatic repertoire. Further diversity is accumulated by multiple rounds of gene conversion events involving a variety of upstream pseudogenes which are used to mutate the rearranged V_HDJ_H and V_LJ_L [84]. These mutations are mainly focused in the CDRs. To create OmniChicken®️, the endogenous Ig heavy and light chain loci were humanised in situ in the cultured primordial germ cells prior to establishing transgenic lines. The humanisation of the heavy chain locus was achieved in a two-step process: (1) a landing pad containing selectable markers and an attP site was utilised to replace the single J_H segment by homologous recombination, producing an IgH knockout, and (2) co-transfection of phiC31 integrase along with a synthetic human locus encompassing 20 V_H pseudogenes followed by a region encoding a rearranged functional V_HDJ_H which consists of V_H3-23, D1 and J_H4, a frequently found combination in the human population [84]. The integrase facilitated the integration of human transgene into the attP site specifically, allowing the V_HDJ_H region to splice precisely to the endogenous chicken Ig constant region gene segments. The pseudogenes were carefully designed to encode the similar framework regions as in the rearranged V_HDJ_H but with very high diversity in CDRs. Using a strategy analogous to heavy chain humanisation, chicken Ig light chain locus was humanised to contain a series of V_K or V_λ pseudogenes followed by a region encoding a rearranged functional V_KJ_K or $V_\lambda J_\lambda$ which is well expressed in humans.

Robust immune responses and production of high-affinity chimeric antibodies have been demonstrated in OmniChicken® [84]. OmniRat®, OmniMouse® and OmniChicken® form the core of the multi-species OmniAb® therapeutic human antibody platforms which can be licensed from Ligand Pharmaceuticals Inc.

5.3.4   VelociMouse®, Kymouse™ and Trianni Mouse®

VelociMouse® was developed by Regeneron Pharmaceuticals from the late 1990s to early 2010s. It is the first transgenic mouse platform with a megabase of endogenous Ig gene loci being precisely humanised in situ using the ES cell-based technology. As illustrated in Fig. 5.2a, the mouse IgH locus was humanised in nine steps [22]. The initial step was to

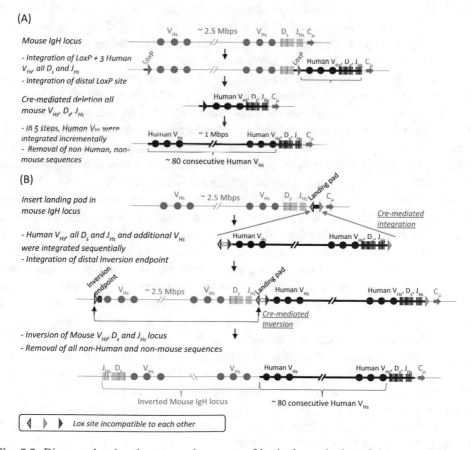

Fig. 5.2 Diagram showing the transgenic process of in situ humanisation of the mouse IgH gene locus when creating (A) VelociMouse® and (B) Kymouse™. Functional V_H genes are represented by solid circles. Regions in continuous sequences but not shown in the diagram are represented by broken lines. Mouse endogenous IgH locus is in blue, and human transgene locus is in black

integrate a 144 kbps transgenic construct derived from a modified BAC into the mouse genome to replace a 16 kbps region including all J_{Hs} via homologous recombination. The construct contains a LoxP site and a human genomic region encompassing 3 V_H, all D and J_H in germline configuration flanked by mouse genomic sequences matching those surrounding the designated insertion sites. This integration ensured the intact mouse intronic enhancer and C_μ switch region to be located immediately downstream of the human J_H segments which in turn prompted the efficient production of chimeric antibodies comprising human variable and mouse Fc region. The second step was to insert a LoxP site along with an appropriate selectable marker upstream of the most distal mouse V_H segment. In the third step, via Cre-LoxP-mediated chromosome deletion, a genomic region spanning ~2.5 Mbps and containing the entire mouse V_H, D and J_H locus was removed in the ES cells. Subsequently, in five steps, additional human V_H carried on modified BACs were inserted upstream of the existing ones through homologous recombination. In the final step, the remaining selectable markers after the final insertion step were removed through FLP-FRT-mediated deletion. As a result, the mouse V_H, D and J_H locus was precisely replaced by ~1 Mbps of human equivalent sequence consisting of 80 consecutive V_H, 27 D and 6 J_H. In addition, through a similar strategy, the mouse V_K and J_K locus spanning ~3 Mbps were deleted and subsequently replaced by a 500 kbps of human genomic region encompassing the entire proximal V_K gene cluster and all J_K segments [22]. Humanisation of the mouse Igλ locus wasn't reported in the published data of VelociMouse®. In this platform, the serum levels of antibodies of all subtypes are similar to those found in wild-type mouse pre- and post-immunisation, and the gene usage of V_H, D and J_H as well as V_K and J_K in the antibody repertoire is comparable to that in humans [54]. To date, four FDA-approved antibody drugs are derived from VelociMouse®.

Kymouse™ was developed by Kymab Ltd. and accomplished later than VelociMouse®. The creation of this platform also involved in situ humanisation of mouse IgH, IgK and, additionally, Igλ gene locus in ES cells [60]. As shown in Fig. 5.2b, in the first stage, ~1 Mbps of human V_H, D_s and J_H locus similar to that described for VelociMouse® was inserted immediately downstream of mouse J_H region in several steps. A technical advance in this work is the inventive use of three types of Lox sites (LoxP and two derivatives) which are incompatible with each other. Firstly, via homologous recombination, a landing pad containing an appropriate selectable marker flanked by two different Lox sites was integrated downstream of the most proximal mouse J_H segment. Subsequently, a series of human V_H, all D and J_H segments flanked by two Lox sites matching those in the initial landing pad were integrated into the genome facilitated by the transiently expressed Cre recombinase. Along with the integrated human transgene, a new landing pad at its 5' end was generated which has a different makeup from the initial one. Additional human V_H gene segments were inserted upstream of existing ones through several rounds of Cre-Lox-mediated integration. The process was designed so that between rounds, non-human and non-mouse sequences except those in the newly generated landing pad were removed through transiently expressed PiggyBac transposase. In the second stage, an 'Inversion endpoint' was integrated upstream of the most distal mouse V_H

segment. The Lox site in the endpoint and the matching one in the most distal landing pad are inverted relative to each other. Using Cre-Lox-mediated inversion, the entire mouse V_H, all D and J_H locus was put into opposite orientation to the mouse Ig constant region genes, resulting in no functional endogenous antibodies being produced. Lastly, all non-human and non-mouse sequences were removed by PiggyBac transposase. The inversion strategy of the endogenous locus differentiates from the large genome deletion used in developing VelociMouse®. Using a similar process, a derivative of human IgK locus spanning ~825 kbps including an inverted distal V_K gene cluster, the proximal V_K gene cluster and all J_K segments was integrated into the mouse genome upstream of the C_K segment, and a ~ 26.3 Mbps mouse genomic region encompassing all V_K and J_K segments was inverted relative to mouse C_K gene [60]. Furthermore, a ~917 kbps human genomic region encompassing all V_λ genes, J_λ-C_λ cassettes and the 3' enhancer region was integrated into the mouse Igλ locus [60]. Robust immune responses and an antibody repertoire with diversities as extensive as in humans were demonstrated in Kymouse™. Trianni Mouse® is a newly accomplished platform by Trianni Inc. In this platform, the in situ humanisation of mouse V_H, D, J_H, V_K, J_K, V_λ and J_λ gene segments was restricted to the coding regions only. All intronic and intergenic regions in mouse genome within and surrounding each of the gene segment were kept intact, preserving all the authentic cis-elements for gene expression control and consequently, eliminating any interspecies barrier between human and mouse. Details of the transgene design haven't been published for this platform.

5.3.5 Transgenic Platforms Expressing Different Formats of Antibodies

5.3.5.1 OmniFlic®and OmniClic®

Several transgenic platforms including OmniFlic®, OmniClic® and MeMo® mouse (Table 5.1) were created to express only a single human IgK light chain, so their antibody diversity was derived primarily from the human IgH transgene. These fixed light chain antibodies are especially useful in formulating bi-specific and multispecific antibody drugs as the undesirable mispairing between heavy and light chains of multiple different antibodies is eliminated. OmniFlic® was developed by OMT Inc. as a transgenic rat platform whose chimeric human-rat IgH transgene locus is the same as in OmniRat®. The fully human IgK transgene locus in this platform contains a genomic region encoding the rearranged IgKV3–15-J_K1 which is commonly found in the human population, followed by the C_K gene and downstream regulatory region [85]. OmniClic® was developed by Ligand Pharmaceuticals Inc. as a transgenic chicken platform which bears the same human IgH transgene as in OmniChicken®. The human genomic region encoding the rearranged IgKV3–15-J_K1 was used to replace the endogenous functional V_L and J_L segment, and all upstream V_L pseudogenes were removed. Upon immunisation, both OmniFlic® and OmniClic® can yield diverse, specific and highly potent antibodies with a common light chain against a variety of antigens ([85] and unpublished data). Both have been included in the OmniAb® collection of therapeutic human antibody platforms.

5.3.5.2 HCAb Transgenic Mice (Harbour Biomed), Humabody® VH Platform and UniRat®

Heavy chain only antibodies (HCAbs) devoid of any Ig light chain occur naturally in Camelidae and cartilaginous fish [86, 87]. In camelids these homodimeric IgG antibodies lack the first constant domain (CH1) [88] and use a diverse but distinct set of V_H region genes. These genes, termed V_{HH}, contain four hydrophilic amino acid substitutions at conserved positions which normally associate with the light chain and so improve single domain solubility [89]. The lack of the CH1 domain removes both a cysteine used in making a disulphide bond with the light chain constant region and a control mechanism ensuring the pairing of heavy and light chain in conventional antibodies. In the absence of light chain, the chaperonin protein BiP binds to the CH1 domain and prevents the heavy chain from being secreted from the ER [90, 91]. The specificity of camelid HCAbs is dependent on the variable domain which can be expressed separately, producing an antibody fragment (nanobody) of 12–14 kDa. These single domain antibodies (sdAb) have good biophysical properties including excellent stability and low levels of aggregation [92, 93]. They also target epitopes not readily accessible to classical antibodies [94, 95] and are predicted to penetrate tissue or tumours better than mAbs [96, 97]. A further feature is they can be easily linked together to generate multivalent or multispecific molecules [96, 98]. Several clinical trials involving nanobodies are ongoing, with the first therapeutic drug, caplacizumab, a humanised bivalent anti-von Willebrand factor nanobody, receiving approval in Europe and the USA recently [99].

The first transgenic animals to make HCAbs used a hybrid llama/human antibody construct. This contained two llama V_{HH} genes and human D, J_H and mu and/or gamma constant region genes missing CH1, in a mouse background lacking endogenous B-cell development (μMT⁻). These animals also gave the first evidence that the mu/delta constant regions were not necessary to produce HCAbs and actually may be detrimental [100]. This hybrid transgenic mouse was modified to make fully human HCAbs by replacing the V_{HH} genes with, in the first instance human VH1–46, VH3–11, VH3–53 and VH3–23 [101]. These HCAb transgenic mice, developed by Harbour Antibodies (Harbour Biomed), have been modified further to include more V_H genes.

A second mouse transgenic platform, the Humabody® VH platform, otherwise known as the Crescendo Mouse, was created by Crescendo Biologics Ltd. using a 339 kbps YAC introduced into fertilised oocytes by pronuclear microinjection. The YAC comprised ten human V_H genes, all human D and J_H genes, the mouse μ enhancer, mouse Cγ1 without CH1 and murine 3′ enhancer region [102]. The background for this mouse is a triple knockout, silenced for endogenous heavy or light chain expression. These animals successfully utilised all V_H and J_H gene segments producing high-affinity VH domains with bio-therapeutic potential.

A platform producing human HCAbs in transgenic rats has been developed by TeneoBio Inc., a spinoff from OMT Inc. Similar to the OmniRat® platform, two UniRat® strains in triple knockout background were generated of which the human V_H, D and J_H loci are identical to both OmniRat® strains. The engineered rat IgH constant gene locus in both UniRat® strains contains the rat intronic enhancer at the $3'$ of J_H locus and μ switch region which is directly linked to Cγ2a-Cγ1-Cγ2b, all lacking CH1 domains, followed by Cε-Cα genes in conjunction with all associated regulatory regions and the $3'$ locus control region [83]. One germline change was introduced, W103R (Kabat numbering) in J_H4, a residue which is constitutively conserved and is an important contact for the light chain [103]. This change decreases the V domain hydrophobicity and was found to reduce the occurrence of lambda light chain association on screening UniRat sequences by ELISA. The UniRat® platform gave robust immune responses to a wide array of antigens, with HCAbs covering nearly the entire productive human heavy chain V(D)J repertoire being identified, producing a large panel of stable sdAbs with high affinity and specificity [83].

5.4   Strategies in Antibody Discovery from Transgenic Platforms

Hybridoma technology is commonly used for the discovery of fully human or chimeric mAbs from immunised transgenic rodents. Antigen-specific single B-cell sorting followed by antibody repertoire cloning from single cells is also widely used [104]. Encapsulating single plasma cells along with detection reagents in microdroplets combined with image analysis is especially useful in identifying antigen-specific mAbs from immunised transgenic chickens [84]. New nanotechnology by placing single plasma cells on a microchip to achieve one cell per nano-well followed by screening and antibody repertoire cloning has also been validated for mAbs discovery from transgenic rodents [105]. Moreover, from immunised transgenic mice, a phage or yeast display library of antibody fragments with native heavy and light chain pairing has been constructed, which involves partitioning single B cells into emulsion droplets containing oligo(dT)-microbeads for synthesising beads-captured cDNA and, subsequently, amplifying linked cognate heavy and light chain transcripts via emulsion PCR from single cDNA-beads [106]. The library screening provides a fast way of identifying antigen-specific mAbs. Alternatively, with platforms producing mAbs with fixed light chain or HCAbs, since the antibody diversity is derived from the heavy chain only, following immunisation, a list of candidate binders can be predicted through next-generation sequencing of the entire heavy chain V(D)J repertoire. Ranking analysis, followed by high-throughput expression of antibody fragments from the candidate list and downstream screening, enables the antigen-specific VH domains to be identified [83, 85]. After the discovery phase, all selected binders can be re-formatted into products with desired formulations and expressed in large scale in a suitable system, such as the CHO-expression system which can routinely yield 5 g/L for mAbs and their derivatives.

A big challenge in antibody discovery from transgenic platforms is to obtain specific binders to antigens that are highly conserved between human and the host species. Many strategies have been adopted to break the immune tolerance, which include the knockout or knockdown of the homolog of a given antigen in the host prior to immunisations. Another solution to this problem is to choose a suitable transgenic host other than mice which have been dominantly used. These species include transgenic rat, rabbit and chicken that were made available commercially in recent years. Moreover, developing transgenic platforms is an expensive and highly time- as well as resource-consuming process, and consequently, each platform is protected through sophisticated patents. End users usually pay a substantial amount of licensing fee in adopting transgenic platforms for their discovery projects. In addition, a hefty royalty charge ensues if an antibody drug developed from a particular platform gains FDA approval. However, the choices of different platforms (multi-species and multiple antibody formats) have been significantly broadened recently, which may in turn drive down the cost in using transgenic platforms for antibody discovery.

Take Home Messages
- In transgenic platforms for human antibody discovery, the host antibody heavy chain VDJ and light chain VJ repertoire are substituted by their counterparts which are encoded exclusively by human transgenes.
- Transgenic platforms take advantage of the host's naturally antigen-driven antibody selection and maturation process which is associated with the primary and secondary immune response following immunisations with a selected target antigen.
- In transgenic platforms, the endogenous antibody production is inactivated via gene knockout and large chromosome deletion or inversion.
- Transgenic constructs in BACs or YACs containing human germline V_{Hs}, D_s and J_{Hs} genes in conjunction with C_H gene segments and all associated regulatory regions derived from human or host species and, separately, constructs containing human IgK and/or Igλ locus can be integrated into the host genome following oocyte microinjection or spheroplast-ES cell fusion.
- Alternatively, human humoral immunity in transgenic animals can be recapitulated from HACs containing human IgH, IgK and/or Igλ locus which are maintained in ES cells or transferred into cultured foetal fibroblast cells prior to establishing transgenic lines.
- In mouse ES cells or chicken primordial germ cells, the variable region gene segments in host IgH, IgK and/or Igλ loci can be humanised in situ.
- Antibody production in transgenic animals carrying full human IgH locus may be suboptimal. This has been attributed to imperfect interaction of the human constant region with endogenous signalling components such as the Igα/β in mouse, rat or cattle. Significant improvements have been obtained when the human IgH variable region genes were linked to the endogenous C_H gene segments and associated regulatory regions.

(continued)

- The antibody discovery process using transgenic platforms can be as short as 6–8 months from the immunisations to obtaining ideal binders, which significantly reduces the development time for antibody-based therapies.

Further Reading

Almagro JC, Daniels-Wells TR, Perez-Tapia SM, Penichet ML. Progress and Challenges in the Design and Clinical Development of Antibodies for Cancer Therapy. Front Immunol 2018; 8:1751

Brüggemann M, Osborn MJ, Ma B, Buelow R. Strategies to obtain diverse and specific Human monoclonal antibodies from transgenic animals. Transplantation 2017; 101 (8):1770–1776

Chen WC, Murawsky CM. Strategies for Generating Diverse Antibody Repertoires Using Transgenic Animals Expressing Human Antibodies. Front Immunol 2018; 9:460.

References

1. Shawler DL, Bartholomew RM, Smith LM, Dillman RO. Human immune response to multiple injections of murine monoclonal IgG. J Immunol. 1985;135(2):1530–5.
2. Hwang WY, Foote J. Immunogenicity of engineered antibodies. Methods. 2005;36(1):3–10.
3. Abramowicz D, Crusiaux A, Goldman M. Anaphylactic shock after retreatment with OKT3 monoclonal antibody. N Engl J Med. 1992;327(10):736.
4. Boulianne GL, Hozum N, Shulman MJ. Production of functional chimaeric mouse/human antibody. Nature. 1984;312:643–6.
5. Morrison SL, Johnson MJ, Herzenberg LA, Oi VT. Chimeric human antibody molecules: mouse antigen-binding domains with human constant region domains. Proc Natl Acad Sci U S A. 1984;81:6851–5.
6. Jones PT, Dear PH, Foote J, Neuberger MS, Winter G. Replacing the complementarity-determining regions in a human antibody with those from a mouse. Nature. 1986;321(6069):522–5.
7. Queen C, Schneider WP, Selick HE, Payne PW, Landolfi NF, Duncan JF, et al. A humanised antibody that binds to the interleukin 2 receptor. Proc Natl Acad Sci U S A. 1989;88(24):10029–33.
8. Clavero-Álvarez A, Di Mambro T, Perez-Gaviro S, Magnani M, Bruscolini P. Humanization of antibodies using a statistical inference approach. Sci Rep. 2018;8(1):14820.
9. Hwang WY, Almagro JC, Buss TN, Tan P, Foote J. Use of human germline genes in a CDR homology-based approach to antibody humanization. Methods. 2005;36:35–42.
10. Torres M, Fernandez-Fuentes N, Fiser A, Casadevall A. Exchanging murine and human immunoglobulin constant chains affects the kinetics and thermodynamics of antigen binding and chimeric antibody autoreactivity. PLoS One. 2007;2:e1310.
11. Garber E, Demarest SJ. A broad range of fab stabilities within a host of therapeutic IgGs. Biochem Biophys Res Commun. 2007;355:751–7.

12. McCafferty J, Griffiths AD, Winter G, Chiswell DJ. Phage antibodies: filamentous phage displaying antibody variable domains. Nature. 1990;348:552–4.
13. de Haard HJ, van Neer N, Reurs A, Hufton SE, Roovers RC, Henderikx P, et al. A large non-immunized human fab fragment phage library that permits rapid isolation and kinetics analysis of high affinity antibodies. J Biol Chem. 1999;274(26):18218–30.
14. Jain T, Sun T, Durand S, Hall A, Houston NR, Nett JH, et al. Biophysical properties of the clinical-stage antibody landscape. Proc Natl Acad Sci U S A. 2017;114(5):944–9.
15. Spencer S, Bethea D, Shantha Raju T, Giles-Komar J. Feng Y. Solubility evaluation of murine hybridoma antibodies mAb. 2012;4:319–25.
16. Brüggemann M, Caskey HM, Teale C, Waldmann H, Williams GT, Surani MA, Neuberger MS. A repertoire of monoclonal antibodies with human heavy chains from transgenic mice. Proc Natl Acad Sci U S A. 1989;86:6709–13.
17. Brüggemann M, Spicer C, Buluwela L, Rosewell I, Barton S, Surani MA, Rabbitts TH. Human antibody production in transgenic mice: expression from 100 kb of the human IgH locus. Eur J Immunol. 1991;21:1323–6.
18. Taylor LD, Carmack CE, Schramm SR, Mashayekh R, Higgins KM, Kuo C, et al. A transgenic mouse that expresses a diversity of human sequence heavy and light chain immunoglobulins. Nucleic Acids Res. 1992;20(23):6287–95.
19. Tuaillon N, Taylor LD, Lonberg N, Tucker PW, Capra JD. Human immunoglobulin heavy-chain minilocus recombination in transgenic mice: gene-segment use in μ and γ transcripts. Proc Natl Acad Sci U S A. 1993;90:3720–4.
20. Lee EC, Yu D, Martinez de Velasco J, Tessarollo L, Swing DA, Court DL, et al. A highly efficient *Escherichia coli*-based chromosome engineering system adapted for recombinogenic targeting and subcloning of BAC DNA. Genomics. 2001;73(1):56–65.
21. Zhang Y, Muyrers JP, Testa G, Stewart AF. DNA cloning by homologous recombination in *Escherichia coli*. Nat Biotechnol. 2000;18(12):1314–7.
22. Macdonald LE, Karow M, Stevens S, Auerbach W, Poueymirou WT, Yasenchak J, et al. Precise and in situ genetic humanization of 6 Mb of mouse immunoglobulin genes. Proc Natl Acad Sci U S A. 2014;111(14):5147–52.
23. Li L, Blankenstein T. Generation of transgenic mice with megabase-sized human yeast artificial chromosomes by yeast spheroplast-embryonic stem cell fusion. Nat Protoc. 2013;8(8):1567–82.
24. Kuroiwa Y, Tomizuka K, Shinohara T, Kazuki Y, Yoshida H, Ohguma A, et al. Manipulation of human minichromosomes to carry greater than megabase-sized chromosome inserts. Nat Biotechnol. 2000;18(10):1086–90.
25. Ishida I, Tomizuka K, Yoshida H, Kuroiwa Y. TransChromo mouse. Biotechnol Genet Eng Rev. 2002;19(1):73–82.
26. Wall RJ. Pronuclear microinjection. Cloning Stem Cells. 2001;3(4):209–20.
27. Vintersten K, Testa G, Stewart AF. Microinjection of BAC DNA into the pronuclei of fertilized mouse oocytes. Methods Mol Biol. 2004;256:141–58.
28. Valenzuela DM, Murphy AJ, Frendewey D, Gale NW, Economides AN, Auerbach W, et al. High-throughput engineering of the mouse genome coupled with high-resolution expression analysis. Nat Biotechnol. 2003;21(6):652–9.
29. Zou X, Xian J, Davies NP, Popov AV, Brüggemann M. Dominant expression of a 1.3 Mb human Ig kappa locus replacing mouse light chain production. FASEB J. 1996;10(10):1227–32.
30. Kuroiwa Y, Kasinathan P, Choi YJ, Naeem R, Tomizuka K, Sullivan EJ, et al. Cloned transchromosomic calves producing human immunoglobulin. Nat Biotechnol. 2002;20 (9):889–94.

31. Kuroiwa Y, Kasinathan P, Sathiyaseelan T, Jiao JA, Matsushita H, Sathiyaseelan J, et al. Antigen-specific human polyclonal antibodies from hyperimmunized cattle. Nat Biotechnol. 2009;27(2):173–81.
32. van de Lavoir MC, Diamond JH, Leighton PA, Mather-Love C, Heyer BS, Bradshaw R, et al. Germline transmission of genetically modified primordial germ cells. Nature. 2006;441 (7094):766–9.
33. Collarini EJ, Leighton PA, Van de Lavoir MC. Production of transgenic chickens using cultured primordial germ cells and Gonocytes. Methods Mol Biol. 1874;2019:403–30.
34. Kitamura D, Roes J, Kühn R, Rajewsky K. A B cell-deficient mouse by targeted disruption of the membrane exon of the immunoglobulin mu chain gene. Nature. 1991;350(6317):423–6.
35. Chen J, Trounstine M, Alt FW, Young F, Kurahara C, Loring JF, et al. Immunoglobulin gene rearrangement in B cell deficient mice generated by targeted deletion of the JH locus. Int Immunol. 1993;5(6):647–56.
36. Ren L, Zou X, Smith JA, Brüggemann M. Silencing of the immunoglobulin heavy chain locus by removal of all eight constant-region genes in a 200-kb region. Genomics. 2004;84(4):686–95.
37. Sanchez P, Drapier AM, Cohen-Tannoudji M, Colucci E, Babinet C, Cazenave PA. Compartmentalization of lambda subtype expression in the B cell repertoire of mice with a disrupted or normal C kappa gene segment. Int Immunol. 1994;6(5):711–9.
38. Chen J, Trounstine M, Kurahara C, Young F, Kuo CC, Xu Y, et al. B cell development in mice that lack one or both immunoglobulin kappa light chain genes. EMBO J. 1993;12(3):821–30.
39. Zou X, Piper TA, Smith JA, Allen ND, Xian J, Brüggemann M. Block in development at the pre-B-II to immature B cell stage in mice without Ig kappa and Ig lambda light chain. J Immunol. 2003;170(3):1354–61.
40. Kuroiwa Y1, Kasinathan P, Matsushita H, Sathiyaselan J, Sullivan EJ, Kakitani M, et al. Sequential targeting of the genes encoding immunoglobulin-mu and prion protein in cattle. Nat Genet. 2004;36(7):775–80.
41. Matsushita H, Sano A, Wu H, Jiao JA, Kasinathan P, Sullivan EJ, et al. Triple immunoglobulin gene knockout transchromosomic cattle: bovine lambda cluster deletion and its effect on fully human polyclonal antibody production. PLoS One. 2014;9(3):e90383.
42. Mendicino M, Ramsoondar J, Phelps C, Vaught T, Ball S, LeRoith T, et al. Generation of antibody- and B cell-deficient pigs by targeted disruption of the J-region gene segment of the heavy chain locus. Transgenic Res. 2011;20(3):625–41.
43. Ramsoondar J, Mendicino M, Phelps C, Vaught T, Ball S, Monahan J, et al. Targeted disruption of the porcine immunoglobulin kappa light chain locus. Transgenic Res. 2011;20(3):643–53.
44. Schusser B, Collarini EJ, Yi H, Izquierdo SM, Fesler J, Pedersen D, et al. Immunoglobulin knockout chickens via efficient homologous recombination in primordial germ cells. Proc Natl Acad Sci U S A. 2013;110(50):20170–5.
45. Geurts AM, Cost GJ, Freyvert Y, Zeitler B, Miller JC, Choi VM, et al. Knockout rats via embryo microinjection of zinc-finger nucleases. Science. 2009;325(5939):433.
46. Flisikowska T, Thorey IS, Offner S, Ros F, Lifke V, Zeitler B, et al. Efficient immunoglobulin gene disruption and targeted replacement in rabbit using zinc finger nucleases. PLoS One. 2011;6(6):e21045.
47. Ménoret S, Iscache AL, Tesson L, Rémy S, Usal C, Osborn MJ, et al. Characterization of immunoglobulin heavy chain knockout rats. Eur J Immunol. 2010;40(10):2932–41.
48. Osborn MJ, Ma B, Avis S, Binnie A, Dilley J, Yang X, et al. High-affinity IgG antibodies develop naturally in Ig-knockout rats carrying germline human IgH/Igκ/Igλ loci bearing the rat CH region. J Immunol. 2013;190(4):1481–90.

49. Giudicelli V, Chaume D, Lefranc MP. IMGT/GENE-DB: a comprehensive database for human and mouse immunoglobulin and T cell receptor genes. Nucleic Acids Res. 2005;33:D256–61.
50. Kidd MJ, Chen Z, Wang Y, Jackson KJ, Zhang L, Boyd SD, et al. The inference of phased haplotypes for the immunoglobulin H chain V region gene loci by analysis of VDJ gene rearrangements. J Immunol. 2012;188:1333–40.
51. Ewert S, Huber T, Honegger A, Plückthun A. Biophysical properties of human antibody variable domains. J Mol Biol. 2003;325:531–53.
52. Tomlinson IM, Walter G, Marks JD, Llewelyn MB, Winter G. The repertoire of human germline VH sequences reveals about fifty groups of VH segments with different hypervariable loops. J Mol Biol. 1992;227:776–98.
53. Green LL. Transgenic mouse strains as platforms for the successful discovery and development of human therapeutic monoclonal antibodies. Curr Drug Discov Technol. 2014;11(1):74–84.
54. Murphy AJ, Macdonald LE, Stevens S, Karow M, Dore AT, Pobursky K, et al. Mice with megabase humanization of their immunoglobulin genes generate antibodies as efficiently as normal mice. Proc Natl Acad Sci U S A. 2014;111(14):5153–8.
55. Matsushita H, Sano A, Wu H, Wang Z, Jiao JA, Kasinathan P, et al. Species-specific chromosome engineering greatly improves fully human polyclonal antibody production profile in cattle. PLoS One. 2015;10(6):e0130699.
56. Hombach J, Tsubata T, Leclercq L, Stappert H, Reth M. Molecular components of the B-cell antigen receptor complex of the IgM class. Nature. 1990;343(6260):760–2.
57. Nimmerjahn F, Ravetch JV. Fc-receptors as regulators of immunity. Adv Immunol. 2007;96:179–204.
58. Shi X, Eckhardt LA. Deletional analyses reveal an essential role for the hs3b/hs4 IgH 3′ enhancer pair in an Ig-secreting but not an earlier-stage B cell line. Int Immunol. 2001;13(8):1003–12.
59. Ma B, Osborn MJ, Avis S, Ouisse LH, Ménoret S, Anegon I, et al. Human antibody expression in transgenic rats: comparison of chimeric IgH loci with human VH, D and JH but bearing different rat C-gene regions. J Immunol Methods. 2013;400-401:78–86.
60. Lee EC, Liang Q, Ali H, Bayliss L, Beasley A, Bloomfield-Gerdes T, et al. Complete humanization of the mouse immunoglobulin loci enables efficient therapeutic antibody discovery. Nat Biotechnol. 2014;32(4):356–63.
61. Weichhold GM, Ohnheiser R, Zachau HG. The human immunoglobulin kappa locus consists of two copies that are organized in opposite polarity. Genomics. 1993;16(2):503–11.
62. Watson CT, Steinberg KM, Graves TA, Warren RL, Malig M, Schein J, et al. Sequencing of the human IG light chain loci from a hydatidiform mole BAC library reveals locus-specific signatures of genetic diversity. Genes Immun. 2015;16(1):24–34.
63. Schaible G, Rappold GA, Pargent W, Zachau HG. The immunoglobulin kappa locus: polymorphism and haplotypes of Caucasoid and non-Caucasoid individuals. Hum Genet. 1993;91(3):261–7.
64. Collins AM, Watson CT. Immunoglobulin light chain gene rearrangements. Receptor editing and the development of a self-tolerant antibody repertoire. Front Immunol. 2018;9:2249.
65. Popov AV, Zou X, Xian J, Nicholson IC, Brüggemann M. A human immunoglobulin lambda locus is similarly well expressed in mice and humans. J Exp Med. 1999;189(10):1611–20.
66. Woloschak GE, Krco CJ. Regulation of kappa/lambda immunoglobulin light chain expression in normal murine lymphocytes. Mol Immunol. 1987;24(7):751–7.
67. Molé CM, Béne MC, Montagne PM, Seilles E, Faure GC. Light chains of immunoglobulins in human secretions. Clin Chim Acta. 1994;224(2):191–7.
68. Nicholson IC, Zou X, Popov AV, Cook GP, Corps EM, Humphries S, et al. Antibody repertoires of four- and five-feature translocus mice carrying human immunoglobulin heavy chain and

kappa and lambda light chain yeast artificial chromosomes. J Immunol. 1999;163(12):6898–906.

69. Pruzina S, Williams GT, Kaneva G, Davies SL, Martín-López A, Brüggemann M, et al. Human monoclonal antibodies to HIV-1 gp140 from mice bearing YAC-based human immunoglobulin transloci. Protein Eng Des Sel. 2011;24(10):791–9.

70. Lonberg N, Taylor LD, Harding FA, Trounstine M, Higgins KM, Schramm SR, et al. Antigen-specific human antibodies from mice comprising four distinct genetic modifications. Nature. 1994;368(6474):856–9.

71. Fishwild DM, O'Donnell SL, Bengoechea T, Hudson DV, Harding F, Bernhard SL, et al. High-avidity human IgG kappa monoclonal antibodies from a novel strain of minilocus transgenic mice. Nat Biotechnol. 1996;14(7):845–51.

72. Strohl WR. Current progress in innovative engineered antibodies. Protein Cell. 2018;9(1):86–120.

73. Green LL, Hardy MC, Maynard-Currie CE, Tsuda H, Louie DM, Mendez MJ, et al. Antigen-specific human monoclonal antibodies from mice engineered with human Ig heavy and light chain YACs. Nat Genet. 1994;7(1):13–21.

74. Mendez MJ, Green LL, Corvalan JR, Jia XC, Maynard-Currie CE, Yang XD, et al. Functional transplant of megabase human immunoglobulin loci recapitulates human antibody response in mice. Nat Genet. 1997;15(2):146–56.

75. Jakobovits A, Amado RG, Yang X, Roskos L, Schwab G. From XenoMouse technology to panitumumab, the first fully human antibody product from transgenic mice. Nat Biotechnol. 2007;25(10):1134–43.

76. Ishida I, Tomizuka K, Yoshida H, Tahara T, Takahashi N, Ohguma A, et al. Production of human monoclonal and polyclonal antibodies in TransChromo animals. Cloning Stem Cells. 2002;4(1):91–102.

77. Sano A, Matsushita H, Wu H, Jiao JA, Kasinathan P, Sullivan EJ, et al. Physiological level production of antigen-specific human immunoglobulin in cloned transchromosomic cattle. PLoS One. 2013;8(10):e78119.

78. Dye JM, Wu H, Hooper JW, Khurana S, Kuehne AI, Coyle EM, et al. Production of potent fully human polyclonal antibodies against Ebola Zaire virus in Transchromosomal cattle. Sci Rep. 2016;6:24897.

79. Beigel JH, Voell J, Kumar P, Raviprakash K, Wu H, Jiao JA, et al. Safety and tolerability of a novel, polyclonal human anti-MERS coronavirus antibody produced from transchromosomic cattle: a phase 1 randomised, double-blind, single-dose-escalation study. Lancet Infect Dis. 2018;18(4):410–8.

80. Luke T, Bennett RS, Gerhardt DM, Burdette T, Postnikova E, Mazur S, et al. Fully Human Immunoglobulin G from Transchromosomic bovines treats nonhuman primates infected with Ebola virus Makona isolate. J Infect Dis 2018; 218(suppl_5):S636-S648.

81. Wu H, Fan Z, Brandsrud M, Meng Q, Bobbitt M, Regouski M, et al. Generation of H7N9-specific human polyclonal antibodies from a transchromosomic goat (caprine) system. Sci Rep. 2019;9(1):366.

82. Buelow R, van Schooten W. The future of antibody therapy. Ernst Schering Found Symp Proc. 2006;4:83–106.

83. Clarke SC, Ma B, Trinklein ND, Schellenberger U, Osborn MJ, Ouisse LH, et al. Multispecific antibody development platform based on human heavy chain antibodies. Front Immunol. 2019;9:3037.

84. Ching KH, Collarini EJ, Abdiche YN, Bedinger D, Pedersen D, Izquierdo S, et al. Chickens with humanized immunoglobulin genes generate antibodies with high affinity and broad epitope coverage to conserved targets. MAbs. 2018;10(1):71–80.

85. Harris KE, Aldred SF, Davison LM, Ogana HAN, Boudreau A, Brüggemann M, et al. Sequence-based discovery demonstrates that fixed light chain human transgenic rats produce a diverse repertoire of antigen-specific antibodies. Front Immunol. 2018;9:889.
86. Hamers-Casterman C, Atarhouch T, Muyldermans S, Robinson G, Hamers C, Songa EB, et al. Naturally occurring antibodies devoid of light chains. Nature. 1993;363(6428):446–8.
87. Könning D, Zielonka S, Grzeschik J, Empting M, Valldorf B, Krah S, et al. Camelid and shark single domain antibodies: structural features and therapeutic potential. Curr Opin Struct Biol. 2017;45:10–6.
88. Nguyen VK, Hamers R, Wyns L, Muyldermans S. Loss of splice consensus signal is responsible for the removal of the entire C(H)1 domain of the functional camel IGG2A heavy-chain antibodies. Mol Immunol. 1999;36:515–24.
89. Nguyen VK, Muyldermans S, Hamers R. The specific variable domain of camel heavy-chain antibodies is encoded in the germline. J Mol Biol. 1998;275:413–8.
90. Hendershot LM. Immunoglobulin heavy chain and binding protein complexes are dissociated in vivo by light chain addition. J Cell Biol. 1990;111:829–37.
91. Feige MJ, Groscurth S, Marcinowski M, Shimizu Y, Kessler H, Hendershot LM, et al. An unfolded CH1 domain controls the assembly and secretion of IgG antibodies. Mol Cell. 2009;34:569–79.
92. Dumoulin M, Conrath K, Van Meirhaeghe A, Meersman F, Heremans K, Frenken LG, et al. Single-domain antibody fragments with high conformational stability. Protein Sci. 2002;11 (3):500–15.
93. van der Linden RH, Frenken LG, de Geus B, Harmsen MM, Ruuls RC, Stok W, et al. Comparison of physical chemical properties of llama VHH antibody fragments and mouse monoclonal antibodies. Biochim Biophys Acta. 1999;1431(1):37–46.
94. Lauwereys M, Arbabi Ghahroudi M, Desmyter A, Kinne J, Holzer W, De Genst E, et al. Potent enzyme inhibitors derived from dromedary heavy-chain antibodies. EMBO J. 1998;17:3512–20.
95. De Genst E, Silence K, Decanniere K, Conrath K, Loris R, Kinne J, et al. Molecular basis for the preferential cleft recognition by dromedary heavy-chain antibodies. Proc Natl Acad Sci U S A. 2006;103(12):4586–91.
96. Tijink BM, Laeremans T, Budde M, Stigter-van Walsum M, Dreier T, de Haard HJ, et al. Improved tumor targeting of anti-epidermal growth factor receptor nanobodies through albumin binding: taking advantage of modular nanobody technology. Mol Cancer Ther. 2008;7(8):2288–97.
97. Oliveira S, van Dongen GA, Stigter-van Walsum M, Roovers RC, Stam JC, Mali W, et al. Rapid visualization of human tumor xenografts through optical imaging with a near-infrared fluorescent anti-epidermal growth factor receptor nanobody. Mol Imaging. 2012;11(1):33–46.
98. Huet HA, Growney JD, Johnson JA, Li J, Bilic S, Ostrom L, et al. Multivalent nanobodies targeting death receptor 5 elicit superior tumor cell killing through efficient caspase induction. MAbs. 2014;6(6):1560–70.
99. Scully M, Cataland SR, Peyvandi F, Coppo P, Knöbl P, Kremer Hovinga JA, et al. Caplacizumab treatment for acquired thrombotic thrombocytopenic purpura. N Engl J Med. 2019;380(4):335–46.
100. Janssens R, Dekker S, Hendriks RW, Panayotou G, Remoortere AV, San JK, et al. Generation of heavy-chain-only antibodies in mice. Proc Natl Acad Sci U S A. 2006;103:15130–5.
101. Drabek D, Janssens R, de Boer E, Rademaker R, Kloess J, Skehel J, et al. Expression cloning and production of human heavy-chain-only antibodies from murine transgenic plasma cells. Front Immunol. 2016;7:619.

102. Teng Y, Young JL, Edwards B, Hayes P, Thompson L, Johnston C, et al. Diverse human VH antibody fragments with bio-therapeutic properties from the crescendo mouse. New Biotechnol. 2020;55:65–76.

103. Chothia C, Novotny J, Bruccoleri R, Karplus M. Domain association in immunoglobulin molecules. The packing of variable domains. The packing of variable domains. J Mol Biol. 1985;186:651–63.

104. Ouisse LH, Gautreau-Rolland L, Devilder MC, Osborn M, Moyon M, Visentin J, et al. Antigen-specific single B cell sorting and expression-cloning from immunoglobulin humanized rats: a rapid and versatile method for the generation of high affinity and discriminative human monoclonal antibodies. BMC Biotechnol. 2017;17(1):3.

105. Winters A, McFadden K, Bergen J, Landas J, Berry KA, Gonzalez A, et al. Rapid single B cell antibody discovery using nanopens and structured light. MAbs. 2019;11(6):1025–35.

106. Adler AS, Bedinger D, Adams MS, Asensio MA, Edgar RC, Leong R, et al. A natively paired antibody library yields drug leads with higher sensitivity and specificity than a randomly paired antibody library. MAbs. 2018;10(3):431–43.

Applications of Antibodies in Therapy, Diagnosis, and Science

6

Laura Rhiel and Stefan Becker

Contents

L. Rhiel · S. Becker (✉)

Protein Engineering and Antibody Technologies, Merck Healthcare KGaA, Darmstadt, Germany

e-mail: laura.rhiel@merckgroup.com; stefan.c.becker@merckgroup.com

© Springer Nature Switzerland AG 2021

F. Rüker, G. Wozniak-Knopp (eds.), *Introduction to Antibody Engineering*,

Learning Materials in Biosciences,

https://doi.org/10.1007/978-3-030-54630-4_6

Keywords

Therapeutic antibodies · Cancer treatment · Diagnostic antibodies · Tool antibodies · Lateral flow assays · Enzyme-linked immunosorbent assay · Fluorescence-activated cell sorting (FACS) · Western blotting · Immunohistochemistry · Immunocytochemistry · Immunoprecipitation

What You Will Learn in This Chapter

This chapter will give an overview on the plethora of applications of antibodies and antibody-like molecules in therapy, diagnosis, and research. In the first section of the chapter, you will learn about the indications, molecular targets, and mode of actions of some selected approved therapeutic antibodies. The second section of this chapter will deal about the utilization of antibodies for disease diagnostics in various indications like infectious diseases, especially regarding tuberculosis and HIV, and in cancer diagnosis with reference to approved and commercially available test kits. The last section of the chapter describes important antibody-based technologies and assay formats like flow cytometry, immunohistochemistry, and ELISA that are commonly deployed in biomedical research. Due to their outstanding specificities, many basic and routine laboratory applications are based on the interactions of antibodies with their respective antigens.

6.1 Application of Antibodies in Therapy

Over the last three decades, monoclonal antibodies (mAbs) and antibody-like molecules (like antibody fragments or bispecifics, as described in detail in other chapters) were extremely successful in becoming effective human therapeutics. The first approved monoclonal antibody was muromonab (Orthoclone OKT3®; Janssen-Cilag) in 1986, targeting CD3 for the prevention of kidney transplant rejection. It binds to the CD3 epsilon chain (a part of the T cell receptor complex) on the surface of circulating T cells, eventually leading to anergy and apoptosis of the T cells, which protects the transplant against a respective immune response [1]. This molecule was still fully murine, posing a severe immunogenicity risk to patients due to the development of anti-drug antibodies. The major advances in molecular biology in the 1990s and the translation of basic biomedical sciences into clinical practice led to the approval of the first chimeric (containing murine and human regions) antibody rituximab (Rituxan®; Genentech) in 1997 for the treatment of low-grade B cell lymphoma. This technology, followed by humanized and eventually fully human antibodies, led to a dramatical increase in the rate of market approvals. The first approved fully human antibody was adalimumab (Humira®; AbbVie) in 2004 for the treatment of rheumatoid arthritis. Nowadays most therapeutic antibodies in development are either humanized or fully human [2]. Antibodies are currently also the fastest growing group of biotechnology-derived molecules in clinical trials [3, 4]. In 2017, the global monoclonal antibody therapeutics market was valued at approximately $95 billion [5]. About 60 antibodies are currently approved by the FDA and under evaluation in various phases of clinical trials for treating various diseases and conditions including cancer, chronic inflammatory diseases, transplantation, infectious diseases, and cardiovascular diseases [2]. The following paragraphs will provide an overview on the different indications where therapeutic antibodies are applied and will discuss some prominent examples.

Oncology is currently the most important indication for the application of therapeutic antibodies. Initially just utilized as antagonists of oncogenic receptor tyrosine kinases, their therapeutic application has expanded significantly in recent years, including the targeted delivery of chemotherapeutic agents (so-called ADCs—antibody-drug conjugates—explicitly described in a separate chapter of this book) and the manipulation of anticancer immune responses. Several clinical trials are ongoing where those targeted cancer therapies are studied for use alone or in combination with chemotherapy or with other targeted therapies [6].

6.1.1 Cetuximab

A prominent representative of the class of receptor tyrosine kinase antagonists is cetuximab (Erbitux®; Merck KGaA/Bristol-Myers Squibb), a chimeric human-murine monoclonal IgG1 antibody targeting EGFR (epidermal growth factor receptor). It acts by inhibiting endogenous ligand binding, cell motility, invasiveness, metastasis, and promoting apoptosis [7]. Cetuximab is approved for squamous cell carcinoma of the head and neck (SCCHN) in combination with radiation therapy for locally or regionally advanced disease, in combination with platinum-based therapy for locoregional or metastatic disease, or as single-agent therapy for recurrent or metastatic disease progressing after platinum-based therapy [8–15]. It is also approved for metastatic colorectal carcinoma in combination with irinotecan in patients who are refractory to irinotecan-based chemotherapy or as monotherapy in patients who have failed irinotecan- and oxaliplatin-based regimens or in patients who are intolerant of irinotecan-based chemotherapy and in combination therapy (with irinotecan, 5-FU, leucovorin) for first-line treatment [16–21].

6.1.2 Trastuzumab

Another example for a highly effective monoclonal antibody directed against a receptor tyrosine kinase is trastuzumab (Herceptin®; Genentech). It binds to the human epidermal growth factor receptor 2 protein (HER2/neu) [22] and causes an immune-mediated response and internalization and downregulation of HER2 as well as upregulation of cell cycle inhibitors [23]. This eventually leads to inhibition of cell proliferation and induction of apoptosis. Trastuzumab also mediates anti-angiogenic effects by induction of anti-angiogenic factors as well as repression of pro-angiogenic factors [7]. It is approved for the adjuvant treatment of HER2-overexpressing node-positive breast cancer as part of a regimen containing doxorubicin, cyclophosphamide, and either paclitaxel or docetaxel or with docetaxel and carboplatin or as a single agent following multimodality anthracycline-based therapy. It is also approved for metastatic breast cancer in which the tumor overexpresses the HER2 protein including first-line treatment in combination with pacli-taxel and single-agent therapy in patients who have received one or more chemotherapy regimens for metastatic disease. Another indication is metastatic gastric cancer in first-line therapy in patients with HER2 overexpressing gastric or gastroesophageal junction adeno-carcinoma in combination with cisplatin and capecitabine or 5-fluorouracil [6, 24–33].

A trastuzumab-derived ADC called ado-trastuzumab emtansine (Kadcyla®; Genentech) is approved for the treatment of patients with HER2-positive metastatic breast cancer. The molecule consists of trastuzumab covalently linked to the cytotoxic agent DM1 (emtansine) [34].

6.1.3   Rituximab

Apart from receptor tyrosine kinases, CD (cluster of differentiation) antigens are also frequently targets for therapeutic antibodies. A prominent example which falls into this category is rituximab (MabThera®; Roche). It is a chimeric monoclonal antibody targeting CD20 found on both normal B cells and most low-grade and some higher-grade B cell lymphomas. It's mode of action consists of induction of antibody-dependent cellular cytotoxicity (ADCC) and complement-dependent cytotoxicity (CDC) as well as induction of apoptosis [35]. Rituximab is approved for non-Hodgkin lymphoma (NHL) in relapsed or refractory low-grade or follicular CD20-positive B cell NHL as a single agent, in combination with first-line chemotherapy for follicular CD20-positive B cell NHL, and in patients achieving a complete or partial response to rituximab in combination with chemo-therapy as single-agent maintenance therapy. Other indications are non-progressive low-grade CD20-positive B cell NHL as single agent after first-line combination chemotherapy and previously untreated diffuse large B cell CD20-positive NHL in combination with CHOP (cyclophosphamide, hydroxydaunorubicin, Oncovin®, prednisone) or other anthracycline-based chemotherapy regimens. It is also approved for chronic lymphocytic leukemia (CLL) in combination with fludarabine and cyclophosphamide in previously untreated and previously treated CD20-positive CLL and in other immunological diseases, such as rheumatoid arthritis and granulomatosis with polyangiitis and microscopic polyangiitis [36–39].

A very important recent discovery in the field of oncology is that of immune checkpoint molecules, which dampen anticancer immune responses. Such proteins include programmed cell death protein-1 (PD-1), its ligands programmed death-ligand 1 (PD-L1) and 2 (PD-L2), and cytotoxic T-lymphocyte-associated protein 4 (CTLA-4), among others.

Inhibitors of immune checkpoint proteins have since been developed to "take the breaks off" of an otherwise impeded anticancer immune response. Such modulation of immune checkpoints became prominent as a means to treat a number of solid malignancies, given the durable response seen in many patients and improved side effect profile compared to conventional chemotherapeutic agents [40].

6.1.4   Ipilimumab

Ipilimumab (Yervoy®; Bristol-Myers Squibb) is currently the only anti-CTLA-4 antibody with FDA approval. The fully human antibody binds and blocks the interaction of CTLA-4 with its ligands, CD80/CD86, leading to activation and proliferation of T cells and is therefore believed to play a critical role in sustaining an active immune response in its attack on cancer cells. Ipilimumab is approved for unresectable or metastatic melanoma (2011) and, in combination with Opdivo, for advanced renal cell carcinoma and metastatic colorectal cancer (2018) [41, 42].

The first phase I trial investigating an anti-PD-1 antibody was reported in 2012, and the field has grown significantly since then [43]. The most well-studied PD-1 inhibitors are represented by nivolumab (Opdivo®; Bristol-Myers Squibb) and pembrolizumab (Keytruda®; Merck Sharp & Dohme), although several other mAbs within this class exist. Both, nivolumab and pembrolizumab, are of the IgG4 subclass, but the epitope binding regions of each differ. Both have high affinity and high specificity for PD-1 and block the binding of PD-L1 [44].

6.1.5   Nivolumab

Nivolumab received its first FDA approval in December 2014 for the treatment of unresectable/metastatic melanoma after failure of ipilimumab and (if BRAF V600 mutation positive) a BRAF inhibitor. Further skin cancer-related indications include combination therapy with ipilimumab for BRAF V600 wild-type, unresectable/metastatic melanoma and adjuvant therapy following complete surgical resection for patients with stage III melanoma. In 2017 nivolumab received approval as second-line monotherapy for metastatic or surgically unresectable urothelial carcinoma that had progressed or recurred despite prior treatment with at least one platinum-based chemotherapy regimen [45]. Further indications are metastatic DNA mismatch repair-deficient/microsatellite instability-high colorectal cancer [46]; recurrent SCCHN that had progressed on or after chemotherapy [47]; hepatocellular carcinoma (HCC) in patients who had failed previous vascular endothelial growth factor inhibition [48]; classical Hodgkin lymphoma (cHL) with progression following autologous stem cell transplantation and post-transplant brentuximab vedotin [49, 50]; metastatic squamous non-small cell lung cancer (NSCLC) and nonsquamous NSCLC that had progressed following platinum-based chemotherapy [51, 52];

advanced renal cell carcinoma (RCC) with history of prior anti-angiogenic therapy; previously untreated advanced renal cell carcinoma in combination with ipilimumab [53]; and small cell lung cancer (SCLC) with disease progression following platinum-based chemotherapy and one other line of therapy [54].

6.1.6 Pembrolizumab

Pembrolizumab was granted its first ever FDA approval in September 2014 for the treatment of melanoma. Indications have expanded since; they include metastatic melanoma with disease progression on ipilimumab and, if BRAF V600 mutation positive, a BRAF inhibitor, as well as adjuvant treatment following resection for stage III disease [55–58]. Further indications are cervical squamous cell carcinoma (CSCC) as second-line treatment [59]; recurrent, advanced gastric or gastroesophageal junction (GEJ) adenocarcinoma with progression on multiple prior therapies and known tumor PD-L1 expression [60]; first-line treatment for metastatic/unresectable recurrent SCCHN, for tumors with known PD-L1 expression alone and in combination with platinum-based chemotherapy and fluorouracil [61]; HCC as monotherapy for patients that had been previously treated with sorafenib [62]; refractory or relapsed cHL after three or more lines of prior therapy [63]; first-line treatment in adult and pediatric locally advanced or metastatic Merkel cell carcinoma (MCC) [64]; metastatic NSCLC with known tumoral PD-L1 expression that had progressed on or after platinum-based chemotherapy; metastatic SCLC that had progressed on or after platinum-based chemotherapy and at least one other line of treatment [65]; relapsed/refractory primary mediastinal B cell lymphoma (rrPMBCL) that had relapsed following two lines of chemotherapy [66]; and RCC as first-line treatment in patients with advanced disease [67, 68]. Apart from that, the FDA approved pembrolizumab in 2017 as the first tissue-agnostic cancer therapy for unresectable or metastatic solid cancers expressing MSI-H or dMMR, marking the first FDA approval based on biomarker expression rather than on specific disease. The approval currently exists for solid tumors that have progressed after treatment with no other current treatment options [69].

6.1.7 Avelumab

While anti-PD-1 therapies paved the way for modulation of the PD-1/PD-L1 axis, research into mAbs directed against PD-L1 followed shortly after. Avelumab (Bavencio®; Merck/Pfizer) is one example of a therapeutic anti-PD-L1 mAb. It was granted its first FDA approval in March 2017 for the treatment of metastatic MCC in patients 12 years and older [70], followed by accelerated FDA approval for treatment of metastatic urothelial carcinoma refractory to 12 months of platinum-based chemotherapy in May 2017 [71]. It was also approved in combination with axitinib for first-line treatment in patients with advanced RCC in May 2019 [72].

6.1.8   Adalimumab

Even though it is apparently the most important indication, oncology is not the only clinical application field for therapeutic antibodies. A prominent example for a mAb approved for the treatment of rheumatological diseases is adalimumab (Humira®; AbbVie). Adalimumab is a fully human mAb that acts by binding to TNF-α and blocking its interaction with the p55 and p75 cell surface TNF receptors, thereby leading to immunosuppression. It was the first human mAb approved by the FDA in December 2002 as a treatment for adult patients with moderately to severely active rheumatoid arthritis. Other indications are psoriatic arthritis (2005), ankylosing spondylitis (2006), Crohn's disease (2007), juvenile idiopathic arthritis (2008), chronic plaque psoriasis (2008), ulcerative colitis (2012), and hidradenitis suppurativa (2015) [4].

Apart from classical monoclonal antibodies, there is a growing interest in the development of bispecific antibodies (BsAbs) for therapeutic applications. In the last decade, the number of articles devoted to BsAbs has been steadily increasing. The journal *Nature Reviews Drug Discovery* called BsAbs in 2014 the "next-generation antibodies" [73]. Bispecific antibodies are explicitly discussed in a separate chapter of this book. However, two examples of BsAbs in clinical applications should briefly be mentioned here: blinatumomab (Blincyto®; Amgen) and emicizumab (Hemlibra®; Roche).

6.1.9   Blinatumomab

Blinatumomab (Blincyto®; Amgen) is a bispecific CD19-directed CD3 "T cell engager." It binds to both CD19 and CD3 expressed on the surface of the cells of B-lineage origin and T cells, respectively, resulting in the activation of endogenous T cells and lysis of CD19-positive benign and malignant B cells by cytotoxic T cells [2]. It is currently approved for B cell precursor acute lymphoblastic leukemia (ALL) in first or second complete remission with minimal residual disease greater than or equal to 0.1% as well as relapsed or refractory B cell precursor ALL.

6.1.10   Emicizumab

Emicizumab (Hemlibra®; Roche) is a BsAb that binds to both activated coagulation factor IX and factor X, thereby mediating the activation of the latter. This is actually the function of coagulation factor VIII in healthy individuals but is missing in hemophilia A patients [74]. Emicizumab was approved for the treatment of patients with hemophilia A who had developed resistance to other treatments in November 2017 and for those who have not developed resistance to other treatments subsequently in April 2018.

An additional overview on the therapeutic antibodies discussed above is also given in Table 6.1.

Table 6.1 Listing of selected and approved therapeutic monoclonal antibodies with reference to the respective antibody type, target, indication and mechanism of action

Name	Type	Target	Approved indications	Mechanism of action
Cetuximab (Erbitux®; Merck KGaA, Bristol-Myers Squibb)	Chimeric human-murine IgG1	EGFR	Head and neck cancer; metastatic colorectal cancer	Inhibition of EGFR signalling and ADCC
Trastuzumab (Herceptin®; Genentech)	Humanized IgG1	Erb2	Adjuvant breast cancer, metastatic breast cancer in patients in which tumor overexpresses the HER2 protein; metastatic gastric cancer	Inhibition of Erb2 signalling and ADCC
Rituximab (MabThera®; Roche)	Chimeric human-murine IgG1	CD20	CD20-positive B cell NHL and CLL, maintenance therapy for untreated follicular CD20-positive NHL	ADCC, direct induction of apoptosis and CDC
Ipilimumab (Yervoy®; Bristol-Myers Squibb)	Human IgG1	CTLA-4	Unresectable or metastatic melanoma	Inhibition of CTLA4 signalling
NivolumabOpdivo®; Bristol-Myers Squibb)	Human IgG4	PD-1	Metastatic melanoma, adjuvant stage III melanoma; urothelial carcinoma; metastatic CRC; recurrent SCCHN; HCC; cHL; metastatic squamous and non-squamous NSCLC; advanced RCC; SCLC	Inhibition of PD-1 signalling
Pembrolizumab (Keytruda®; Merck Sharp and Dohme)	Humanized IgG4	PD-1	Metastatic melanoma, adjuvant stage III melanoma; CSCC; GEJ adenocarcinoma; metastatic SCCHN; HCC; refractory or relapsed cHL; metastatic MCC; metastatic NSCLC; metastatic SCLC; relapsed/refractory PMBCL; RCC metastatic solid cancers expressing MSI-H or dMMR	Inhibition of PD-1 signalling
Avelumab (Bavencio®; Merck KGaA, Pfizer)	Human IgG1	PD-L1	Metastatic MCC; metastatic urothelial carcinoma; RCC in combination with axitinib	ADCC, inhibition of PD-1 signalling
Adalimumab (Humira®; AbbVie)	Human IgG1	TNF-α	Rheumatoid arthritis; psoriatic arthritis; ankylosing spondylitis; Crohn's disease;	Inhibition of TNF-α signalling

(continued)

Table 6.1 (continued)

Name	Type	Target	Approved indications	Mechanism of action
			juvenile idiopathic arthritis; chronic plaque psoriasis; ulcerative colitis; hidradenitis suppurativa	
Blinatumomab (Blincyto®; Amgen)	Bispecific scFv fusion	CD19/ CD3	B-cell precursor ALL	Engaging T cells toward tumor cells
Emicizumab (Hemlibra®; Roche)	Bispecific IgG	Factor IX/ factor X	Hemophilia A	Crosslinking of coagulation factor IX and factor X

6.2   Applications of Antibodies in Diagnosis

As antibodies display high affinities and high antigen specificities, they are perfect candidates for noninvasive in vivo imaging and in vitro immunoassays in the context of disease diagnostics and prognosis evaluation. Antibodies are applied in various disciplines like oncology, immunology, microbiology, hematology, and pathology. In addition, they are useful tools for the identification of infectious diseases through viruses, bacteria, or parasites, to diagnose metabolic disorders and for cancer imaging.

The unequivocal identification of a given infectious pathogen is not only relevant for the specific patient but for public health as well. In this context it is critical for national health institutes to discriminate between pandemic and only local infections in human populations [75]. The early recognition of a pandemic bacterial strain or virus, particularly when infections can proceed asymptomatic or latent, helps to take the required actions in time to prevent the spreading of the disease. One proven procedure to detect an infection with *Mycobacterium tuberculosis*, for example, which causes tuberculosis (TB), is a specific nested PCR (polymerase chain reaction) approach using the commercial test kit Xpert® MTB/RIF (Cepheid) [76] or to culture the bacteria, which can take weeks to obtain reliable results [77]. The Xpert® MTB/RIF test is a fully automated nucleic acid amplification test. It strongly contributes to the goal to control global TB infections and reduce disease prevalence, because it is a rapid and sensitive diagnostic tool [78]. Compared to the procedure of culturing bacteria, it delivers fast results but is rather cost intensive, and a special processing device (GeneXpert, Cepheid) is needed to provide results [79]. There are already improved methods regarding early TB detection available like an interferon-gamma release assay called QuantiFERON®-TB Gold-Plus (QTF-Plus) from Qiagen [80]. QTF-Plus is a two-step assay that consists of an interferon γ (IFNγ) in vitro stimulation step in whole blood samples and a subsequently performed ELISA (enzyme-linked immunosorbent assay) using murine monoclonal anti-INFγ antibodies to quantify the amount of IFNγ release that is associated with *M. tuberculosis* infection [81]. QTF-Plus needs a skilled operator and laboratory infrastructure to be applied. Newly developed point-of-care tests utilize mono- or bispecific antibodies to bind mycobacterial lipoarabinomannan (LAM) from swab or urine samples in a sandwich-type, one-step assay [82, 83]. Commercial immunological test kits that are based on this method are, e.g., Determine®-TB Alere (Abott) and FujiLAM (Fujifilm) [84, 85] (www.who.int).

Another example is to test patients for an infection with the respiratory syncytial virus (RSV). RSV is a common respiratory virus that causes mild symptoms but can be serious for children under 1 year of age where it is the main cause of bronchiolitis and pneumonia [86]. Next to the gold standard of testing RSV infection using a real-time PCR (RT-PCR) approach [87], the qualitative test kit CerTest RSV (CerTest Biotec S.L.) is available and less complex. It is a one-step colored chromatographic immunoassay that gives very rapid information about the infection status of an individual by detection of RSV antigen in nasal swab, nasopharyngeal wash, or aspirate specimens [88]. Despite the advantage of having fast results available, this test is much less sensitive in specific subpopulations of patients [89]. CerTest is based on a chromatographic immunoassay method for the detection of RSV antigen from patient respiratory secretion material. The technology behind this and the above described tests to detect TB infection like Determine®-TB Alere is called lateral flow immunoassay (LFIA) [90]. LFIAs are not only useful in medical applications but also for the detection of drug abuse, agriculture safety, animal health, and food quality. One of the first tests was generated to detect human chorionic gonadotropin (hCG) in female urine and therefore to confirm or exclude pregnancy [91]. One advantage of LFIA is the potential direct application by patients and the simple, rapid, and low-cost technology [90]. The test can even be conducted without skilled personnel or a laboratory facility which makes it attractive for the application in developing countries without adequate health infrastructure. LFIAs are according to WHO characteristics described as ASSURED: affordable, sensitive, specific, user-friendly, rapid and robust, equipment free, and deliverable to end users [92]. This section focuses on LFIAs where antibodies are used for recognition. Tests utilizing nucleic acid probes (NALFIA, nucleic acid lateral immunoassay) are being developed as well [93, 94]. For NALFIA the analyte is an amplicon that is generated via PCR prior to the analysis with the test strip. Specific primers are used that attach two different tags onto the amplicon which are afterward detected by tag-specific monoclonal antibodies on the strip. Compared to conventional PCR analysis, this combined method is much faster because the amplicon does not need to be separated by electrophoresis anymore. The amplicon in the PCR mixture can, e.g., be labeled with biotin and FITC through specific primers and is after addition onto the sample pad of NALFIA bound by a gold-nanoparticle-labeled monoclonal anti-FITC antibody. This complex migrates through the membrane and is at the test line bound by immobilized anti-biotin monoclonal antibodies whereby the gold nanoparticles are concentrated and a colorimetric signal is generated with an intensity that is proportional to the amplicon concentration in the mixture. Pecchia and Da Lio described this method and used it to detect the fungal pathogenic *Macrophomina phaseolina* in soil samples [95].

LFIAs can be used to detect antibodies and other antigens like hormones, peptides, or toxins from blood, serum, or other liquid and complex samples [96]. The methodology relies on a combination of chromatography and immune staining. In the beginning of LFIA technology, tests were designed to give only qualitative results like appearance or absence of signals that translated into yes or no results. Nowadays tests are being developed to provide semi-quantitative results and are combined with automated reader devices [97]. Figure 6.1 shows the typical construction of a test strip. The strip consists of many different elements and is often more complex than it seems. The liquid sample containing the analyte is applied onto the sample pad where it is mixed with rehydrated buffer, salts, and surfactants or other liquids. This step is important for sample preparation and ensures a constant flow rate and controlled distribution of the sample to the conjugate pad and is critical for antigen-antibody interaction because protein solubility is influenced by pH. The isoelectric point (pI) of many antibodies is between pH 5.5 and 7.5. Thus, a pH between 7.0 and 7.5 can be optimal, and the buffer conditions should be adjusted in this sense [98]. The flow rate on the other hand is determined by the material of the membrane itself. Nitrocellulose is a commonly used material. Others are, e.g., polyvinylidene fluoride, charge-modified nylon, and polyethersulfone [99]. A favorable membrane material has advantageous capillary forces and constant flow rates and allows an easy binding and immobilization of the antibody at the test and control lines. Binding mechanisms are mainly electrostatic and hydrophobic interactions. The immobilization capability is defined by

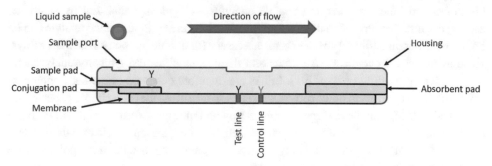

Fig. 6.1 Schematic representation of a standard lateral flow assay (side view). The test strip is typically localized inside a plastic housing with one or in case of a multiplex format a varying number of openings for sample administration and a window to observe the results of test and control lines. The liquid sample is admitted onto the sample pad through the sample port where it is mixed with rehydrated buffer components and chemicals like detergents, surfactants, or blocking agents and others depending on the specific application of the test. The sample migrates to the overlapping conjugate pad where the detector complex is located. The complex is rehydrated and moves with the sample onto the membrane where it passes test and control zones. At the end of the strip, the absorbent pad soaks up the liquid and enhances the overall capacity of the test strip. The absorbent pad in this way increases assay sensitivity. At the test and control line assay, specific capture antibodies are immobilized that capture the analyte-detector or control complexes

the amount of polymer surface area, porosity, thickness, and structural characteristics of the material itself. Following application of the liquid sample, it migrates to the conjugate pad. This holds the detector complex and delivers sample and detector complex to the membrane and to the test and control lines. The recognition antibody conjugated to colloidal gold nanoparticles is a commonly used detector complex localized at the conjugate pad. Cellulose nanobeads, latex beads, paramagnetic beads, or fluorescent molecules are usable, too [100]. The selection of the recognition/capture antibody is the most important part in designing a specific LFIA since binding characteristics like the antibody's binding constant (K_D) and specificity have a huge impact on test sensitivity and background signal [90]. The signal is in its simplest form a colorimetric signal which is generated by the concentration of analyte-associated aggregation (sandwich assay) or disaggregation (competitive assay). The recognition antibody is absorbed to the gold particles and can bind to the analyte in complex samples. Other labels can, e.g., be fluorescent dyes, luminescent molecules, and quantum dots [101]. The utilization of these labels may lead to an improvement in sensitivity of the LFIA but cannot be detected directly by eye of the operator anymore and need an additional device like a fluorescence reader. Shi et al. used a superparamagnetic nanoparticle and a magnetic reader and improved the assay sensitivity by tenfold [102]. Colorimetric signals can be enhanced by applying an enzymatic signal amplification reaction. Horseradish peroxidase (HRP) or alkaline phosphatase (AP) are both working well in improving sensitivity in ELISA and are although applied for LFIAs [103]. An absorbent pad is localized at the end of the test strip. It's main function is to soak up the fluid and therefore to enhance the volume of the analyzed sample.

In general, LFIAs can be distinguished into direct and indirect assays. Tests for hormones and other proteins and peptide analytes are typically analyzed by sandwich assays, because this type of analyte has often more than one epitope and the analyte can be bound by different antibody species that is necessary for double antibody sandwich assays. Haptens and small chemicals like drugs and their metabolites are typically analyzed by competitive or inhibition assays [91]. LFIAs are therefore designed like other immunoassay formats. In contrast to other complex formats like ELISA, they are simple paper-based analytical tests and can be designed in microarray and multiplex formats and deliver rapid results without the need of a trained and skilled operator. Despite their pitfalls like less quantitative results and often minor sensitivity, they greatly contribute to the point-of-care and near patient diagnostic sector [93].

Enzyme-linked immunosorbent assays [104, 105] are very specific and sensitive to detect and quantify very low analyte concentrations in simple or complex samples. The principle is based on the very specific interaction of antigen and antibody. Its application is among many others in science and diagnostics, for example, the detection of autoimmune disease biomarkers. These can be auto-antibodies in patient serum samples [106]. Commonly three types of ELISAs are utilized that are schematically displayed in Fig. 6.2: indirect ELISA [107, 108], sandwich ELISA [109, 110], and competitive ELISA [111]. In indirect ELISA the sample (e.g., patient's serum) with the analyte is coated to the well of a microtiter plate. After blocking of the remaining binding sites, the detection antibody and

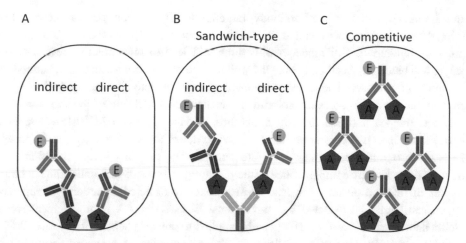

Fig. 6.2 Three different ELISA formats are commonly used for diagnostic applications. (**a**) shows the setup for a direct ELISA where the antigen of interest (analyte) is coated onto the well surface of an ELISA plate. The antigen is bound by an enzyme-coupled detection antibody or for signal amplification by a detection antibody followed by a polyclonal enzyme-coupled secondary antibody. In (**b**) the sandwich-type ELISA assay is displayed schematically. The analyte is attached to the well surface through binding of a capture antibody and is via another epitope bound by the enzyme-coupled detection antibody or again for signal amplification through utilization of a polyclonal enzyme-coupled secondary antibody. An example for the competitive ELISA setup is shown in (**c**). The analyte to be detected is immobilized onto the well surface and can be bound by the enzyme-coupled detection antibody. If the concentration of the analyte in the sample increases, the detection antibody will more likely bind to the soluble analyte which translates into a decreasing signal intensity

the enzyme conjugated secondary antibody that is specific for the primary detection antibody are added. After addition of the substrate, the conversion to the colored product can be measured using a specified ELISA reader device. Chemiluminescent- and fluorescent-based reporter molecules can be developed as well and may improve assay sensitivity. Special equipment is needed in this regard. The signal intensity correlates directly with the concentration of the analyte in the sample and is compared to a gage row. It is indispensable to have a purified analyte sample available in order to prepare serial dilutions for analyzing the gage row simultaneously to the sample dilutions. The linear relationship between analyte concentration and signal intensity is limited through to the narrow dynamic range [112]. Indirect ELISAs are commonly used to identify and quantify analytes like antibodies or cytokine levels in human serum samples, for example, to diagnose endocrine diseases (human IL-13 Quantikine ELISA KIT, R&D Systems, Inc.) [113–115]. The indirect ELISA method is prone to matrix effects that can occur through inhibition of full analyte recovery if the analyte is part of a complex matrix like blood or serum. Adverse matrix effects can be limited by dilution of the serum samples [116]. By using a specific capture antibody to immobilize the analyte on the surface of the plate well, the ELISA can be highly specific. In sandwich ELISAs the immobilized capture antibody binds the analyte from the matrix, and a primary antibody, targeting a different epitope, can afterward be detected with an enzyme-coupled detection antibody [117]. Inefficient blocking of remaining binding sites or inadequate washing will lead to false-positive results as the analyte will bind non-specifically onto the well surface or as unbound detection antibody is remaining in solution. The enzyme coupled secondary detection antibody binds the captured analyte and enables the colorimetric readout after addition of the substrate. The analyte is trapped between the two antibodies. Alliance HIV-1 p24 ELISA test from PerkinElmer, Inc., is a commonly used test kit to detect the viral marker capsid protein p24 [118] in serum, blood, and cell culture samples. The very low levels of p24 at early infection times and the complexation of free p24 antigen and anti-p24 antibodies in blood samples of infected persons can lead to inconsistently detected infections. Complexed antigen can hardly be detected by conventional ELISA tests. A photochemical signal amplification method based on HRP can be applied to enhance sensitivity of that ELISA up to 40-fold [119]. During competitive ELISA the complex of detection antibody and bound antigen in the sample is added to a well in which the recombinant purified antigen is coated to the well surface. The sample antigen (the analyte) and the coated antigen compete for the binding to the antibody. In this setup the decrease in signal intensity correlates with analyte concentration [117].

Antibodies can furthermore be helpful in the recognition of abnormal cells in neoplastic diseases and the identification of the originating cell lineage, which is critical for the diagnosis of the type of cancer and the determination of the following therapy. Antibodies are, for example, utilized in classification of lymphoma subtypes from lymph node fine needle aspirates, tissue samples, and other body fluids by immunophenotyping using flow cytometry [120]. In this approach multiple biomarkers, such as surface antigens on cells, can be assayed in parallel through the simultaneous application of several specific mono-clonal antibodies conjugated with different fluorophores [121]. Malignant cells can, e.g., be identified through the detection of a modified expression pattern of these antigens [122]. During flow cytometry, a patient sample containing viable cells to be assessed is pre-incubated with antigen-specific fluorescence-labeled monoclonal antibodies. Within the flow cytometer, the cells are scattered and pass a laser beam one by one in suspension. Labeled cells are excited by the laser and emit light in a specific wavelength corresponding to the fluorophore's spectrum that is measured by the detector. Using a suitable set of labeled antibodies and laser sources, various physical parameters of each cell in a complex population can be analyzed that give rise about expression pattern of surface antigens and granularity and size of the cell [123]. This contrasts with IHC, where only one or two markers can be assessed at a time [124]. To subtype lymphomas, it is advantageous to analyze the phenotypic expression of leukocyte markers [125]. The expression pattern of these markers gives notice about the cell type origin and stage of differentiation, which can predict the response of the neoplastic cell clone in therapy. Flow cytometry in this regard enables the simultaneous, multiparameter, and quantitative analysis of several surface markers and helps to distinguish between, e.g., CLL (chronic lymphatic leukemia) and more aggressive leukemia types [122]. After therapy, flow cytometry is used to analyze the amount of minimal remaining malignant cells that were not eliminated during treatment and that can initiate cancer recurrence. This is called minimal residue disease and is a reliable prognostic marker for chronic lymphocytic leukemia [122].

During therapy it is advantageous and often required to monitor the response to a treatment and the progression and severity of the disease. An important example in this regard is the classification of breast cancer tumors. Prognosis and treatment decisions are mainly guided by tumor stage, tumor grade, hormone receptor status, and HER2 status [126]. Only HER2-positive tumors are being treated using the anti-HER2 antibody trastuzumab. Two different methods are commonly used to analyze the HER2 status of a breast cancer tumor. FISH (fluorescence in situ hybridization) analyzes HER2 gene amplification [127], and IHC (immunohistochemistry) analyzes HER2 expression and distribution in tissue samples. HER2 overexpression is often correlated to a poor prognosis [128]. FDA-approved test kits are, e.g., HercepTest™ [129], InSite® (BioGenex Laboratories Inc.), Bond Oracle™ (Leica Biosystems), and Ventana HER2 Pathway® System (Ventana Medical Systems, Inc.) [130]. It utilizes a rabbit monoclonal anti-HER2 antibody (clone 4B5) and a peroxidase-conjugated secondary antibody to semi-quantify the HER2 status of formalin-fixed and paraffin-embedded breast cancer tissue samples using an immunohistochemical assay. Unfortunately, variables in test procedure and sample handling affect the outcome of the test [131]. These are the type of fixative, length of fixation, the storage of paraffin blocks, and the thickness of the section (manual Agilent/DAKO). The epitope recognition by the primary detection antibody is frequently affected detrimentally by formalin treatment, and this effect can have a large impact on the staining pattern. It turned out that analysis of *HER2* gene amplification by FISH is the more reliable diagnostic technology in this regard.

A special category of detection reagents regarding immunoassays represents bispecific antibody molecules. The utilization of bispecific antibodies in diagnostic assays can reduce assay complexity by reducing the total number of procedure steps and enhancing sensitivity over mono-specific antibodies [77]. This is achieved through their inherent ability to bind two distinct epitopes. In an immunoassay, a bispecific detection antibody can bind to the analyte with one arm and to the reporter molecule with the other arm. This arrangement has the advantage that the reporter molecule, e.g., FITC (fluorescein) or an enzyme, and the antibody are no longer covalently linked to each other which can deleteriously affect their functionalities [132].

An application in cancer imaging is immuno-PET (immune-based positron emission tomography). Immuno-PET is used to analyze tumor antigen expression and distribution in vivo and on this basis can be beneficial to choose the most suitable targeted therapy [133]. Immuno-PET probes are typically made of positron-emitting radioisotope-labeled monoclonal antibodies specific for a cancer-associated antigen, a surface molecule on tumor cells or biomarker, and a PET scanner. One critical aspect in this regard is the biological half-life of the antibody molecule and the physical half-life of the radioisotope that needs to be balanced. As antibodies have a prolonged retention time in the body due to their big size (150 kDa) and FcRn-(neonatal Fc-receptor) recycling [134], they have also a poorer tumor penetration compared to smaller molecules like recombinant antibody fragments [135]. For this reason, they need to be coupled to a radioisotope with a suitable half-life as imaging is typically performed 1 day post-injection to enable biodistribution of the probe [136]. The antibody probe is applied to the circulation and binds to the tumor-associated antigen for specific labeling of the lesion. The patient is exposed to a certain amount of radiation during that time that is increasing with the retention time of the probe in the body. An improvement of this method was, e.g., to engineer a hapten specificity of one paratope of a bispecific antibody. The hapten itself was labeled with the radioisotope, for example, with ^{68}Ga- or ^{18}F. The other paratope was specific for the tumor-associated antigen CEA (carcinoembryonic antigen). The CEA × hapten bispecific antibody showed specific tumor targeting and minor targeting to CEA-negative tissue [137]. This pre-targeting approach using a bispecific anti-hapten antibody showed a faster clearance of the radioiso-tope from circulation with no unspecific labeling and good and fast visualization of the tumor. It reduces the overall exposure time to the isotope which can be a great benefit for the patients [138]. Another approach is the utilization of smaller antibody fragments like F(ab)$_2'$ or scFv fragments with an improved clearance rate for targeting [139].

6.3   Application of Antibodies in Science

The extremely high sensitivity and specificity of antibodies toward their particular antigens make them also ideal reagents for many research applications. In fact, with the facilitation of large-scale production by modern biotechnology, antibodies have become major workhorses in biomedical research. Both monoclonal (homogeneous isotype and antigen specificity) and polyclonal (heterogeneous isotype and antigen specificity) antibodies are available from industrial vendors. Polyclonal antibodies are isolated from the sera of animals that have been immunized against a target antigen. They contain a mix of different antibodies against different epitopes of the same antigen and may also contain antibodies against unrelated proteins, which can affect assay results. Additionally, a polyclonal antibody supply is dependent on the source animal, and thus no two batches of polyclonal antibodies against a particular antigen are identical. In contrast, monoclonal antibodies are obtained from hybridomas or made recombinantly from expression vectors, both of which ensure continuous supply of homogeneous antibody [140].

6.3.1   Immunoprecipitation

Immunoprecipitation (IP) is a method where antibodies are applied to precipitate or "pull down" their respective target antigens from aqueous solutions like cell lysates or biological fluids. The process of precipitation is facilitated by the utilization of special beads that bind to the Fc region of an antibody (usually via covalent conjugation of an anti-Fc antibody or protein A/G). The beads itself typically consist of agarose, sepharose, or other synthetic polymers. Separation of antibody-antigen complexes on the beads is achieved by centrifugation, magnetic force, or other mechanical methods. IP assays are commonly used in many biochemical, cellular, or molecular biology applications. A basic application is the simple purification of a target protein for further analysis in subsequent assays like electrophoresis or Western blotting (see below). IP could also be utilized to analyze the interactions of multiple proteins. In those cases an antibody is used to precipitate the target protein, and subsequent assays (like mass spectroscopy) are used to determine if other proteins were pulled down with the complex [140].

6.3.2   Chromatin Immunoprecipitation

Chromatin immunoprecipitation (ChIP) is a special variant of IP to analyze interactions of proteins with nucleic acids. The methodology was originally developed for the evaluation of RNA polymerase II with its target gene nucleotide sequence [141]. In this procedure, cells were fixed to crosslink the DNA with any bound proteins. After lysis of the cells and fragmentation of the DNA-protein complexes into fragments of 100–300 base pairs, immunoprecipitation is performed with an antibody specific for the protein of interest (e. g., transcription factor) according to standard IP protocols. The DNA-protein complexes are finally dissociated by resolving the crosslinks, and precipitated DNA fragments are analyzed by PCR amplification. ChIP could also be used to analyze other proteins that co-precipitated with the DNA via mass spectrometry [142]. Sometimes chemical crosslinking could change the structure of the antigen epitope which leads to disruption of antibody binding. For such cases NChIP could be performed, a variation of ChIP where the crosslinking step is omitted and IP is performed on native antigen. However, this strategy is limited to target proteins that strongly bind to DNA [143]. Recent advances in ChIP are ChIP-chip, genome-wide microarray analysis of ChIP-isolated DNA, and ChIP-seq, where ChIP is followed by high-throughput sequencing of the immunoprecipitated DNA [144, 145].

6.3.3   Western Blotting

Western blotting (WB) is a method, where the resolution of gel electrophoresis is combined with the specificity of antibody detection. In this procedure proteins are separated using sodium dodecyl sulfate-polyacrylamide gel electrophoresis (SDS-PAGE), followed by a transfer to an absorbent blotting membrane by an electric charge. For antibodies that do not recognize their respective target antigen after protein denaturation, native PAGE (without SDS) could also be applied. After protein transfer the blotting membrane is blocked to reduce non-specific protein binding and subsequently incubated with the target protein specific primary antibody. Since primary antibodies are often unlabeled, a labeled secondary antibody specific for the Fc portion of the primary antibody is utilized for detection.

Respective labels are most often enzymes for colorimetric or chemiluminescent reactions or fluorophores, depending on the following type of analysis [140]. WB can be applied to detect the presence and quantity of an antigen, its molecular weight, and the efficiency of antigen extraction. This method is especially helpful when dealing with antigens that are insoluble, difficult to label, or easily degraded and thus not amenable to procedures such as immunoprecipitation [144].

6.3.4   Enzyme-Linked Immunosorbent Assay (ELISA)

The ELISA method was already described in the section "Antibodies in Diagnosis" as it is an established and robust method to detect and quantify very low amounts of disease-related analytes in complex biological samples like human serum. In general, the enzyme-linked immunosorbent assay is a method to identify and analyze the concentration of molecules in fluids. It relies on the very specific and high affinity interaction of an antibody and its specific antigen. Enzymes used for ELISA are, e.g., beta-galactosidase, glucose oxidase, peroxidase, and alkaline phosphatase. If the specific soluble substrate is added, the enzyme-substrate reaction occurs, and a colorimetric product signal is generated that is read on a spectrophotometer at 400–600 nm and gives notice about the concentration of the analyte, dependent on a standard curve [146].

ELISAs are commonly used in peptide and protein analytics. Besides other very sensitive and more modern methods, it is still a robust, easy, and frequently applied assay system that does not require very much automation [147]. Anyhow, complete automated systems containing pipetting robots, microplate washer, and absorbance reader are available (see, e.g., Hamilton Company). During monoclonal antibody production using hybridoma technology that was introduced by Köhler and Milstein in 1975 [148], ELISA is used to analyze serum antibody titers of immunized animals. After fusion of antibody producing B cells and immortal myeloma cells, the corresponding hybridoma cells produce the monoclonal antibody. Cell supernatants are screened using ELISA to identify antibody producing hybridoma cells [149]. During recombinant antibody discovery using phage display, a phage-ELISA is applied for biochemical characterization of enriched phage pools after panning [150].

6.3.5   Immunohistochemistry/Immunocytochemistry

Immunohistochemistry (IHC) and immunocytochemistry (ICC) are generally both methods for the in situ detection of the presence and the location of proteins within tissues and cells. While ICC is performed on intact cells that have been removed from their surrounding extracellular matrix, IHC is performed on fixed whole tissue sections (Fig. 6.3). Both methods utilize monoclonal or polyclonal antibodies for the detection of target structures. However, it is important to consider that the latter is a heterogeneous mix of

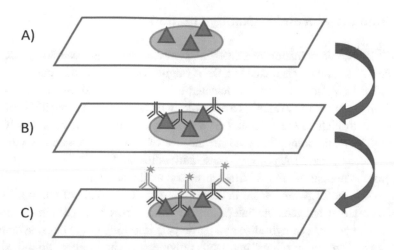

Fig. 6.3 Schematic representation of the immunohistochemistry procedure. Tissue sample is mounted on a slide (**a**). Incubation of tissue sample with target-specific primary antibody (**b**). Incubation of tissue sample with reporter-labelled secondary antibody specific for the constant region of the primary antibody (**c**)

antibodies that recognize several epitopes which also might lead to unspecific binding. Staining can either be performed via direct labeling of the primary antibody or by utilization of a labeled secondary antibody which is specific for the constant part of the primary antibody. Different staining reporters can be used depending on the subsequent method of detection. Colorimetric staining with enzyme-linked antibodies and colorimetric substrates is commonly used. This method is relatively easy, and the slides can be analyzed by standard microscopy. Disadvantages are potentially higher signal backgrounds due to endogenous enzymatic activity as well as the fact that only one or two antigens can be targeted per sample. An alternative approach is the utilization of fluorochrome-tagged antibodies (immunofluorescence), which allows the simultaneous labeling and detection of several antigens [140]. This can be achieved for IHC and ICC by using primary antibodies from different species or subclasses together with the respective corresponding conjugated secondary antibodies.

For high-resolution analyses, immunogold staining (IGS) can also be applied. This method uses antibodies that are linked to gold particles of varied sizes, which allows the detection of different antigens in a single sample. Sample analysis is eventually performed in electron microscope scans at high resolution with high sensitivity, permitting a very precise localization of target antigens within cells and tissues [151]. IHC and ICC are widely used in basic research to understand the distribution and localization of biomarkers and differentially expressed proteins in different cells or parts of a biological tissue.

6.3.6   Immuno-PCR and Proximity Ligation Assay

Immuno-PCR is a combination of ELISA and polymerase chain reaction (PCR). In this case the detection antibody is linked to a DNA oligonucleotide which is amplified by PCR. The amplified DNA molecule can be detected using standard fluorescent methods [152]. The proximity ligation assay (PLA) is a technology that extends the capabilities of classical immuno-PCR to include the detection of protein interactions with high specificity and sensitivity. The methodology is based on the use of secondary antibodies labeled with oligonucleotides (called PLA probes). These antibodies bind

the constant regions of the different primary antibodies directed against the target antigens. When the PLA probes are in close proximity (about 20–40 nm, e.g., when the two analyzed target antigens are interacting or form a part of a protein complex), the oligonucleotide probes can hybridize and participate in rolling circle DNA synthesis after addition of two "connector oligos" and DNA polymerase. The resulting amplified circular structure remains covalently attached to one of the PLA probes and can again be detected via standard fluorescent methods [153].

6.3.7   Flow Cytometry

Flow cytometry is a highly sophisticated method to analyze multiple physical parameters of single cells, including bacterial and microbiological cells, from samples containing a large number of these cells in high throughput. Artificial particles can be analyzed as well, like polymeric beads that can, e.g., be loaded with proteins and fluorescent molecules. The method is for that reason applied in clinics to support the diagnosis of diseases on a cellular level, e.g., immunophenotyping (see "Antibodies in Diagnosis"), and in science to screen large peptide and protein-based libraries in the context of directed evolution [154]. The principle behind flow cytometry is the scattering and emission of light through the analyzed cells and particles. Morphological structure of a given cell and fluorophore loading correlates with light scattering and fluorescence emission. Important components of a flow cytometry device are fluidics, optics, detectors, photomultipliers, and a computer. They all work together to focus a cell containing liquid stream and a laser beam to the flow chamber where the cells pass the laser one by one, in order to detect the emitted and scattered light through a detector. Special filters are required for the detection of fluorescence signals that rely on the wavelength of the emitted fluorescence. A photomultiplier downstream of the detector converts the light signal into an amplified electronic signal [123].

Especially for scientific applications, the possibility to couple flow cytometry with a sorting unit is attractive. Typically, special FACS (fluorescence activated cell sorting) machines are in use. FACS enables the separation of single cells with defined parameters from a large, heterogeneous cell population. The sorting of single cells occurs through the deflection of charged cell-containing droplets that pass a strong electrical field generated by two deflection plates. For directed evolution approaches in the field of antibody engineering, binders with improved properties, like specificity or binding constant (K_D), can be isolated from large antibody-based libraries that are, for example, displayed on microbial cells [155].

Important for flow cytometry was the discovery of quantum dots that display superior properties regarding spectral resolution and reduce the issue of overlapping spectra known from conventional fluorochromes like FITC (fluorescein) or PE (R-phycoerythrin). Quantum dots do not bleach and are very photostable. Anyhow, they did not succeed in establishing a robust range of tool products so far in the context of secondary staining molecules as this area is still dominated by antibody-fluorophore conjugates.

6.3.8   Mass Spectrometry

Mass spectrometry (MS) is an analytical method that measures the mass-to-charge (m/z) ratio of ions in order to identify and quantify molecules both in simple and complex matrices. MS is a workhorse for a broad range of scientific areas like proteomics, drug discovery, environmental analysis, and biomedical research. However, a common issue in MS is the presence of high-abundance proteins that mask some proteins of interest. A strategy to overcome this issue is the combination of MS with antibody-based affinity methods. In immunoprecipitation-based mass spectrometry (IP-MS), a molecule of interest is captured and isolated using IP, followed by analysis by MS [156]. There are several other quantitative and qualitative applications that have been developed to improve the sensitivity and specificity of MS by combining it with antibody-based immunoassays. Introduced in 2004, stable isotope standards and capture by anti-peptide antibodies (SISCAPA) is a method in which peptides of interest (after proteolytic digestion with, e.g., the enzyme trypsin) are enriched by using anti-peptide antibodies, immobilized on flow-through columns or magnetic beads. By selecting target peptides whose sequences only occur in the selected target proteins, the target peptides can serve as direct quantitative surrogates for the target proteins. Synthetic versions of the target peptides containing a stable isotope label are added in known amounts to the digested sample to serve as internal standards. Since the target peptides and standards can be measured separately due to the mass difference of the stable isotope label, their ratio provides the desired quantitative estimate of the target peptide amount [157]. SISCAPA is used in a variety of studies in the field of proteomics, as well as in clinical blood tests, e.g., for the validation and quantification of biomarkers like the human serum transferrin receptor (sTfR) in breast cancer patients [140, 158]. A methodology similar to SISCAPA is immuno-MALDI (iMALDI), where proteolytic peptides are also captured with anti-peptide antibodies, but the enriched sample is subsequently spotted onto a specific matrix and eventually analyzed by matrix-assisted laser desorption/ionization (MALDI) MS. The methodology has been applied for the diagnosis of different pathologies such as hypertension and Alzheimer's disease [159, 160]. Another similar approach is immuno-MRM, where multiple reaction monitoring mass spectrometry (MRM-MS) is combined with antibody-based enrichment of the peptide mixture. In MRM-MS, ions of a particular mass are selected in the first stage of a tandem mass spectrometer, and ion products of a fragmentation reaction of precursor ions are selected in the second stage for detection [161]. The mass spectrometric immunoassay (MSIA) is a top-down MS-based approach that combines immunoaffinity separation with mass spectrometric detection. Antibodies toward targeted proteins are attached to porous micro-columns fitted at the entrance of a pipette tip and enable for protein affinity extraction directly from a biological sample. After the affinity capture, proteins are eluted either directly onto a target plate for MALDI MS analysis or with a small volume of elution solution for subsequent liquid chromatography-MS (LC-ESI MS) analysis [162]. Mass cytometry is a fusion of two experimental platforms: flow cytometry and elemental mass spectrometry. In this approach target proteins on cells are labeled with antibodies conjugated to unique stable heavy metal isotopes. The labeled cells are subsequently nebulized, followed by ionization of the metal-conjugated antibody via passage through argon plasma. The mass-to-charge ratio is then analyzed by a time-of-flight (TOF) mass spectrometer. The methodology overcomes limitations of spectral overlap in flow cytometry, which are based on the broad emission spectra of classical fluorophores. In contrast, elemental mass spectrometry is able to discriminate isotopes of different atomic weights with high accuracy. This fundamental difference enables significantly more cellular features to be assayed simultaneously using a mass-based platform [163]. However, current limitations of mass cytometry are low flow rates and high operation costs (compared to classical flow cytometry).

> **Take Home Message**
> Antibodies are extremely versatile tools and have become the most important class of biomolecules in therapy, diagnosis, and biomedical sciences.

References

1. Norman DJ. Mechanisms of action and overview of OKT3. Ther Drug Monit. 1995;17(6):615–20.
2. Singh S, Kumar NK, Dwiwedi P, Charan J, Kaur R, Sidhu P, et al. Monoclonal antibodies: a review. Curr Clin Pharmacol. 2018;13(2):85–99.
3. Liu JK. The history of monoclonal antibody development – progress, remaining challenges and future innovations. Ann Med Surg (Lond). 2014;3(4):113–6.
4. Nelson AL, Dhimolea E, Reichert JM. Development trends for human monoclonal antibody therapeutics. Nat Rev Drug Discov. 2010;9(10):767–74.
5. marketwatch.com. Global Monoclonal Antibody Therapeutics Market to Surpass US$ 174. 2 Billion by 2026 [Press release]. www.marketwatch.com2019. Available from: https://www.marketwatch.com/press-release/global-monoclonal-antibody-therapeutics-market-to-surpass-us-174-2-billion-by-2026-2019-01-18

6. Gasser M, Waaga-Gasser AM. Therapeutic antibodies in cancer therapy. Adv Exp Med Biol. 2016;917:95–120.

7. Kalofonos HP, Grivas PD. Monoclonal antibodies in the management of solid tumors. Curr Top Med Chem. 2006;6(16):1687–705.

8. Thienelt CD, Bunn PA Jr, Hanna N, Rosenberg A, Needle MN, Long ME, et al. Multicenter phase I/II study of cetuximab with paclitaxel and carboplatin in untreated patients with stage IV non-small-cell lung cancer. J Clin Oncol. 2005;23(34):8786–93.

9. Modi S, D'Andrea G, Norton L, Yao TJ, Caravelli J, Rosen PP, et al. A phase I study of cetuximab/paclitaxel in patients with advanced-stage breast cancer. Clin Breast Cancer. 2006;7 (3):270–7.

10. Hofheinz RD, Horisberger K, Woernle C, Wenz F, Kraus-Tiefenbacher U, Kahler G, et al. Phase I trial of cetuximab in combination with capecitabine, weekly irinotecan, and radiotherapy as neoadjuvant therapy for rectal cancer. Int J Radiat Oncol Biol Phys. 2006;66(5):1384–90.

11. Belani CP, Schreeder MT, Steis RG, Guidice RA, Marsland TA, Butler EH, et al. Cetuximab in combination with carboplatin and docetaxel for patients with metastatic or advanced-stage nonsmall cell lung cancer: a multicenter phase 2 study. Cancer. 2008;113(9):2512–7.

12. Bonner JA, Harari PM, Giralt J, Azarnia N, Shin DM, Cohen RB, et al. Radiotherapy plus cetuximab for squamous-cell carcinoma of the head and neck. N Engl J Med. 2006;354(6):567–78.

13. Bourhis J, Rivera F, Mesia R, Awada A, Geoffrois L, Borel C, et al. Phase I/II study of cetuximab in combination with cisplatin or carboplatin and fluorouracil in patients with recurrent or metastatic squamous cell carcinoma of the head and neck. J Clin Oncol. 2006;24 (18):2866–72.

14. Burtness B, Goldwasser MA, Flood W, Mattar B, Forastiere AA. Eastern cooperative oncology G. phase III randomized trial of cisplatin plus placebo compared with cisplatin plus cetuximab in metastatic/recurrent head and neck cancer: an eastern cooperative oncology group study. J Clin Oncol. 2005;23(34):8646–54.

15. Curran D, Giralt J, Harari PM, Ang KK, Cohen RB, Kies MS, et al. Quality of life in head and neck cancer patients after treatment with high-dose radiotherapy alone or in combination with cetuximab. J Clin Oncol. 2007;25(16):2191–7.

16. Di Nicolantonio F, Martini M, Molinari F, Sartore-Bianchi A, Arena S, Saletti P, et al. Wild-type BRAF is required for response to panitumumab or cetuximab in metastatic colorectal cancer. J Clin Oncol. 2008;26(35):5705–12.

17. Jonker DJ, O'Callaghan CJ, Karapetis CS, Zalcberg JR, Tu D, Au HJ, et al. Cetuximab for the treatment of colorectal cancer. N Engl J Med. 2007;357(20):2040–8.

18. Karapetis CS, Khambata-Ford S, Jonker DJ, O'Callaghan CJ, Tu D, Tebbutt NC, et al. K-ras mutations and benefit from cetuximab in advanced colorectal cancer. N Engl J Med. 2008;359 (17):1757–65.

19. Lenz HJ, Van Cutsem E, Khambata-Ford S, Mayer RJ, Gold P, Stella P, et al. Multicenter phase II and translational study of cetuximab in metastatic colorectal carcinoma refractory to irinotecan, oxaliplatin, and fluoropyrimidines. J Clin Oncol. 2006;24(30):4914–21.

20. Saltz LB, Lenz HJ, Kindler HL, Hochster HS, Wadler S, Hoff PM, et al. Randomized phase II trial of cetuximab, bevacizumab, and irinotecan compared with cetuximab and bevacizumab alone in irinotecan-refractory colorectal cancer: the BOND-2 study. J Clin Oncol. 2007;25 (29):4557–61.

21. Tabernero J, Van Cutsem E, Diaz-Rubio E, Cervantes A, Humblet Y, Andre T, et al. Phase II trial of cetuximab in combination with fluorouracil, leucovorin, and oxaliplatin in the first-line treatment of metastatic colorectal cancer. J Clin Oncol. 2007;25(33):5225–32.

22. Slamon DJ, Leyland-Jones B, Shak S, Fuchs H, Paton V, Bajamonde A, et al. Use of chemotherapy plus a monoclonal antibody against HER2 for metastatic breast cancer that overexpresses HER2. N Engl J Med. 2001;344(11):783–92.
23. Bange J, Zwick E, Ullrich A. Molecular targets for breast cancer therapy and prevention. Nat Med. 2001;7(5):548–52.
24. Arnould L, Arveux P, Couturier J, Gelly-Marty M, Loustalot C, Ettore F, et al. Pathologic complete response to trastuzumab-based neoadjuvant therapy is related to the level of HER-2 amplification. Clin Cancer Res. 2007;13(21):6404–9.
25. Baselga J, Carbonell X, Castaneda-Soto NJ, Clemens M, Green M, Harvey V, et al. Phase II study of efficacy, safety, and pharmacokinetics of trastuzumab monotherapy administered on a 3-weekly schedule. J Clin Oncol. 2005;23(10):2162–71.
26. Belkacemi Y, Gligorov J, Ozsahin M, Marsiglia H, De Lafontan B, Laharie-Mineur H, et al. Concurrent trastuzumab with adjuvant radiotherapy in HER2-positive breast cancer patients: acute toxicity analyses from the French multicentric study. Ann Oncol. 2008;19(6):1110–6.
27. Burstein HJ, Keshaviah A, Baron AD, Hart RD, Lambert-Falls R, Marcom PK, et al. Trastuzumab plus vinorelbine or taxane chemotherapy for HER2-overexpressing metastatic breast cancer: the trastuzumab and vinorelbine or taxane study. Cancer. 2007;110(5):965–72.
28. Hussain MH, MacVicar GR, Petrylak DP, Dunn RL, Vaishampayan U, Lara PN Jr, et al. Trastuzumab, paclitaxel, carboplatin, and gemcitabine in advanced human epidermal growth factor receptor-2/neu-positive urothelial carcinoma: results of a multicenter phase II National Cancer Institute trial. J Clin Oncol. 2007;25(16):2218–24.
29. Sato N, Sano M, Tabei T, Asaga T, Ando J, Fujii H, et al. Combination docetaxel and trastuzumab treatment for patients with HER-2-overexpressing metastatic breast cancer: a multicenter, phase-II study. Breast Cancer. 2006;13(2):166–71.
30. Schaller G, Fuchs I, Gonsch T, Weber J, Kleine-Tebbe A, Klare P, et al. Phase II study of capecitabine plus trastuzumab in human epidermal growth factor receptor 2 overexpressing metastatic breast cancer pretreated with anthracyclines or taxanes. J Clin Oncol. 2007;25(22):3246–50.
31. Suter TM, Procter M, van Veldhuisen DJ, Muscholl M, Bergh J, Carlomagno C, et al. Trastuzumab-associated cardiac adverse effects in the herceptin adjuvant trial. J Clin Oncol. 2007;25(25):3859–65.
32. Viani GA, Afonso SL, Stefano EJ, De Fendi LI, Soares FV. Adjuvant trastuzumab in the treatment of her-2-positive early breast cancer: a meta-analysis of published randomized trials. BMC Cancer. 2007;7:153.
33. Vogel CL, Cobleigh MA, Tripathy D, Gutheil JC, Harris LN, Fehrenbacher L, et al. Efficacy and safety of trastuzumab as a single agent in first-line treatment of HER2-overexpressing metastatic breast cancer. J Clin Oncol. 2002;20(3):719–26.
34. Lewis Phillips GD, Li G, Dugger DL, Crocker LM, Parsons KL, Mai E, et al. Targeting HER2-positive breast cancer with trastuzumab-DM1, an antibody-cytotoxic drug conjugate. Cancer Res. 2008;68(22):9280–90.
35. Seyfizadeh N, Seyfizadeh N, Hasenkamp J, Huerta-Yepez S. A molecular perspective on rituximab: a monoclonal antibody for B cell non Hodgkin lymphoma and other affections. Crit Rev Oncol Hematol. 2016;97:275–90.
36. Byrd JC, Peterson BL, Morrison VA, Park K, Jacobson R, Hoke E, et al. Randomized phase 2 study of fludarabine with concurrent versus sequential treatment with rituximab in symptomatic, untreated patients with B-cell chronic lymphocytic leukemia: results from cancer and leukemia group B 9712 (CALGB 9712). Blood. 2003;101(1):6–14.

37. Coiffier B, Lepage E, Briere J, Herbrecht R, Tilly H, Bouabdallah R, et al. CHOP chemotherapy plus rituximab compared with CHOP alone in elderly patients with diffuse large-B-cell lymphoma. N Engl J Med. 2002;346(4):235–42.
38. Ghielmini M, Schmitz SF, Cogliatti SB, Pichert G, Hummerjohann J, Waltzer U, et al. Prolonged treatment with rituximab in patients with follicular lymphoma significantly increases event-free survival and response duration compared with the standard weekly x 4 schedule. Blood. 2004;103(12):4416–23.
39. Marcus R, Imrie K, Belch A, Cunningham D, Flores E, Catalano J, et al. CVP chemotherapy plus rituximab compared with CVP as first-line treatment for advanced follicular lymphoma. Blood. 2005;105(4):1417–23.
40. Gravbrot N, Gilbert-Gard K, Mehta P, Ghotmi Y, Banerjee M, Mazis C, et al. Therapeutic monoclonal antibodies targeting immune checkpoints for the treatment of solid tumors. Antibodies (Basel). 2019;8(4).
41. Weber J. Review: anti-CTLA-4 antibody ipilimumab: case studies of clinical response and immune-related adverse events. Oncologist. 2007;12(7):864–72.
42. Weber JS, O'Day S, Urba W, Powderly J, Nichol G, Yellin M, et al. Phase I/II study of ipilimumab for patients with metastatic melanoma. J Clin Oncol. 2008;26(36):5950–6.
43. Topalian SL, Hodi FS, Brahmer JR, Gettinger SN, Smith DC, McDermott DF, et al. Safety, activity, and immune correlates of anti-PD-1 antibody in cancer. N Engl J Med. 2012;366 (26):2443–54.
44. Zhang N, Tu J, Wang X, Chu Q. Programmed cell death-1/programmed cell death ligand-1 checkpoint inhibitors: differences in mechanism of action. Immunotherapy. 2019;11(5):429–41.
45. Sharma P, Retz M, Siefker-Radtke A, Baron A, Necchi A, Bedke J, et al. Nivolumab in metastatic urothelial carcinoma after platinum therapy (CheckMate 275): a multicentre, single-arm, phase 2 trial. Lancet Oncol. 2017;18(3):312–22.
46. Overman MJ, McDermott R, Leach JL, Lonardi S, Lenz HJ, Morse MA, et al. Nivolumab in patients with metastatic DNA mismatch repair-deficient or microsatellite instability-high colorectal cancer (CheckMate 142): an open-label, multicentre, phase 2 study. Lancet Oncol. 2017;18(9):1182–91.
47. Ferris RL, Blumenschein G Jr, Fayette J, Guigay J, Colevas AD, Licitra L, et al. Nivolumab for recurrent squamous-cell carcinoma of the head and neck. N Engl J Med. 2016;375(19):1856–67.
48. El-Khoueiry AB, Sangro B, Yau T, Crocenzi TS, Kudo M, Hsu C, et al. Nivolumab in patients with advanced hepatocellular carcinoma (CheckMate 040): an open-label, non-comparative, phase 1/2 dose escalation and expansion trial. Lancet. 2017;389(10088):2492–502.
49. Ansell SM, Lesokhin AM, Borrello I, Halwani A, Scott EC, Gutierrez M, et al. PD-1 blockade with nivolumab in relapsed or refractory Hodgkin's lymphoma. N Engl J Med. 2015;372 (4):311–9.
50. Younes A, Santoro A, Shipp M, Zinzani PL, Timmerman JM, Ansell S, et al. Nivolumab for classical Hodgkin's lymphoma after failure of both autologous stem-cell transplantation and brentuximab vedotin: a multicentre, multicohort, single-arm phase 2 trial. Lancet Oncol. 2016;17(9):1283–94.
51. Brahmer J, Reckamp KL, Baas P, Crino L, Eberhardt WE, Poddubskaya E, et al. Nivolumab versus docetaxel in advanced squamous-cell non-small-cell lung cancer. N Engl J Med. 2015;373(2):123–35.
52. Borghaei H, Paz-Ares L, Horn L, Spigel DR, Steins M, Ready NE, et al. Nivolumab versus docetaxel in advanced nonsquamous non-small-cell lung cancer. N Engl J Med. 2015;373 (17):1627–39.
53. Motzer RJ, Escudier B, McDermott DF, George S, Hammers HJ, Srinivas S, et al. Nivolumab versus Everolimus in advanced renal-cell carcinoma. N Engl J Med. 2015;373(19):1803–13.

54. Antonia SJ, Lopez-Martin JA, Bendell J, Ott PA, Taylor M, Eder JP, et al. Nivolumab alone and nivolumab plus ipilimumab in recurrent small-cell lung cancer (CheckMate 032): a multicentre, open-label, phase 1/2 trial. Lancet Oncol. 2016;17(7):883–95.
55. Robert C, Ribas A, Wolchok JD, Hodi FS, Hamid O, Kefford R, et al. Anti-programmed-death-receptor-1 treatment with pembrolizumab in ipilimumab-refractory advanced melanoma: a randomised dose-comparison cohort of a phase 1 trial. Lancet. 2014;384(9948):1109–17.
56. Ribas A, Puzanov I, Dummer R, Schadendorf D, Hamid O, Robert C, et al. Pembrolizumab versus investigator-choice chemotherapy for ipilimumab-refractory melanoma (KEYNOTE-002): a randomised, controlled, phase 2 trial. Lancet Oncol. 2015;16(8):908–18.
57. Robert C, Schachter J, Long GV, Arance A, Grob JJ, Mortier L, et al. Pembrolizumab versus Ipilimumab in advanced melanoma. N Engl J Med. 2015;372(26):2521–32.
58. Eggermont AMM, Blank CU, Mandala M, Long GV, Atkinson V, Dalle S, et al. Adjuvant Pembrolizumab versus placebo in resected stage III melanoma. N Engl J Med. 2018;378 (19):1789–801.
59. Chung HC, Ros W, Delord JP, Perets R, Italiano A, Shapira-Frommer R, et al. Efficacy and safety of Pembrolizumab in previously treated advanced cervical cancer: results from the phase II KEYNOTE-158 study. J Clin Oncol. 2019;37(17):1470–8.
60. Fuchs CS, Doi T, Jang RW, Muro K, Satoh T, Machado M, et al. Safety and efficacy of Pembrolizumab monotherapy in patients with previously treated advanced gastric and gastro-esophageal junction cancer: phase 2 clinical KEYNOTE-059 trial. JAMA Oncol. 2018;4(5): e180013.
61. Seiwert TY, Burtness B, Mehra R, Weiss J, Berger R, Eder JP, et al. Safety and clinical activity of pembrolizumab for treatment of recurrent or metastatic squamous cell carcinoma of the head and neck (KEYNOTE-012): an open-label, multicentre, phase 1b trial. Lancet Oncol. 2016;17 (7):956–65.
62. Zhu AX, Finn RS, Edeline J, Cattan S, Ogasawara S, Palmer D, et al. Pembrolizumab in patients with advanced hepatocellular carcinoma previously treated with sorafenib (KEYNOTE-224): a non-randomised, open-label phase 2 trial. Lancet Oncol. 2018;19(7):940–52.
63. Chen R, Zinzani PL, Fanale MA, Armand P, Johnson NA, Brice P, et al. Phase II study of the efficacy and safety of Pembrolizumab for relapsed/refractory classic Hodgkin lymphoma. J Clin Oncol. 2017;35(19):2125–32.
64. Nghiem P, Bhatia S, Lipson EJ, Sharfman WH, Kudchadkar RR, Brohl AS, et al. Durable tumor regression and overall survival in patients with advanced Merkel cell carcinoma receiving Pembrolizumab as first-line therapy. J Clin Oncol. 2019;37(9):693–702.
65. Garon EB, Rizvi NA, Hui R, Leighl N, Balmanoukian AS, Eder JP, et al. Pembrolizumab for the treatment of non-small-cell lung cancer. N Engl J Med. 2015;372(21):2018–28.
66. Armand P, Rodig S, Melnichenko V, Thieblemont C, Bouabdallah K, Tumyan G, et al. Pembrolizumab in relapsed or refractory primary mediastinal large B-cell lymphoma. J Clin Oncol. 2019;37(34):3291–9.
67. Atkins MB, Plimack ER, Puzanov I, Fishman MN, McDermott DF, Cho DC, et al. Axitinib in combination with pembrolizumab in patients with advanced renal cell cancer: a non-randomised, open-label, dose-finding, and dose-expansion phase 1b trial. Lancet Oncol. 2018;19(3):405–15.
68. Rini BI, Plimack ER, Stus V, Gafanov R, Hawkins R, Nosov D, et al. Pembrolizumab plus Axitinib versus Sunitinib for advanced renal-cell carcinoma. N Engl J Med. 2019;380 (12):1116–27.
69. Stenger M. Pembrolizumab in MSI-H or dMMR solid tumors: 'First Tissue/Site-Agnostic' approval by FDA The ASCO Post; 2018. Available from: https://www.ascopost.com/issues/

february-10-2018/pembrolizumab-in-msi-h-or-dmmr-solid-tumors-first-tissuesite-agnostic-approval-by-fda/

70. Kaufman IIL, Russell J, Hamid O, Bhatia S, Terheyden P, D'Angelo SP, et al. Avelumab in patients with chemotherapy-refractory metastatic Merkel cell carcinoma: a multicentre, single-group, open-label, phase 2 trial. Lancet Oncol. 2016;17(10):1374–85.

71. Apolo AB, Infante JR, Balmanoukian A, Patel MR, Wang D, Kelly K, et al. Avelumab, an anti-programmed death-ligand 1 antibody, in patients with refractory metastatic urothelial carcinoma: results from a multicenter, Phase Ib study. J Clin Oncol. 2017;35(19):2117–24.

72. Motzer RJ, Penkov K, Haanen J, Rini B, Albiges L, Campbell MT, et al. Avelumab plus Axitinib versus Sunitinib for advanced renal-cell carcinoma. N Engl J Med. 2019;380 (12):1103–15.

73. Sedykh SE, Prinz VV, Buneva VN, Nevinsky GA. Bispecific antibodies: design, therapy, perspectives. Drug Des Devel Ther. 2018;12:195–208.

74. Shima M, Hanabusa H, Taki M, Matsushita T, Sato T, Fukutake K, et al. Factor VIII-mimetic function of humanized bispecific antibody in hemophilia a. N Engl J Med. 2016;374(21):2044–53.

75. Wu JC, Chen CH, Fu JW, Yang HC. Electrophoresis-enhanced detection of deoxyribonucleic acids on a membrane-based lateral flow strip using avian influenza H5 genetic sequence as the model. Sensors (Basel). 2014;14(3):4399–415.

76. Blakemore R, Story E, Helb D, Kop J, Banada P, Owens MR, et al. Evaluation of the analytical performance of the Xpert MTB/RIF assay. J Clin Microbiol. 2010;48(7):2495–501.

77. Byrne H, Conroy PJ, Whisstock JC, O'Kennedy RJ. A tale of two specificities: bispecific antibodies for therapeutic and diagnostic applications. Trends Biotechnol. 2013;31(11):621–32.

78. WHO. Global tuberculosis report, executive summary. 2019.

79. Boehme CC, Nabeta P, Hillemann D, Nicol MP, Shenai S, Krapp F, et al. Rapid molecular detection of tuberculosis and rifampin resistance. N Engl J Med. 2010;363(11):1005–15.

80. Doan TN, Eisen DP, Rose MT, Slack A, Stearnes G, McBryde ES. Interferon-gamma release assay for the diagnosis of latent tuberculosis infection: a latent-class analysis. PLoS One. 2017;12(11):e0188631.

81. Qiagen_Group. QuantiFERON®-TB Gold Plus (QFT®-Plus) ELISA Package Insert. The whole blood IFN-γ test measuring responses to ESAT-6 and CFP-10 peptide antigens02/2016.

82. Sigal GB, Pinter A, Lowary TL, Kawasaki M, Li A, Mathew A, et al. A novel sensitive immunoassay targeting the 5-methylthio-d-xylofuranose-lipoarabinomannan epitope meets the WHO's performance target for tuberculosis diagnosis. J Clin Microbiol. 2018;56(12).

83. Sarkar S, Tang XL, Das D, Spencer JS, Lowary TL, Suresh MR. A bispecific antibody based assay shows potential for detecting tuberculosis in resource constrained laboratory settings. PLoS One. 2012;7(2):e32340.

84. Broger T, Sossen B, du Toit E, Kerkhoff AD, Schutz C, Ivanova Reipold E, et al. Novel lipoarabinomannan point-of-care tuberculosis test for people with HIV: a diagnostic accuracy study. Lancet Infect Dis. 2019;19(8):852–61.

85. Drain PK, Heichman KA, Wilson D. A new point-of-care test to diagnose tuberculosis. Lancet Infect Dis. 2019;19(8):794–6.

86. Collins PL, Graham BS. Viral and host factors in human respiratory syncytial virus pathogenesis. J Virol. 2008;82(5):2040–55.

87. Eboigbodin KE, Moilanen K, Elf S, Hoser M. Rapid and sensitive real-time assay for the detection of respiratory syncytial virus using RT-SIBA(R). BMC Infect Dis. 2017;17(1):134.

88. CerTest_Biotech. CERTEST RSV. One step respiratory syncytial virus card test.

89. Nam HH, Ison MG. Respiratory syncytial virus infection in adults. BMJ. 2019;366:l5021.

90. Koczula KM, Gallotta A. Lateral flow assays. Essays Biochem. 2016;60(1):111–20.

91. Posthuma-Trumpie GA, Korf J, van Amerongen A. Lateral flow (immuno)assay: its strengths, weaknesses, opportunities and threats. A literature survey. Anal Bioanal Chem. 2009;393 (2):569–82.

92. Wu G, Zaman MH. Low-cost tools for diagnosing and monitoring HIV infection in low-resource settings. Bull World Health Organ. 2012;90(12):914–20.

93. Ngom B, Guo Y, Wang X, Bi D. Development and application of lateral flow test strip technology for detection of infectious agents and chemical contaminants: a review. Anal Bioanal Chem. 2010;397(3):1113–35.

94. Ngom B, Guo Y, Wang X, Bi D. Correction to: development and application of lateral flow test strip technology for detection of infectious agents and chemical contaminants: a review. Anal Bioanal Chem. 2018;410(11):2859.

95. Pecchia S, Da Lio D. Development of a rapid PCR-nucleic acid lateral flow immunoassay (PCR-NALFIA) based on rDNA IGS sequence analysis for the detection of Macrophomina phaseolina in soil. J Microbiol Methods. 2018;151:118–28.

96. Sajid M. Designs, formats and applications of lateral flow assay: a literature review. J Saudi Chem Soc. 2015;19(6):689–705.

97. Li Z, Chen H, Wang P. Lateral flow assay ruler for quantitative and rapid point-of-care testing. Analyst. 2019;144(10):3314–22.

98. Jia Li DM, Macdonald J. Enhancing the signal of lateral flow immunoassays by using different developing methods. Sensors Mater. 2015;27(7):549–61.

99. Millipore E. Rapid lateral flow test strips: considerations for product development. 2013.

100. Hsieh HV, Dantzler JL, Weigl BH. Analytical tools to improve optimization procedures for lateral flow assays. Diagnostics (Basel). 2017;7(2).

101. Ren M, Xu H, Huang X, Kuang M, Xiong Y, Xu H, et al. Immunochromatographic assay for ultrasensitive detection of aflatoxin B(1) in maize by highly luminescent quantum dot beads. ACS Appl Mater Interfaces. 2014;6(16):14215–22.

102. Shi L, Wu F, Wen Y, Zhao F, Xiang J, Ma L. A novel method to detect *Listeria monocytogenes* via superparamagnetic lateral flow immunoassay. Anal Bioanal Chem. 2015;407(2):529–35.

103. Panferov VG, Safenkova IV, Varitsev YA, Zherdev AV, Dzantiev BB. Enhancement of lateral flow immunoassay by alkaline phosphatase: a simple and highly sensitive test for potato virus X. Mikrochim Acta. 2017;185(1):25.

104. Engvall E, Perlmann P. Enzyme-linked immunosorbent assay (ELISA). Quantitative assay of immunoglobulin G. Immunochemistry. 1971;8(9):871–4.

105. Engvall E, Jonsson K, Perlmann P. Enzyme-linked immunosorbent assay. II. Quantitative assay of protein antigen, immunoglobulin G, by means of enzyme-labelled antigen and antibody-coated tubes. Biochim Biophys Acta. 1971;251(3):427–34.

106. Castro C, Gourley M. Diagnostic testing and interpretation of tests for autoimmunity. J Allergy Clin Immunol. 2010;125(2 Suppl 2):S238–47.

107. Lindstrom P, Wager O. IgG autoantibody to human serum albumin studied by the ELISA-technique. Scand J Immunol. 1978;7(5):419–25.

108. Kohl TO, Ascoli CA. Indirect immunometric ELISA. Cold Spring Harb Protoc. 2017;2017(5).

109. Kato K, Hamaguchi Y, Okawa S, Ishikawa E, Kobayashi K. Use of rabbit antiboty IgG bound onto plain and aminoalkylsilyl glass surface for the enzyme-linked sandwich immunoassay. J Biochem. 1977;82(1):261–6.

110. Kato K, Hamaguchi Y, Okawa S, Ishikawa E, Kobayashi K, Katunuma N. Use of rabbit antibody IgG-loaded silicone pieces for the sandwich enzymoimmunoassay of macromolecular antigens. J Biochem. 1977;81(5):1557–66.

111. Yorde DE, Sasse EA, Wang TY, Hussa RO, Garancis JC. Competitive enzyme-linked immunoassay with use of soluble enzyme/antibody immune complexes for labeling. I. Measurement of human choriogonadotropin. Clin Chem. 1976;22(8):1372–7.
112. Hornbeck PV. Enzyme-linked immunosorbent assays. Curr Protoc Immunol. 2015;110:2 1–2 1 23.
113. Matyjaszek-Matuszek B, Pyzik A, Nowakowski A, Jarosz MJ. Diagnostic methods of TSH in thyroid screening tests. Ann Agric Environ Med. 2013;20(4):731–5.
114. Kohl TO, Ascoli CA. Immunometric double-antibody sandwich enzyme-linked immunosorbent assay. Cold Spring Harb Protoc. 2017;2017(6):pdb prot093724.
115. Haapakoski R, Karisola P, Fyhrquist N, Savinko T, Lehtimaki S, Wolff H, et al. Toll-like receptor activation during cutaneous allergen sensitization blocks development of asthma through IFN-gamma-dependent mechanisms. J Invest Dermatol. 2013;133(4):964–72.
116. Kragstrup TW, Vorup-Jensen T, Deleuran B, Hvid M. A simple set of validation steps identifies and removes false results in a sandwich enzyme-linked immunosorbent assay caused by anti-animal IgG antibodies in plasma from arthritis patients. Springerplus. 2013;2(1):263.
117. Sakamoto S, Putalun W, Vimolmangkang S, Phoolcharoen W, Shoyama Y, Tanaka H, et al. Enzyme-linked immunosorbent assay for the quantitative/qualitative analysis of plant secondary metabolites. J Nat Med. 2018;72(1):32–42.
118. Alexander TS. Human immunodeficiency virus diagnostic testing: 30 years of evolution. Clin Vaccine Immunol. 2016;23(4):249–53.
119. Bystryak S, Santockyte R. Increased sensitivity of HIV-1 p24 ELISA using a photochemical signal amplification system. J Acquir Immune Defic Syndr. 2015;70(2):109–14.
120. Hulett HR, Bonner WA, Barrett J, Herzenberg LA. Cell sorting: automated separation of mammalian cells as a function of intracellular fluorescence. Science. 1969;166(3906).747–9.
121. El-Sayed AM, El-Borai MH, Bahnassy AA, El-Gerzawi SM. Flow cytometric immunophenotyping (FCI) of lymphoma: correlation with histopathology and immunohistochemistry. Diagn Pathol. 2008;3:43.
122. Craig FE, Foon KA. Flow cytometric immunophenotyping for hematologic neoplasms. Blood. 2008;111(8):3941–67.
123. Adan A, Alizada G, Kiraz Y, Baran Y, Nalbant A. Flow cytometry: basic principles and applications. Crit Rev Biotechnol. 2017;37(2):163–76.
124. Demurtas A, Stacchini A, Aliberti S, Chiusa L, Chiarle R, Novero D. Tissue flow cytometry immunophenotyping in the diagnosis and classification of non-Hodgkin's lymphomas: a retrospective evaluation of 1,792 cases. Cytometry B Clin Cytom. 2013;84(2):82–95.
125. Karawajew L, Dworzak M, Ratei R, Rhein P, Gaipa G, Buldini B, et al. Minimal residual disease analysis by eight-color flow cytometry in relapsed childhood acute lymphoblastic leukemia. Haematologica. 2015;100(7):935–44.
126. Yamauchi H, Stearns V, Hayes DF. When is a tumor marker ready for prime time? A case study of c-erbB-2 as a predictive factor in breast cancer. J Clin Oncol. 2001;19(8):2334–56.
127. Press MF, Bernstein L, Thomas PA, Meisner LF, Zhou JY, Ma Y, et al. HER-2/neu gene amplification characterized by fluorescence in situ hybridization: poor prognosis in node-negative breast carcinomas. J Clin Oncol. 1997;15(8):2894–904.
128. Andrulis IL, Bull SB, Blackstein ME, Sutherland D, Mak C, Sidlofsky S, et al. neu/erbB-2 amplification identifies a poor-prognosis group of women with node-negative breast cancer. Toronto breast cancer study group. J Clin Oncol. 1998;16(4):1340–9.
129. Roche PC, Ingle JN. Increased HER2 with U.S. Food and Drug Administration-approved antibody. J Clin Oncol. 1999;17(1):434.
130. Perez EA, Cortes J, Gonzalez-Angulo AM, Bartlett JM. HER2 testing: current status and future directions. Cancer Treat Rev. 2014;40(2):276–84.

131. Sauter G, Lee J, Bartlett JM, Slamon DJ, Press MF. Guidelines for human epidermal growth factor receptor 2 testing: biologic and methodologic considerations. J Clin Oncol. 2009;27 (8):1323–33.
132. Takai H, Kato A, Nakamura T, Tachibana T, Sakurai T, Nanami M, et al. The importance of characterization of FITC-labeled antibodies used in tissue cross-reactivity studies. Acta Histochem. 2011;113(4):472–6.
133. Zalutsky MR. Potential of immuno-positron emission tomography for tumor imaging and immunotherapy planning. Clin Cancer Res. 2006;12(7 Pt 1):1958–60.
134. Roopenian DC, Akilesh S. FcRn: the neonatal Fc receptor comes of age. Nat Rev Immunol. 2007;7(9):715–25.
135. Thurber GM, Schmidt MM, Wittrup KD. Antibody tumor penetration: transport opposed by systemic and antigen-mediated clearance. Adv Drug Deliv Rev. 2008;60(12):1421–34.
136. Bailly C, Clery PF, Faivre-Chauvet A, Bourgeois M, Guerard F, Haddad F, et al. Immuno-PET for clinical theranostic approaches. Int J Mol Sci. 2016;18(1).
137. Schoffelen R, Sharkey RM, Goldenberg DM, Franssen G, McBride WJ, Rossi EA, et al. Pretargeted immuno-positron emission tomography imaging of carcinoembryonic antigen-expressing tumors with a bispecific antibody and a 68Ga- and 18F-labeled hapten peptide in mice with human tumor xenografts. Mol Cancer Ther. 2010;9(4):1019–27.
138. McBride WJ, Zanzonico P, Sharkey RM, Noren C, Karacay H, Rossi EA, et al. Bispecific antibody pretargeting PET (immunoPET) with an 124I-labeled hapten-peptide. J Nucl Med. 2006;47(10):1678–88.
139. Batra SK, Jain M, Wittel UA, Chauhan SC, Colcher D. Pharmacokinetics and biodistribution of genetically engineered antibodies. Curr Opin Biotechnol. 2002;13(6):603–8.
140. Oliver J. Antibody applications. Mater Methods. 2013;3:182.
141. Carey MF, Peterson CL, Smale ST. Chromatin immunoprecipitation (ChIP). Cold Spring Harb Protoc. 2009;2009(9):pdb prot5279.
142. Eberl HC, Mann M, Vermeulen M. Quantitative proteomics for epigenetics. Chembiochem. 2011;12(2):224–34.
143. O'Neill LP, Turner BM. Immunoprecipitation of native chromatin: NChIP. Methods. 2003;31 (1):76–82.
144. Mutneja M, Mohan C, Long KD, Das C. An introduction to antibodies and their applications. 2013.
145. Landt SG, Marinov GK, Kundaje A, Kheradpour P, Pauli F, Batzoglou S, et al. ChIP-seq guidelines and practices of the ENCODE and modENCODE consortia. Genome Res. 2012;22 (9):1813–31.
146. Aydin S. A short history, principles, and types of ELISA, and our laboratory experience with peptide/protein analyses using ELISA. Peptides. 2015;72:4–15.
147. Engvall E. The ELISA, enzyme-linked immunosorbent assay. Clin Chem. 2010;56(2):319–20.
148. Kohler G, Milstein C. Continuous cultures of fused cells secreting antibody of predefined specificity. Nature. 1975;256(5517):495–7.
149. Saeed AF, Wang R, Ling S, Wang S. Antibody engineering for pursuing a healthier future. Front Microbiol. 2017;8:495.
150. Ledsgaard L, Kilstrup M, Karatt-Vellatt A, McCafferty J, Laustsen AH. Basics of antibody phage display technology. Toxins (Basel). 2018;10(6).
151. Faulk WP, Taylor GM. An immunocolloid method for the electron microscope. Immunochemistry. 1971;8(11):1081–3.
152. Chang L, Li J, Wang L. Immuno-PCR: an ultrasensitive immunoassay for biomolecular detection. Anal Chim Acta. 2016;910:12–24.

153. Fredriksson S, Gullberg M, Jarvius J, Olsson C, Pietras K, Gustafsdottir SM, et al. Protein detection using proximity-dependent DNA ligation assays. Nat Biotechnol. 2002;20(5):473–7.
154. Doerner A, Rhiel L, Zielonka S, Kolmar H. Therapeutic antibody engineering by high efficiency cell screening. FEBS Lett. 2014;588(2):278–87.
155. Rhiel L, Krah S, Gunther R, Becker S, Kolmar H, Hock B. REAL-select: full-length antibody display and library screening by surface capture on yeast cells. PLoS One. 2014;9(12):e114887.
156. Federspiel JD, Cristea IM. Considerations for identifying endogenous protein complexes from tissue via immunoaffinity purification and quantitative mass spectrometry. Methods Mol Biol. 1977;2019:115–43.
157. Anderson NL, Anderson NG, Haines LR, Hardie DB, Olafson RW, Pearson TW. Mass spectrometric quantitation of peptides and proteins using stable isotope standards and capture by anti-peptide antibodies (SISCAPA). J Proteome Res. 2004;3(2):235–44.
158. Xu Q, Zhu M, Yang T, Xu F, Liu Y, Chen Y. Quantitative assessment of human serum transferrin receptor in breast cancer patients pre- and post-chemotherapy using peptide immunoaffinity enrichment coupled with targeted proteomics. Clin Chim Acta. 2015;448:118–23.
159. Mason DR, Reid JD, Camenzind AG, Holmes DT, Borchers CH. Duplexed iMALDI for the detection of angiotensin I and angiotensin II. Methods. 2012;56(2):213–22.
160. Nakamura A, Kaneko N, Villemagne VL, Kato T, Doecke J, Dore V, et al. High performance plasma amyloid-beta biomarkers for Alzheimer's disease. Nature. 2018;554(7691):249–54.
161. Schoenherr RM, Saul RG, Whiteaker JR, Yan P, Whiteley GR, Paulovich AG. Anti-peptide monoclonal antibodies generated for immuno-multiple reaction monitoring-mass spectrometry assays have a high probability of supporting Western blot and ELISA. Mol Cell Proteomics. 2015;14(2):382–98.
162. Trenchevska O, Sherma ND, Oran PE, Reaven PD, Nelson RW, Nedelkov D. Quantitative mass spectrometric immunoassay for the chemokine RANTES and its variants. J Proteome. 2015;116:15–23.
163. Spitzer MH, Nolan GP. Mass cytometry: single cells, many features. Cell. 2016;165(4):780–91.

Further Reading

Vaughan T, Osbourn J, Jallal B. Protein therapeutics. Wiley-VCH Verlag GmbH & Co. KGaA; 2017.
George AJT, Urch CE. Diagnostic and therapeutic antibodies. Totowa, NJ: Humana Press; 2000. xiv, 477 pp.

Bispecific Antibodies

7

Gordana Wozniak-Knopp

Contents

G. Wozniak-Knopp (✉)

Christian Doppler Laboratory for Innovative Immunotherapeutics, Institute of Molecular Biotechnology, Department of Biotechnology, University of Natural Resources and Life Sciences, Vienna (BOKU), Vienna, Austria

e-mail: gordana.wozniak@boku.ac.at

© Springer Nature Switzerland AG 2021

F. Rüker, G. Wozniak-Knopp (eds.), *Introduction to Antibody Engineering*,
Learning Materials in Biosciences,
https://doi.org/10.1007/978-3-030-54630-4_7

Keywords

Obligate bispecific antibodies · Novel modes of action · T-cell engagement · Molecular Trojan horses · Fragment-based bispecific antibodies

What You Will Learn in This Chapter

After the introduction to the history and contemporary aspects of bispecific antibodies, this chapter reveals the background of the great enthusiasm of the protein engineering community and pharmaceutical industry for development of this class of molecules into therapeutic agents. You will be able to elaborate on their unique mode of action as well as the underlying mechanistic and spatiotemporal aspects of their activity. The insights in diverse molecular architectures of bispecific formats will enable you to evaluate their suitability in a particular biological situation. Specific challenges in their manufacturing and quality control will trigger your interest in implementing the acquired knowledge in design of novel solutions to improve this promising family of therapeutics for the future.

Please note that this chapter only describes immunoglobulin-based bispecific molecules, and refer to the chapter dedicated to bispecific binders based on alternative scaffolds to find out more on this subject.

7.1 Introduction

Inherently, bispecific antibodies (bsAbs) combine specificities for two epitopes or two antigens in a single molecule, and even from this simple point of view, their application appears advantageous over having to produce and characterize two distinct molecules with single designated targets. Such combinatorial bsAbs are invaluable in clinical situations when the redundancy of disease-mediating ligands has to be overcome. Moreover, during the past three decades, a substantial knowledge base has been built on significant biological effects that can only be achieved by the physical linkage of the two specificities in a single molecule and not by their mixture, which has raised the popularity of such obligate bsAbs, resulting in 2 clinically approved molecules and 85 further that are tested in clinical trials during 2019 [1]. These unique effects can be spatial, where binding with both arms is critical for the positioning of the antibody: notable examples include the molecules that can connect two different cell types or link an enzyme to its substrate [2]. Another group of obligate bsAbs is temporal, where binding of the two domains proceeds in sequential manner and antigen interaction of one arm depends on binding of the other one, either due to kinetic or steric reasons [2].

As depicted in this chapter, the bsAbs often tend to incorporate a complete native-like IgG and/or fragments thereof in a single molecule, as well as antibody fragments whose antigen-engaging properties are retained by a designed linker-mediated arrangement in a single polypeptide chain. Although they are composed of amino acid strands that are postulated to build identical domains as present in a complete antibody, they may be different in their antigen engagement and functional properties, when reformatted to a full-length IgG [3]. Library selections can however be designed to steer toward directed evolution of antibody fragments with attractive properties such as high antigen affinity [4], but also high expression level, favorable biophysical properties, and a particular binding mode [5]. For this feature to be utilized, strategies to incorporate screening of the full-length antibodies early in the discovery stage are being intensely explored [6].

Questions

1.1 Define bispecific antibodies.

1.2 What are obligate bispecific antibodies?

7.2   History of Bispecific Antibodies

Almost 50 years ago, scientist Alfred Nisonoff, working in one of Linus Pauling's groups, concluded on an important discovery on the structure of antibody molecules: every antibody molecule consists of two combining sites with identical specificity to the target antigen [7]. Continuing the work of the Nobel Laureate Rodney Porter, who used papain to cleave an antibody into two Fab fragments and an Fc fragment [8], Nisonoff used another enzyme, pepsin, to achieve proteolytic cleavage of the antibody into a single bivalent F(ab')$_2$ fragment and a number of peptide fragments derived from the Fc fragment. He reasoned that the two enzymes cleave on the opposite side of disulfide bridges of the heavy chains. These results changed the previous paradigm of the positioning of the two Fab fragments at each end of the Fc fragment in favor of the four-chain model, where the Fab fragments are placed at one side of the Fc fragment. The idea of man-made bsAbs quickly followed this discovery: using pepsin cleavage, mild reduction, and oxidation of two F(ab') fragments of different specificities, Nisonoff and Rivers produced a bivalent F(ab')$_2$ of mixed specificity [9]. Their original approach was applied for preparation of a bispecific reagent directed against mouse immunoglobulin (Ig) and ferritin, which was used for detection of Ig on the surface of mouse lymphocytes using electron microscopy [10].

In the following 20 years, the period in which the hybridoma technique developed by Köhler and Milstein became widely used and enabled efficient sourcing of monoclonal antibodies [11], the yield of bispecific F(ab')$_2$ could be improved by using chemical coupling with dithiol-complexing and thiol-activating agents [12] or thioether linkage [13]. In fact, Milstein himself extended the concept of hybridoma into "hybrid hybridoma" envisioning a single "quadroma" [14] cell harboring heavy and light chain-encoding genes

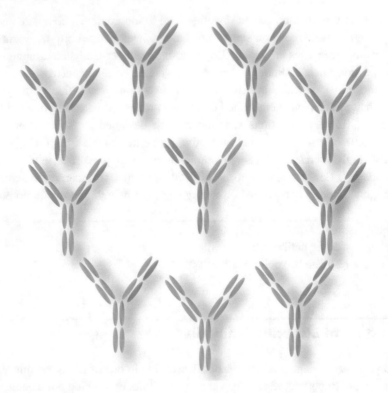

Fig. 7.1 Chain association issue. Upon expression of 4 different antibody chains in one cell (here the chains corresponding to each antibody are depicted in blue and green, respectively), a mixture of 10 potential products can be expected. Several engineering methods aim toward steering the pairing preference toward the desired bispecific product (center)

of two different antibodies [15] and was able to demonstrate the secretion of a bsAb, albeit as one of several molecular species (termed "waste" by Milstein) formed upon promiscuous pairing of the heavy and light chains (Fig. 7.1).

Approximately from that time dates the first report of a bsAb that can induce potent cellular cytotoxicity by redirection of T cells to attack tumor cells: engagement of T-cell receptor (TCR) was used to activate T-cell clones, and Thy-1 antigen, expressed on a lymphoma cell line, was chosen as a tumor-associated antigen (TAA) [16]. Within years followed the evidence of redirection of T-cell activity toward tumor cells, mediated by bsAbs that consisted of chemically coupled $F(ab')_2$ fragments only [17, 18]. The independency of MHC restriction on one hand and the extreme potency of induced cell lysis on the other hand were the two features that contributed to the fact that bsAbs with such mode of action have remained in the focus of clinical development until the present time.

The next decade's bsAb engineering was strongly influenced by the development of recombinant DNA techniques. In 1988, the first report of a single-chain variable fragment (scFv), where the variable domains of antibody's heavy and light chains are connected with a flexible linker into a minimalistic antigen-binding entity [19, 20], initiated a new era in the design of multispecific binding molecules. To circumvent random antibody chain association, multiple antigen-binding molecules could now be combined in one polypeptide chain, and the absence of glycosylated Fc fragment enabled high-titer inexpensive production in *E. coli*. Additionally, the valency of such multispecific agents could be chosen at will, further increasing their versatile use in various biological situations. Nevertheless, the lack of the Fc fragment results in a much shorter half-life in vivo and has to be compensated by introduction of a human serum albumin (HSA) – or an Fc – fusion partner for the optimal biological activity. It is therefore not surprising that the first published symmetric format features a tetravalent bispecific molecule, a monoclonal antibody with a scFv fused to the C-terminus of each of the heavy chains [21]. Further, the issues of aggregation and low stability often required re-optimization of the final scFv-based products [22, 23].

In 1996, another pioneering approach was used to tackle the chain association issue, the term used to describe random pairing of simultaneously expressed heavy and light chains, resulting in a mixture of ten different molecular species (Fig. 7.1). Following the observation that heavy chain dimerization is initiated through dimerization of the C_H3-domains with an affinity over 10–15 nM, Carter and co-workers proposed a "Knobs-into-Holes" engineering strategy that enables guided heterodimerization of two antibody halves by the replacement of an amino acid with a small side chain for one with a bulky side chain in one of the heavy chains ("knob"), and opposite steps were used to modify the other heavy chain ("hole"), which resulted in over 95% of heterodimeric product upon co-expression [24] (Fig. 7.2a). This steric hindrance approach to produce asymmetric antibodies has been complemented with further mutagenesis to increase the yield of the heterodimer [25], and additional engineering steps, such as introduction of interchain disulfide bonds [26] as well as point mutagenesis [27], successfully improved the lowered thermal stability of the resulting molecule (Fig. 7.2b). Alternatively, heterodimerization can be governed by the introduction of oppositely charged pairs of amino acids on the interface of the heterologous C_H3 domains [28] (Fig. 7.2c) or a combination of steric hindrance and electrostatic repulsion (Fig. 7.2d). More recent methods aim at a higher efficiency of heterodimerization by applying a more extensive mutagenesis or even by using heterodimerization motifs naturally present in heterodimeric molecules, such as the one introduced by BEAT body platform [29] (Fig. 7.2e). In the first decade of the new millennium, a number of ingenious solutions to tackle the light chain pairing were proposed, such as the use of a common light chain [26, 30] or light chain with species-specific pairing; [31] however first orthogonal interfaces that can promote selective pairing of the cognate light to its heavy chain partner were not described until 2014 [32]. An efficient solution to this question was presented with CrossMab, a format with swapped C_H1 and C_L domains at each arm of the heterodimer [33]. Yet another solution of solving the light chain pairing problem in heterodimeric antibodies consists in exchanging the C_H1/C_L constant domain pair in one of the two Fabs by a pair of C_H3 domains [34].

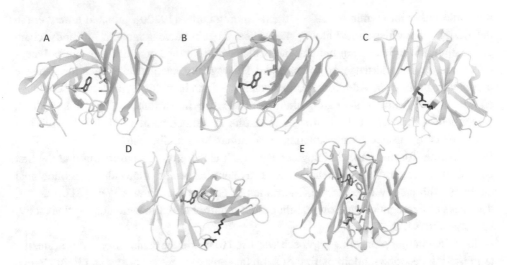

Fig. 7.2 Engineering methods to introduce preferential pairing of two heterologous heavy chains. The heterologous C_H3-domain backbones are presented in a cartoon diagram with depicted mutated amino acid residues in the interface between C_H3 domains. (**a**), "Knobs-into-Holes" heterodimer T366W + T366S/L368A/Y407V (PDB: 5DI8); (**b**), disulfide-linked "Knobs-into-Holes" heterodimer (PDB: 5HY9); (**c**), heterodimer formed by introducing oppositely charged residues E356K/D399K + K392D/K409D (PDB: 5DK2); (**d**), heterodimer with a combination of sterically and electrostatically interfering residues "EW-RVT," including mutations K360E/K409W + Q347R/D399V/F405T (PDB: 4X98); (**e**), BEAT body-based heterodimer featuring heterodimerizing residues based on TCR sequences (PDB: 6G1E) (EU numbering scheme is used)

In parallel to the explosion of human efforts to produce bsAbs, there is a much older naturally occurring antibody species with a great potential for development into a bispecific agent, whose particular biological properties have raised early attention; however the structural reasons behind them remained unclear for a long time. The inability of serum-derived IgG4 to crosslink antigens and generate large immune complexes was in contrast with the recombinant IgG4, which was well able to crosslink identical antigens. Elevated IgG4 levels, discovered in the sera of beekeepers with chronical exposure to bee toxin phospholipase-A2 (PLA2), triggered the interest in the study of these full-sized "function-ally monovalent" antibodies that could block the activity of other PLA2-specific antibodies: they consisted of two heavy and light chain pairs, each derived from a different antibody, and had exchanged Fab arms [35]. It was later shown that arm exchange occurs post-translationally by chance: an excess of irrelevant IgG4 prevented in vivo formation of hybrid antibodies [36], and using a sensitive real-time FRET technique, it was demonstrated that Fab arm exchange can occur in vivo under specific redox conditions [37]. Harvest of this phenomenon resulted in one of the very successful techniques for construction of bsAbs of today, the "DuoBody" platform, where the IgG1-halves are mutated in their C_H3 domain to undergo controlled Fab-arms exchange upon mild reduc-tion and re-oxidation [38]. Alternatively to the expression of different antibody chains in one cell, bsAbs of the IgG4 type could even be produced in bacterial co-culture expressing two distinct half-antibodies [39].

To sum up, technical innovations in engineering, manufacturing, as well as downstream processing and quality control of bsAbs not only delivered a variety of over 100 formats available today but have permanently succeeded in propelling the bsAbs to the top interesting compounds among antibody-based therapeutics. In the following paragraphs, we will further address the diverse modes of action that significantly differentiate bsAbs as therapeutics, delineate their format categories in the respect of their valency and the resulting variability in multi-targeting applications, and discuss their opportunities and limitations.

Questions

2.1. Of what molecular format was the first bispecific antibody, and what was the method of its preparation?

2.2. What is the quadroma technique?

2.3. What are the reasons for the popular use of the scFv fragment in bispecific antibodies?

2.4. Which particular class of antibodies is naturally bispecific, and what is the molecular background of this feature?

7.3   Mode of Action of Bispecific Antibodies Today

Historically, certain biological functions that can only be exerted by antibodies joining two target specificities in one molecule have attracted the attention of the scientific community to develop and improve these advantageous therapeutic agents (Fig. 7.3). Examples of therapeutic antibodies that have reached approval in clinical setting are listed in Table 7.1. Apart from T-cell recruitment strategies, as a subset of bridging different cell types and

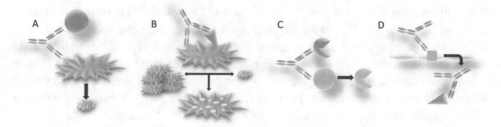

Fig. 7.3 Mechanisms of action of obligate bsAbs. (**a**), bsAb bridges effector and target cell and mediates cell killing; (**b**) bsAb binds to two different antigen epitopes on one cell and causes proliferation, metabolic activation, or apoptosis; (**c**) bsAb acts as a link in an enzyme cascade and enables enzyme activation; (**d**) bsAb as a molecular transfer mediator

Table 7.1 Clinically approved bispecific antibodies

Specificities	Format	Indication	Year of approval
Catumaxomab (Removab®)			
Ep-CAMxCD3x CD16a via Fc	Mouse IgG2b – rat IgG2a hybrid	Malignant ascites	2009, withdrawn in 2017
Blinatumomab (Blincyto®)			
CD19 x CD3	2 linker-connected scFvs ("BiTE")	Hematological malignancies	2014
Emicizumab (Hemlibra®)			
FXIa x FX	Heterodimeric IgG4; humanized	Hemophilia A	2017

allowing cell-to-cell association ("*trans*" interaction) mentioned in Sect. 7.3.1 (Fig. 7.3a), bsAbs are also able to bridge two receptors on the same cell ("*cis*" interaction) (Fig. 7.3b and Sect. 7.3.2), target several ligands of the same signaling pathway to overcome their redundancy (Sect. 7.3.3), or engage two epitopes of the same antigen in biparatopic binding (Sect. 7.3.4). In some cases, the activity of so-called agonistic bsAbs can potentiate metabolic processes and promote regeneration (Fig. 7.3b), or they act as a chain link in an enzyme cascade (Fig. 7.3c and Sect. 7.3.5). Particular biological situations harness their activity as molecular transfer mediators as they can facilitate enrichment or escape of a molecule or induce the activity of a functional binding arm in particular organs, tissues, and cellular compartments (Fig. 7.3d and Sect. 7.3.6).

7.3.1 BsAbs for Bridging Different Cell Types

Historically, this approach introduces a bsAb that can redirect the cytotoxic activity of T cells to eliminate tumor cells. The bsAb acts as a physical linker between T cells and tumor cells, with one arm binding to T cells and the other to tumor cells [18]. Most of the known bsAbs acting in this way bind to the CD3ε component of the TCR and elicit the activation of T cells independently of the epitope specificity of the TCR and of MHC restriction and in addition without the necessity to engage accessory molecules such as CD4 or CD8 [40]. In spite of the encouraging potency of such agents in in vitro tests, the property of these molecules to induce an uncontrolled cytokine release in vivo [41], accompanied with their difficult production, slowed down their development for clinical use. Only after the amazing data of the clinical efficiency of blinatumomab (Blincyto®) were presented [42, 43], the interest in T-cell-engaging bsAb has taken off intensely, and as a result, more than 50% of bsAbs in clinical trials today work per T-cell engagement as the primary mode of action.

Blinatumomab is a bsAb, composed of two scFvs, specific to CD19 and CD3ε, devoid of the Fc fragment and hence only 55 kDa in size [44]. As its serum half-life in vivo is approximately 1 h, it is typically delivered by continuous intravenous infusion using portable pumps and has first proven efficient in treatment of patients with non-Hodgkin lymphoma [45]. An impressive complete response rate of 43% was achieved in patients with relapsed/refractory acute lymphoblastic leukemia (ALL) [46], which lead to its regulatory approval in 2014 in the USA.

How do the T-cell engagers function? Such bsAbs connect T cells and tumor cells to form an immune synapse, which results in TCR activation and release of granzyme and perforin from the T cells, and eventually target cell lysis and death via apoptosis [47]. The results of activity of blinatumomab were so impressive that they prompted clinical development of several bispecific T-cell engagers (BiTEs), which all share its architecture and CD3-engaging property but however differ in the tumor-asociated-antigen engaging arms [48]. Factors that critically influence the potency of T-cell-engaging bsAb include the expression level of the target antigen on the tumor cell surface, its mobility in the membrane, and the epitope distance to the target cell membrane. It has also been suggested that the multivalent engagement of the TAA can improve the efficacy of the T-cell-engaging agent by allowing avid binding to the tumor cell surface, such as in multivalent TandAbs, engaging two CD3 molecules and two TAAs [49]. On the other hand, the CD3-specific arm is typically of relatively low affinity (about 200 nM), which enables the bsAb molecule to engage in several sequential contacts with the CD3 complex and induce a phenomenon described as "serial killing," where multiple target cells can be attacked by a single bsAb-armed T cell [50].

Further, the T-cell-engaging reagents devoid of or equipped with an Fc fragment that cannot engage effector cells appear advantageous in comparison with their functional Fc counterparts. Even before the rise of bispecific reagents targeting CD3, silencing of the effector functions of the Fc fragment improved the safety of OKT3 (Muromonab), a monospecific anti-CD3 antibody used as immunosuppressant in transplantation medicine [51]. With obliterated Fcγ-receptor binding, the unrestricted clustering of CD3 on the surface of the immune cells was prevented, and the antibody variant modified in this way induced significantly lower cytokine release. In contrast to blinatumomab, the first approved bsAb catumaxomab was composed of a heterodimer of mouse and rat chains, specific to CD3 and to the TAA Ep-CAM, and equipped with a functional Fc [52]. The severe adverse events, such as strong local cytokine release and T-cell-mediated hepatotoxicity that were induced upon intravenous administration, were attributed to on-target/off-tumor binding of its Fc region to Fcγ receptor-expressing Kupffer cells in the liver [53].

The majority (67%) of the CD3-targeting T-cell-engaging bsAbs are intended for the treatment of hematological malignancies, and to a great extent, their targets include CD19, CD20, CD33, CD38, and CD123 (B-cell maturation antigen, BCMA) [2]. The development of T-cell-engaging bsAbs for solid tumors is slower due to the fact that many solid tumor antigens are also expressed at low levels in healthy tissues, which can lead to undesired toxicity. Other factors include low tumor penetration of the antibody and the effector cells and the immunosuppressive tumor environment. About 14 T-cell-engaging bsAbs currently in clinical trials target solid tumors and are directed against Her2, EGFR, PSMA, and Ep-CAM [2].

Long-term treatment with T-cell-engaging bsAbs may be rendered inefficient primarily due to loss of the targeted tumor antigen. A prominent example is the proliferation of the CD19-negative clones in ALL patients. Other mechanisms of resistance include suppression of T-cell activity due to the activity of regulatory T cells or expression of immune checkpoint molecules, such as PD-1 and PD-L1. In accord, the concomitant inhibition of the PD-1/PD-L1 contact can enhance the therapeutic effect of T-cell-engaging bsAbs. To this end, bsAb-based therapies enhancing T-cell activity by using targeting of immune checkpoint inhibitors along the PD-1/PD-L1 axis and simultaneously blocking CTLA-4, LAG-3, or TIM-3 antigens are now in early clinical trials [2]. A rationale for such treatments was provided by encouraging results of studies using Ab combinations: the treatment of melanoma patients with ipilimumab (Yervoy®) (anti-CTLA4) and nivolumab (Keytruda®) (anti-PD-1) induced improved long-term progression-free survival outcomes compared with treatment with ipilimumab alone [54]. In the design of a bispecific agent, Fc-silent molecules have been envisioned to block the PD-1 activity with high affinity binding while suppressing CTLA-4 with a low-affinity binding arm on the double-positive tumor-infiltrating lymphocytes, the latter at the same time reducing binding to CTLA-4 positive peripheral T cells [55]. This constellation is intended to increase safety and tolerability of the compound.

Clinical progress in the fight against cancer with use of T-cell-engaging bsAbs has initiated an ambition to expand the spectrum of potentially activated effector cells to other suitable cell types. Vγ9Vδ2 T cells, a class implicated in immune surveillance and reactive with phospho-antigens derived from infectious agents or from tumor cells as a result of metabolic dysregulation [56], were recruited via a bsAb to EGFR-positive tumors and induced lysis of patient-derived colorectal carcinoma cells [57]. Another eligible group of T cells are the natural killer (NK) cells, which have been addressed by targeting CD16a receptor in a trispecific scFv-based format with incorporated IL-15 crosslinker in a Her2-specific construct [58]. TandAb AFM 13, a linear array of four antibody variable fragments coupled via linkers, with two binding sites for CD30 situated between two binding sites for CD16a, could also trigger NK cell-mediated killing of CD30-positive non-Hodgkin lymphoma cells [49].

Besides oncology, T-cell redirection is a valuable concept in the treatment of infectious disease, and the activity of virus-specific bsAbs has been demonstrated in redirecting cytotoxic T cells to infected cells expressing surface antigens of hepatitis B virus [59], cytomegalovirus, [60] and HIV-1 [61]. Such bsAbs were able to induce killing of HIV-infected cells and inhibit the HIV replication ex vivo, as well as to reactivate virus replication in latently infected cells via the engagement of CD3. Recent experiments hold promise for successful use of bsAbs in regenerative medicine: the targeted delivery of endothelial progenitor cells engaged by the stem cell marker CD34 to CD41-positive platelets resulted in effective heart repair in a mouse model of acute myocardial infarction

[62]. The delivery of a subset of PBMCs to the site of injury through a bsAb targeting stem cell antigen 1 and the activated conformation of GPIIb/IIIa on platelets resulted in a significant decrease in infiltrating inflammatory cells and decreased fibrosis, increased capillary density, and restored cardiac function [63].

7.3.2 Signaling Inhibition by Bridging Receptor Molecules

Major successful examples of bsAbs that bridge receptor molecules pertain to the molecules reactive with the tyrosine kinase receptor family. This may be due to their combinatorial properties, but there are examples where the beneficial properties of the therapeutic molecule are attributed to its obligate bispecificity. The stunning feature of all receptor bridging molecules successful in in vitro and in vivo assays until now is that the features of bispecific agents cannot be predicted from the properties of their parental monospecific agents. For example, the superior functional activity of the bispecific anti-met/anti-EGFR antibody was discovered in an extensive screen of a panel of bsAbs and was the only candidate not inducing agonistic activity on met/EGFR-positive cells [64]. As bivalent anti-met antibodies can cause met-crosslinking and the induction of a tumorigenic phenotype, a bispecific asymmetric antibody combining an anti-met and an anti-EGFR arm was designed to block met- and EGFR-induced signaling [64]. Another example is a bsAb with monovalent specificities for Her2 and Her3, which by high affinity binding to Her2 provides a high local concentration of anti-Her3 Fabs, which can then prevent binding of the natural Her3-ligand HRG and inhibit signaling of both receptors [65].

Receptor targeting bsAbs can also exhibit agonistic activity. Activation of the fibroblast growth factor 21 (FGF21) metabolic pathway has been reported to counter obesity and diabetes, but this molecule could not be targeted directly due to its adverse effects [66]. Therefore, activation of this metabolic pathway was achieved by bsAb BFKB8488A (Roche) specific for the broadly expressed fibroblast growth factor receptor 1C and β-klotho receptor with expression restricted to liver, adipose tissue, and pancreas, and its effect was only exerted in the tissues that express both antigens [67].

7.3.3 Overcoming Ligand Redundancy

Inhibition of cytokines that support cancer cell growth was often not sufficient to stop the proliferation of cancer cells, as the signaling pathways involved are dependent on redundant ligands that can step in for the targeted molecule. The bsAbs with this mode of action, able to show a competitive advantage in comparison with agents of single specificity, are those interfering with the process of angiogenesis. Such a molecule of dual variable domain-Ig (DVD-Ig) format with combinatorial inhibitory effect, targeting VEGF and delta-like ligand 4 (DLL-4), is currently being evaluated for clinical use for patients with colorectal cancer [68]. A human IgG1-based CrossMab targeting VEGF and Ang2 is developed for treatment of ophthalmic conditions [69]. The additional engineering of its Fc-region for abolishment of the binding of Fc receptors to reduce the effector functions and to increase the systemic clearance should optimize its suitability for use in patients with diabetes-related macular edema [70].

7.3.4   Targeting Two Epitopes of the Same Antigen: Biparatopic bsAbs

Biparatopic BsAbs target two different epitopes of the same antigen and act as crosslinking agents, often leading to antigen aggregation. In this aspect their activity can be compared with mixtures of antibodies. An example of such an antibody is the anti-Her2 bsAb ZW25, which acts with avid binding to achieve a higher binding affinity, and sequesters Her2 from the cell surface to inhibit Her2-mediated signaling and to overcome Her2 addiction [71]. Its Fc region is engineered to be more potent in eliciting effector functions. In clinical tests involving patients with advanced gastroesophageal and breast cancer, this molecule proved more potent than Herceptin. In addition, this molecule is the basis of the antibody-drug conjugate (ADC) molecule ZW49 that has been launched into clinical trials due to the expectation that the enhanced internalization should enable a more efficient and specific cleavage of the toxin moiety and hence a higher efficacy.

7.3.5   Enzyme Mimetics

A prominent example of a bsAb that acts as a part of an enzyme complex is emicizumab (Hemlibra®), the second bsAb currently approved for human use. These molecules act as a mimetic of Factor VIIIa, the activated form of FVIII, deficient in hemophilia A patients, by simultaneous binding to Factor IXa and Factor X with micromolar affinity such that in plasma only small amounts of the active complex are generated. As a prophylactic treatment, this bsAb can effectively inhibit bleeding in patients with hemophilia A, either with or without Factor VIIIa inhibitors. The discovery of this molecule started with an extensive activity screen of 40,000 clones resulting from combination of 200 antibodies directed against Factor IXa and 200 directed against Factor X, expressed as two heterodimeric IgG4 heavy chains with a common light chain to reduce the chain association problem [72]. A total of 0.3% of those clones tested positive. The light chain of the best candidate was optimized for binding by framework and CDR shuffling. Further optimization rounds included humanization to reduce immunogenicity, charge optimization to reduce aggregation issues, and isoelectric point optimization to enable ion-exchange-chromatography-based isolation of the bispecific product from the contaminants. The final candidate reached approval in USA and Europe in 2018 [1].

7.3.6   BsAbs Enabling Transport: Piggyback Molecules

This concept relates to obligate BsAbs that enable transport and hence delivery or escape of molecules into certain tissues or cellular compartments. Pioneering examples include a construct composed of an antibody as a targeting moiety and toxic ricin A that could induce specific toxicity upon its internalization [73]. Agents containing an antibody and mutant pseudomonas toxin are in clinical development [74].

Currently, bsAbs that act as molecular "Trojan horses" as they can cross the brain-blood barrier and deliver active compounds into this compartment are considered important and are being intensely studied. Most commonly, they follow the transcytosis pathway by binding with one binding arm to the transferrin receptor. Such a bsAb, with the other arm specific for ß-secretase 1 (BACE1), was able to reduce the activity of this enzyme in both mouse and monkey models, resulting in a lower concentration of ß-amyloid plaques in the brain and cerebrospinal fluid [75].

The endosomal compartment appears as another attractive target for the modulation of activities of compounds that can be mediated by bsAb transfer function. A bsAb that could target an extracellularly exposed conserved epitope of Ebola virus glycoprotein mediated its access to the endosomal compartment [76]. After the proteolytic cleavage, cryptic receptor binding sites for the viral intracellular receptor, required to enter the cytoplasm, were revealed. Through this mechanism, the bsAb was able to neutralize several known Ebola virus strains in vitro and confer virus protection in mice [76]. The bsAb that targeted the late-endosomal protein CD63 enabled improved delivery of an antibody-drug conjugate targeting Her2 and reduced peripheral toxicity of the compound [77]. Bispecific binding can also endow antigen binding fragments, devoid of Fc, with a longer serum half-life in vivo by mediating a non-covalent fusion to the Fc or to HSA, hence providing an entry into the neonatal Fc receptor (FcRn) rescue pathway and avoiding lysosomal degradation. This is exemplified in H chain-only variable domain (V_{HH})-based bsAbs vobarilizumab [78] and ozoralizumab [79], directed against the IL-6 receptor and tumor necrosis factor, respectively, which are currently in Phase II studies as fusion proteins with an HSA binding moiety, as well as von Willebrand factor-targeting caplicizumab (Cablivi®), a FDA-approved drug for thrombotic thrombocytopenic purpura.

Questions

3.1. List the modes of action of bispecific antibodies of today.

3.2. Describe the activity of T-cell-engaging bsAbs: describe their biological effect; refer to their specific antigens, preferred valency, and affinity of the binding sites.

3.3. How can the resistance to T-cell-engaging bsAbs due to T-cell suppression be overcome?

3.4. What is a "cis" interaction of a bsAb?

3.5. What types of biological effects can be achieved by crosslinking of the antigens on the same cell with a bsAb?

3.6. Which concept underlies targeting ligands, redundant in a signaling pathway? Which examples are translated in a bsAb application?

3.7. How can bsAbs act as molecular "Trojan horses"?

7.4   Formats of Bispecific Antibodies

The success of bsAbs in overcoming tasks not manageable by monospecific antibodies has raised interest in their development for use in ever more complex biological situations. Their diverse functions, which require different valencies of binding and sometimes even restricted spatial positioning of the two paratopes, have induced motivation in the protein engineering community to design a number of ingenious solutions to cover the unmet requirements for their diverse architectures. As a result, there are over 100 different formats available today (Fig. 7.4).

In natural bivalent antibodies of the IgG class, the two antigen-binding sites are identical, and both heavy and light chain variable domains contribute to antigen binding. Chemical coupling as well as co-expression of two different heavy and light chains leads to a mixture of products, which poses the chain association issue: this lowers the yield of the desired species, requires tagging of expressed chains to facilitate additional purification steps, and poses additional demands on analytical procedures for the end product. Several strategies have been developed to increase the yield of the desired homogeneous bispecific molecule. Apart from their basic architecture, the engineering methods applied can be used to classify the colorful collection of various bispecific formats.

7.4.1   Fragment-Based Bispecific Antibodies

The introduction of the single-chain Fv fragment (scFv), a construct of linker-bound variable domains of the heavy and light chain, as a minimal antibody-derived fragment that preserves the antigen specificity of the full IgG and consists of a single polypeptide chain [19], has become a popular format of producing bispecific and multispecific molecules (Fig. 7.4A2). In this format, the correct spatial arrangement of heavy and light chains can be achieved by choosing adequate linker length, while the lack of the Fc fragment prevents undesired homodimerization (Fig. 7.4A). The latter property also makes them amenable for expression in the periplasm of prokaryotic microorganisms or lower eukaryotes, assuring an inexpensive and simple production. Their modular nature allows the choice of the valency for the particular binding specificity at will [80]: at present, formats with one or two binding sites for each specificity are being evaluated in clinical trials. At the same time, its modular properties offer the possibility to choose an optimal domain arrangement and alleviate the potential issues in solubility and stability of fragment-based bsAbs.

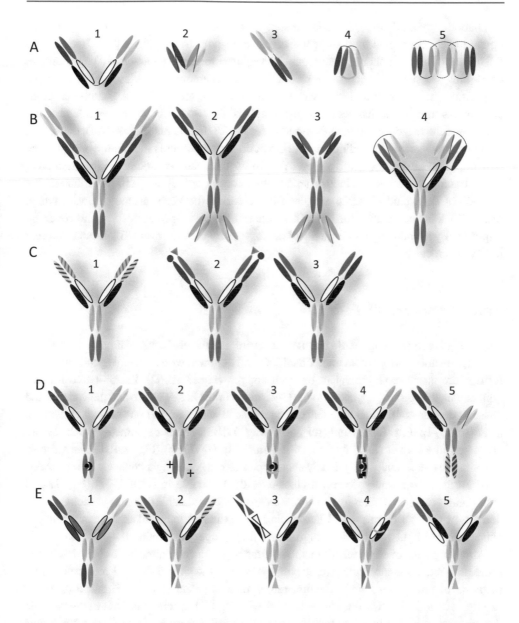

Fig. 7.4 Selection of bsAb formats. Antibody domains follow the color code: red, V_L1; blue, V_H1; yellow, V_L2; green, V_H2; black, C_L; white, C_H1; grey, C_H2; light blue, C_H3. (**A**) Fragment-based formats: A1 F(ab)$_2$, A2 bispecific T-cell engager (BiTE), A3 diabody, A4 dual affinity retargeting (DART), A5 tandem diabody (TandAb); (**B**) symmetric fusion formats, B1 dual-variable domain-Ig (DVD-Ig), B2 IgG-(scFv)$_2$, B3 (scFv)$_4$-Fc, B4 cross-over dual variable Ig (CODV-Ig); (**C**) symmetric formats, C1 Two-in-One Ab, C2 DutaMab, C3 mAb [2] (mutated CH3 domains in pink); (**D**) asymmetric formats, D1 "Knobs-into-Holes" heterodimeric IgG, D2 heterodimeric IgG formed by electrostatic interactions, D3 sterically formed heterodimeric IgG stabilized by disulfide bond, D4 sterically formed heterodimeric IgG stabilized by point mutations, D5 SEED-body (heterodimerized C_H3-halves in stripes); (**E**) asymmetric formats addressing the correct pairing of the light chain, E1

Diabodies (Fig. 7.4A3), for comparison, comprise two chains with one V_H and V_L domain each, connected by a short linker that is usually five residues [81]. Because the linker length is much shorter than would be required to allow an assembly of an antigen combining site such as in scFv, two chains dimerize in an opposite orientation and form a bispecific heterodimeric molecule when expressed from the same cell. Additional engineering has shown that changes in the positioning of the two binding sites can influence the potency of the reagent. One such example is the DART fragments (Fig. 7.4A4), composed of two diabodies with an additional cysteine bond that constraints the binding sites for the target and the effector cell into a closer proximity than can be anticipated for alternative linker-based designs and can induce a more potent target cell killing [82]. Nevertheless, the absence of Fc excludes such molecules from entering the FcRn rescue pathway, which results in a short serum half-life. Their pharmacokinetic properties can however be improved by introducing an Fc fragment de novo or by including an HSA-binding moiety in the fusion construct.

7.4.2 Symmetric Ab Formats

An alternative to omitting the Fc fragment to overcome the chain association is to choose a basic symmetric scaffold of an unmodified IgG and endow it with a second specificity. Such formats are hence usually bivalent for each antigen (Fig. 7.4B, C). Fusion-based formats typically include a scFv, appended to the C-terminus of the heavy chain, and the variable and constant domains of the Fab fragment can be replaced with a scFv at one or both arms, such as in $(scFv)_2$-Ig (Fig. 7.4B2) and $(scFv)_4$-Fc (Fig. 7.4B3), another symmetric bsAb format. However, in a format called dual-variable domain Ig (DVD-Ig) (Fig. 7.4B1), the variable domains are N-terminally extended by another pair of heavy and light chain variable domains [83]. Further, the recently described cross-over dual variable Ig (CODV-Ig) (Fig. 7.4B4) incorporates two specificities by appending an array of two variable domains to each constant domain of the Fab fragment and introducing an optimized long linker sequence between either heavy or light variable domain to assure their proper positioning [84].

Another group of engineering methods aims to produce symmetric bsAbs that at the same time retain the molecular architecture as well as the size of the IgG molecule, using methods of mutagenesis and selection rather than covalent linking (Fig. 7.4C). These examples include the "Two-in-One antibody" (Fig. 7.4C1), where the CDRs of the variable domains are modified in a way that they can specifically recognize two different antigens [85], DutaMab, where the CDRs both of the heavy and the light chains can accommodate

Fig. 7.4 (continued) rat-mouse hybrid IgG (bordered magenta, mouse C_L; bordered pink, mouse C_H1; magenta, mouse C_H3; bordered dark grey, rat C_L; bordered grey, rat C_H1; dark grey, rat C_H3), E2 common light chain heterodimeric IgG, E3 heterodimeric IgG with mutated C_H1/C_L and V_H1/V_L interface, E4 DuetMab, E5 CrossMab

two different antigens (Fig. 7.4C2), and mAb [2] with a library-derived antigen-binding Fc that can specifically confer antigen binding via mutated residues at the tip of the C_H3 domain (Fig. 7.4C3) [86, 87].

7.4.3   Asymmetric bsAb Formats

Bispecific formats designed in this class are based on the premise of introducing a preferential pairing between two heterologous antibody chains of the heterodimer upon co-expression of four polypeptide chains (Fig. 7.4D, E). Apart from the native-like architecture and size of the asymmetric antibody formats, their huge advantage is in minimal use of potentially immunogenic non-native antibody sequences, such as tags and linkers. First strategies to achieve heterodimerization were based on steric effects (Fig. 7.4D1) [88] or electrostatic steering (Fig. 7.4D2) [28], achieved with mutagenesis methods, and the mutated Fc constructs were later optimized to restore molecular stability (Fig. 7.4D3, 4D4) [25]. Recent approaches aimed at efficient heterodimerization employ a more extensive mutagenesis, such as in strand-exchanged engineered domain (SEED)-bodies, which utilize preferential pairing of Ig chains composed of segments of IgG3 and IgA (Fig. 7.4D5) [89], or BEAT bodies based on bispecific engagement by antibodies based on the T-cell receptor—a technology platform, where the residues that mediate the pairing of heterodimeric TCR chains are designed onto the interface of the Fc chains (Fig. 7.2E) [29]. A recent alternative, currently applicable to C_H3-domain adapted IgG1 format, employs post-expression assembly of two antibody half-molecules, known also as controlled Fab-arms exchange [38].

The correct pairing of the light chain in asymmetric constructs was first assured by expressing one of the antigen-combining sites in a scFv format (Fig. 7.4D5) or using antibody domains from multiple species (Fig. 7.4E1). Later this issue was solved by a neat concept of using a common light chain for both binding partners (Fig. 7.4E2) [26], but as this cannot be applied for the expression of pre-existing binding entities, complementary platforms embodied in suitable phage libraries [26] and transgenic animals [90] were implemented to support the discovery of compatible molecules. Alternatively, the interface between the heavy and the light chain was modified to guide preferential formation of a certain Fab species (Fig. 7.4E3) [32]. Correct pairing of the light chain could also be achieved by obliterating the native disulfide bond between the constant domains of the Fab fragment and introducing an alternative one, which gave rise to a format called DuetMab (Fig. 7.4E4) [91]. In CrossMab, the position of the light chain and the Fd (V_H-C_H1)-fragment is swapped in one of the two asymmetric antibody half-molecules (Fig. 7.4E5) [33]. BsAbs whose bispecificity is based on light chains were isolated with a purification-based protocol: in the "κλ-body" format, two unique light chains of different subfamilies (kappa and lambda) are selected, and chromatography employing sequential affinity-based purification steps with matrices such as KappaSelect and LambdaFabSelect is used to selectively capture asymmetric bsAbs that contain elements with both light chain classes

[92]. Alternatively, two anti-idiotypic ligands, binding specifically to the two different Fab fragments in a bsAb, have been applied in chromatography matrices [93].

In fact, certain designs of asymmetric bsAbs are trimmed to support the achievement of optimal yield of the desired homogenous end product by downstream methods. Such protocols are based on fine differentiation of the target molecule from by-products and are based on properties such as differential protein A binding, unique charge that allows isolation with ion exchange chromatography, or size difference enabling isolation through gel filtration columns. The complexity of bsAbs raised a corresponding demand for sensitive analytical tools to monitor and control the isolated correctly paired variant. Their analytics consists of complementary methods, the first of which is usually a liquid chromatography-mass spectrometry (LC-MS)-based procedure to measure the mass of the intact antibody for fast analysis of correct pairing and the second step includes a hydrophobic interaction chromatography (HIC)-based method that detects mispairing based on the unique hydrophobic properties of the potential by-products [94].

Notably, most contemporary asymmetric formats feature a single monovalent binding site for each specificity, and the introduction of potential multiple uniform contact points with the antigen remains a challenge for protein engineers of the future times: the avidity effects achieved could untap the potentials yet unexplored in enhancing the number and hence the strength of the interaction with the target cell displaying the cognate surface antigen.

Questions

4.1. Explain the chain association issue in bsAbs.

4.2. List the advantages and the shortcomings of fragment-based bsAbs.

4.3. Which two classes of engineering methods underlie the construction of symmetric bsAbs?

4.4. Describe the engineering principles required to design an asymmetric bsAb.

4.5. List the methods used to achieve the correct pairing of the light chain in asymmetric bsAbs.

4.6. List the analytical methods used in quality control of a bsAb.

7.5 Outlook for Bispecific Antibodies

After learning to know several valuable bispecific formats, we can conclude that most candidate molecules intended for future therapy contain an Fc fragment, primarily to endow them with a long in vivo serum half-life by enrolling them in the FcRn rescue pathway. Additionally, all Fc-based modifications that previously proved beneficial for monoclonal antibodies are also amenable to derivate bsAbs: half-life extension can be achieved by using suitable Fc variants, and effector functions can be enhanced using point mutagenesis or altered glycosylation or obliterated with few point mutations that for steric reasons or absence of glycosylation render them refractory to binding of FcγR molecules. Alternatively, Fc fragments of IgG2 or IgG4 class can be used if the effector functions are not desired. The use of IgG4 as naturally heterodimerizing agent appears particularly suitable in acting as a backbone for bsAbs in the future.

Multiple valencies for a particular antigen appear an advantageous and developable feature in several antibody formats, although the molecules presently performing in clinical trials feature a maximum of two identical binding sites. Future engineering steps may utilize the naturally multivalent antibody molecules and render feasible tetravalent formats based on IgA or decavalent formats based on IgM.

7.6 Conclusion

In this chapter, we have followed the development of bsAbs through the decades and encountered their multiple formats as well as their multifaceted modes of action. Each format presents a particular strength in assigning the valency in encounter of the cognate antigen and delimiting the physicochemical parameters of distance and concentration in the future encounter of the two antigens addressed. The plethora of formats further differs in their physicochemical properties, manufacturing requirements, post-production analytics and particularities in formulation, as well as pharmacokinetic properties in vivo. Nevertheless, the prerequisite for structural and functional homogeneity and consistent biological performance, similarly to the one applied for classical monospecific members of the antibody family, remains critical for their progress up to the level of "in-man" application and widely approved therapeutic use.

> **Take Home Messages**
> - BsAbs can be, according to their biological effect, combinatorial by exerting their effect as a mixture of two specificities or obligate where the incorporation of two specificities is a prerequisite for the particular unique functionality of the bsAb.
> - Obligate effects of bsAbs are either spatial, where the positioning of the antigen-binding arms occurs simultaneously, or temporal, where the binding of one arm is a prerequisite to achieve the binding activity of the other arm.
> - Most known modes of activity of bsAbs include bridging of different cell types, bridging of different antigens, engaging two epitopes of an antigen in biparatopic binding, targeting two antigens of the same signaling pathway to overcome redundancy, and transfer of functionality into desired organs and cellular compartments.
> - Formats of bsAbs include fragment-based formats devoid of an Fc fragment, symmetric formats with a homodimeric Fc that utilize either fusion or mutagenesis methods to achieve bispecificity, and asymmetric formats that aim to specifically combine two heterologous Fc-based half-molecules by steric or electrostatic effects.

(continued)

- Asymmetric formats tend to introduce a chain association motif that can be established by mutagenesis to induce preferential pairing of two heterodimeric half-antibody heavy chain molecules. Two different antigen combining sites can utilize a common light chain, or proper light chain pairing can be steered by engineering an orthogonal surface for the interaction with the desired heavy chain, introducing an engineered disulfide bond, by domain swapping, or by constant domain exchange. Alternatively, purification-based isolation protocols can be applied for antibodies with light chains from kappa and lambda classes.
- The valency of antigen recognition sites in various formats is different and can be chosen according to the particular biological situation. Even multivalent formats are at reach of contemporary antibody engineering techniques.
- Bispecific formats that include an IgG1-Fc fragment are amenable to tuning of the effector functions, as well as the use of IgG2- and IgG4-based scaffolds.

Answers

1.1. Bispecific antibodies combine specificities for two epitopes or two antigens in a single molecule.
1.2. Obligate bsAbs display unique biological effects that can only be achieved by the physical linkage of the two specificities in a single molecule and not by their mixture.
2.1. The first bsAb was a bivalent F(ab')$_2$, prepared by using pepsin cleavage, mild reduction, and oxidation of two F(ab') fragments of different specificities.
2.2. "Quadroma" technique describes fusing two antibody-producing hybridoma cells. Such quadroma cells harbor genes for two different sets of heavy and light chains four genomes and can secrete antibodies of double specificity.
2.3. As the heavy and light variable domains of a scFv are present on one polypeptide chain, one can avoid the random antibody chain association. Further, multiple antigen-binding molecules could now be combined in one polypeptide chain and their position chosen at will. The absence of glycosylated Fc fragment enables inexpensive production in *E. coli* at a high yield.
2.4. Naturally occurring IgG4 is functionally monovalent, although of the size of a complete antibody. The molecular background of this phenomenon is Fab-arms exchange.
3.1. Modes of action of modern bsAbs include cell bridging, receptor bridging, biparatopic binding, targeting redundant ligands in signaling pathways, acting as enzyme cofactor, and transport of molecules or activities into or out of certain organs and subcellular compartments.
3.2. T-cell-engaging bsAbs bridge the tumor cell or virus-infected cell with the effector T cells and induce their activation. T cells respond with secretion of perforin and granzyme, which results in specific cytotoxicity and apoptotic cell death of the targeted cell. The most common molecular target of the T cells is CD3ε, and the molecular targets on the tumor cells should be unique for the cell type or strongly

overexpressed to reduce on-target/off-tumor effects. Binding to TAA should be bi- or multivalent to profit from avid binding but monovalent for the CD3 to reduce the likelihood of severe side effects, such as cytokine release storm. The affinity to the TAA should be high, but the affinity to CD3 low to allow serial killing.

3.3. T-cell suppression that causes the resistance to T-cell-engaging bsAbs can be overcome by inhibiting the signaling of the immune checkpoint molecules along the PD-1/PD-L1 axis.

3.4. "*Cis*" interaction of a bsAb describes the mode of action when the two arms of the bsAb contact two different molecules or two epitopes on the same cell.

3.5. Crosslinking of the receptors on the same cell by a bsAb can cause signaling inhibition, internalization, or agonistic effects such as proliferation or potentiation of metabolic processes.

3.6. The signaling pathways of a tumor cell can be dependent on redundant ligands that can step in for the molecule neutralized by a monospecific antibody. The bsAbs with this mode of action hence show a competitive advantage, particularly those interfering with the process of angiogenesis (DVD-Ig targeting VEGF and DLL-4, CrossMab targeting VEGF and Ang2).

3.7. Molecular Trojan horse mechanism is exhibited by obligate bsAbs that enable transport and hence delivery or escape of molecules into certain tissues or cellular compartments. For example, bsAb can hijack the transcytosis pathway to cross the blood-brain barrier by binding to human transferrin receptor with one arm and block the activity of an enzyme in the brain with another arm.

4.1. Chain association issue refers to random pairing of simultaneously expressed heavy and light chains, which results in a mixture of ten different molecular species.

4.2. For fragment-based formats, the absence of the Fc fragment prevents undesired homodimerization, and correct positioning of heavy and light chains can be determined by choosing adequate linker length. Their modular nature allows the choice of the number of binding sites for a particular binding specificity at will. Lack of glycosylated Fc fragment makes such molecules amenable to expression in the periplasm of prokaryotic microorganisms, assuring an inexpensive and simple production. This however at the same time results in a short serum half-life in vivo, which can be extended by producing a fusion with an Fc fragment or HSA.

4.3. Symmetric bsAbs are produced either using a fusion of scFvs or variable domains to a homodimeric antibody or by methods involving mutagenesis and selection of novel binding sites that are either dual specific or show a novel antigen specificity.

4.4. In asymmetric bsAbs the association of homodimeric Fc is obstructed using mutagenesis that introduces either steric hindrance or electrostatic repulsion. Alternatively, more extensive mutagenesis can be applied to create novel selectively interacting interfaces of the Fc fragment. Chromatography methods can be used to purify the correctly paired half-antibody molecules on the bases of differential binding to protein A or their different charge.

4.5. Correct pairing of the light chain was first achieved by using light chains from different species, later by creating binding interfaces that interact selectively with the heavy chain, or by using light chains from different subfamilies and applying a sequential (kappa-lambda) chromatography steps over anti-kappa and anti-lambda matrices. An alternative is the use of common light chain for both binding partners, which requires antigen selections de novo.

4.6. The analytical methods for quality control of bsAbs include liquid chromatography-mass spectrometry (LC-MS)-based procedure to measure the mass of intact antibody and detect the potential undesired by-products and hydrophobic interaction chromatography (HIC) that resolves the mispaired species based on their different hydrophobic properties.

Acknowledgments The financial support by the Christian Doppler Society, Austrian Federal Ministry for Digital and Economic Affairs, and National Foundation for Research, Technology and Development is gratefully acknowledged.

References

1. Kaplon H, Reichert JM. Antibodies to watch in 2019. MAbs. 2019;11(2):219–38. https://doi.org/10.1080/19420862.2018.1556465.
2. Labrijn AF, Janmaat ML, Reichert JM, Parren PWHI. Bispecific antibodies: a mechanistic review of the pipeline. Nat Rev Drug Discov. 2019;18(8):585–608. https://doi.org/10.1038/s41573-019-0028-1.
3. Steinwand M, Droste P, Frenzel A, Hust M, Dübel S, Schirrmann T. The influence of antibody fragment format on phage display based affinity maturation of IgG. MAbs. 2014;6(1):204–18. https://doi.org/10.4161/mabs.27227.
4. Li D, Wang L, Maziuk BF, Yao X, Wolozin B, Cho YK. Directed evolution of a picomolar-affinity, high-specificity antibody targeting phosphorylated tau. J Biol Chem. 2018;293 (31):12081–94. https://doi.org/10.1074/jbc.RA118.003557.
5. Tiller KE, Tessier PM. Advances in antibody design. Annu Rev Biomed Eng. 2015;17 (1):191–216. https://doi.org/10.1146/annurev-bioeng-071114-040733.
6. Xiao X, Douthwaite JA, Chen Y, et al. A high-throughput platform for population reformatting and mammalian expression of phage display libraries to enable functional screening as full-length IgG. MAbs. 2017;9(6):996–1006. https://doi.org/10.1080/19420862.2017.1337617.
7. Nisonoff A, Wissler FC, Lipman LN. Properties of the major component of a peptic digest of rabbit antibody. Science. 1960;132(3441):1770–1. https://doi.org/10.1126/science.132.3441.1770
8. Porter RR. The hydrolysis of rabbit y-globulin and antibodies with crystalline papain. Biochem J. 1959;73:119–26. https://doi.org/10.1042/bj0730119.
9. Nisonoff A, Rivers MM. Recombination of a mixture of univalent antibody fragments of different specificity. Arch Biochem Biophys. 1961;93(2):460–2. https://doi.org/10.1016/0003-9861(61)90296-X.
10. Hämmerling U, Aoki T, Wood HA, Old LJ, Boyse EA, De Harven E. New visual markers of antibody for electron microscopy [17]. Nature. 1969;223(5211):1158–9. https://doi.org/10.1038/2231158a0.

11. Köhler G, Milstein C. Continuous cultures of fused cells secreting antibody of predefined specificity. Nature. 1975;256(5517):495–7. https://doi.org/10.1038/256495a0.
12. Brennan M, Davison PF, Paulus H. Preparation of bispecific antibodies by chemical recombination of monoclonal immunoglobulin G1 fragments. Science. 1985;229(4708):81–3. https://doi.org/10.1126/science.3925553
13. Glennie MJ, McBride HM, Worth AT, Stevenson GT. Preparation and performance of bispecific F(ab' gamma)2 antibody containing thioether-linked fab' gamma fragments. J Immunol. 1987;139(7):2367–75.
14. De Lau WBM, Van Loon AE, Heije K, Valerio D, Bast BJEG. Production of hybrid hybridomas based on HATs-neomycin double mutants. J Immunol Methods. 1989;117:1. https://doi.org/10.1016/0022-1759(89)90111-7.
15. Milstein C, Cuello AC. Hybrid hybridomas and their use in immunohistochemistry. Nature. 1983;305(5934):537–40. https://doi.org/10.1038/305537a0.
16. Staerz UD, Bevan MJ. Hybrid hybridoma producing a bispecific monoclonal antibody that can focus effector T-cell activity. Proc Natl Acad Sci U S A. 1986;83(5):1453–7. https://doi.org/10.1073/pnas.83.5.1453.
17. Karpovsky B, Titus JA, Stephany DA, Segal DM. Production of target-specific effector cells using hetero-cross-linked aggregates containing anti-target cell and anti-Fcγ, receptor antibodies. J Exp Med. 1984;160(6):1686–701. https://doi.org/10.1084/jem.160.6.1686.
18. Perez P, Hoffman RW, Shaw S, Bluestone JA, Segal DM. Specific targeting of cytotoxic T cells by anti-T3 linked to anti-target cell antibody. Nature. 1985;316(6026):354–6. https://doi.org/10.1038/316354a0.
19. Skerra A, Plückthun A. Assembly of a functional immunoglobulin Fv fragment in *Escherichia coli*. Science. 1988. https://doi.org/10.1126/science.3285470
20. Bird RE, Hardman KD, Jacobson JW, et al. Single-chain antigen-binding proteins. Science. 1988;242(4877):423–6. https://doi.org/10.1126/science.3140379
21. Coloma MJ, Morrison SL. Design and production of novel tetravalent bispecific antibodies. Nat Biotechnol. 1997;15(2):159–63. https://doi.org/10.1038/nbt0297-159.
22. Arndt KM, Müller KM, Plückthun A. Factors influencing the dimer to monomer transition of an antibody single-chain Fv fragment. Biochemistry. 1998;37(37):12918–26. https://doi.org/10.1021/bi9810407.
23. Kortt AA, Malby RL, Caldwell JB, et al. Recombinant anti-sialidase single-chain variable fragment antibody: characterization, formation of dimer and higher-molecular-mass multimers and the solution of the crystal structure of the single-chain variable fragment/sialidase complex. Eur J Biochem. 1994;221(1):151–7. https://doi.org/10.1111/j.1432-1033.1994.tb18724.x.
24. Ridgway JBB, Presta LG, Carter P. "Knobs-into-holes" engineering of antibody C H 3 domains for heavy chain heterodimerization. Protein Eng. 1996;9(7):617–21. https://doi.org/10.1016/1380-2933(96)80685-3.
25. Atwell S, Ridgway JBB, Wells JA, Carter P. Stable heterodimers from remodeling the domain interface of a homodimer using a phage display library. J Mol Biol. 1997;270(1):26–35. https://doi.org/10.1006/jmbi.1997.1116.
26. Merchant AM, Zhu Z, Yuan JQ, et al. An efficient route to human bispecific IgG. Nat Biotechnol. 1998;16(7):677–81. https://doi.org/10.1038/nbt0798-677.
27. Von Kreudenstein TS, Escobar-Carbrera E, Lario PI, et al. Improving biophysical properties of a bispecific antibody scaffold to aid developability: quality by molecular design. MAbs. 2013;5(5):646–54. https://doi.org/10.4161/mabs.25632.
28. Gunasekaran K, Pentony M, Shen M, et al. Enhancing antibody fc heterodimer formation through electrostatic steering effects: applications to bispecific molecules and monovalent IgG. J Biol Chem. 2010;285(25):19637–46. https://doi.org/10.1074/jbc.M110.117382.

29. Skegro D, Stutz C, Ollier R, et al. Immunoglobulin domain interface exchange as a platform technology for the generation of Fc heterodimers and bispecific antibodies. J Biol Chem. 2017;292(23):9745–59. https://doi.org/10.1074/jbc.M117.782433.
30. Krah S, Schröter C, Eller C, et al. Generation of human bispecific common light chain antibodies by combining animal immunization and yeast display. Protein Eng Des Sel. 2017;30(4):291–301. https://doi.org/10.1093/protein/gzw077.
31. Lindhofer H, Mocikat R, Steipe B, Thierfelder S. Preferential species-restricted heavy/light chain pairing in rat/mouse quadromas. Implications for a single-step purification of bispecific antibodies. J Immunol. 1995;155(1):219–25.
32. Lewis SM, Wu X, Pustilnik A, et al. Generation of bispecific IgG antibodies by structure-based design of an orthogonal fab interface. Nat Biotechnol. 2014;32(2):191–8. https://doi.org/10.1038/nbt.2797.
33. Schaefer W, Regula JT, Bahner M, et al. Immunoglobulin domain crossover as a generic approach for the production of bispecific IgG antibodies. Proc Natl Acad Sci. 2011;108 (27):11187–92. https://doi.org/10.1073/pnas.1019002108.
34. Dietrich S, Gross AW, Becker S, et al. Constant domain-exchanged Fab enables specific light chain pairing in heterodimeric bispecific SEED-antibodies. Biochim Biophys Acta – Proteins Proteomics. 2020;1868(1). https://doi.org/10.1016/j.bbapap.2019.07.003
35. Aalberse RC, Stapel SO, Schuurman J, Rispens T. Immunoglobulin G4: An odd antibody. Clin Exp Allergy. 2009;39(4):469–77. https://doi.org/10.1111/j.1365-2222.2009.03207.x.
36. Kolfschoten MVDN, Schuurman J, Losen M, et al. Anti-inflammatory activity of human IgG4 antibodies by dynamic Fab arm exchange. Science. 2007;317(5844):1554–7. https://doi.org/10.1126/science.1144603
37. Rispens T, Ooijevaar-De Heer P, Bende O, Aalberse RC. Mechanism of immunoglobulin G4 fab-arm exchange. J Am Chem Soc. 2011;133(26):10302–11. https://doi.org/10.1021/ja203638y.
38. Labrijn AF, Meesters JI, De Goeij BECG, et al. Efficient generation of stable bispecific IgG1 by controlled fab-arm exchange. Proc Natl Acad Sci U S A. 2013;110(13):5145–50. https://doi.org/10.1073/pnas.1220145110.
39. Spiess C, Merchant M, Huang A, et al. Bispecific antibodies with natural architecture produced by co-culture of bacteria expressing two distinct half-antibodies. Nat Biotechnol. 2013;31 (8):753–8. https://doi.org/10.1038/nbt.2621.
40. Baeuerle PA, Reinhardt C. Bispecific T-cell engaging antibodies for cancer therapy. Cancer Res. 2009;69(12):4941–4. https://doi.org/10.1158/0008-5472.CAN-09-0547.
41. Shimabukuro-Vornhagen A, Gödel P, Subklewe M, et al. Cytokine release syndrome. J Immunother Cancer. 2018;6(1). https://doi.org/10.1186/s40425-018-0343-9
42. Bargou R, Leo E, Zugmaier G, et al. Tumor regression in cancer patients by very low doses of a T cell-engaging antibody. Science. 2008;321(5891):974–7. https://doi.org/10.1126/science.1158545
43. Nuñez-Prado N, Compte M, Harwood S, et al. The coming of age of engineered multivalent antibodies. Drug Discov Today. 2015;20(5):588–94. https://doi.org/10.1016/j.drudis.2015.02.013.
44. Rogala B, Freyer CW, Ontiveros EP, Griffiths EA, Wang ES, Wetzler M. Blinatumomab: enlisting serial killer T-cells in the war against hematologic malignances. Expert Opin Biol Ther. 2015;15(6):895–908. https://doi.org/10.1517/14712598.2015.1041912.
45. Hijazi Y, Klinger M, Kratzer A, et al. Pharmacokinetic and Pharmacodynamic relationship of Blinatumomab in patients with non-Hodgkin lymphoma. Curr Clin Pharmacol. 2018;13 (1):55–64. https://doi.org/10.2174/1574884713666180518102514.

46. Le Jeune C, Thomas X. Potential for bispecific T-cell engagers: role of blinatumomab in acute lymphoblastic leukemia. Drug Des Devel Ther. 2016;10:757–65. https://doi.org/10.2147/DDDT. S83848.
47. Offner S, Hofmeister R, Romaniuk A, Kufer P, Baeuerle PA. Induction of regular cytolytic T cell synapses by bispecific single-chain antibody constructs on MHC class I-negative tumor cells. Mol Immunol. 2006;43(6):763–71. https://doi.org/10.1016/j.molimm.2005.03.007.
48. Ross SL, Sherman M, McElroy PL, et al. Bispecific T cell engager (BiTE®) antibody constructs can mediate bystander tumor cell killing. PLoS One. 2017;12(8). https://doi.org/10.1371/journal. pone.0183390
49. Reusch U, Duell J, Ellwanger K, et al. A tetravalent bispecific TandAb (CD19/CD3), AFM11, efficiently recruits T cells for the potent lysis of CD19+tumor cells. MAbs. 2015;7(3):584–604. https://doi.org/10.1080/19420862.2015.1029216.
50. Hoffmann P, Hofmeister R, Brischwein K, et al. Serial killing of tumor cells by cytotoxic T cells redirected with a CD19-/CD3-bispecific single-chain antibody construct. Int J Cancer. 2005;115 (1):98–104. https://doi.org/10.1002/ijc.20908.
51. Alegre ML, Peterson LJ, Xu D, et al. A non - activating "humanized" anti - CD3 monoclonal antibody retains immunosuppressive properties in vivo. Transplantation. 1994;57(11):1537–43. https://doi.org/10.1097/00007890-199457110-00001.
52. Sebastian M, Kuemmel A, Schmidt M, Schmittel A. Catumaxomab: a bispecific trifunctional antibody. Drugs of Today. 2009;45(8):589–97. https://doi.org/10.1358/dot.2009.45.8.1401103.
53. Borlak J, Länger F, Spanel R, Schöndorfer G, Dittrich C. Immune-mediated liver injury of the cancer therapeutic antibody catumaxomab targeting EpCAM, CD3 and Fcγ receptors. Oncotarget. 2016;7(19):28059–74. https://doi.org/10.18632/oncotarget.8574
54. Larkin J, Chiarion-Sileni V, Gonzalez R, et al. Combined nivolumab and ipilimumab or monotherapy in untreated melanoma. N Engl J Med. 2015;373(1):23–34. https://doi.org/10. 1056/NEJMoa1504030.
55. Dovedi SJ, Mazor Y, Elder M, et al. Abstract 2776: MEDI5752: a novel bispecific antibody that preferentially targets CTLA-4 on PD-1 expressing T-cells. In: AACR; 2018, p. 2776. https://doi. org/10.1158/1538-7445.am2018-2776
56. Tanaka Y, Sano S, Nieves E, et al. Nonpeptide ligands for human γδ T cells. Proc Natl Acad Sci U S A. 1994;91(17):8175–9. https://doi.org/10.1073/pnas.91.17.8175.
57. de Bruin RCG, Veluchamy JP, Lougheed SM, et al. A bispecific nanobody approach to leverage the potent and widely applicable tumor cytolytic capacity of Vγ9Vδ2-T cells. Onco Targets Ther. 2017;7:e1375641. https://doi.org/10.1080/2162402X.2017.1375641.
58. Vallera DA, Felices M, McElmurry R, et al. IL15 Trispecific killer engagers (TriKE) make natural killer cells specific to CD33+ targets while also inducing persistence, in vivo expansion, and enhanced function. Clin Cancer Res. 2016;22(14):3440–50. https://doi.org/10.1158/1078-0432. CCR-15-2710.
59. Kruse RL, Shum T, Legras X, et al. In situ liver expression of HBsAg/CD3-bispecific antibodies for HBV immunotherapy. Mol Ther - Methods Clin Dev. 2017;7:32–41. https://doi.org/10.1016/ j.omtm.2017.08.006.
60. Meng W, Tang A, Ye X, et al. Targeting human-cytomegalovirus-infected cells by redirecting T cells using an anti-CD3/anti-glycoprotein B bispecific antibody. Antimicrob Agents Chemother. 2018;62(1). https://doi.org/10.1128/AAC.01719-17
61. Huang Y, Yu J, Lanzi A, et al. Engineered bispecific antibodies with exquisite HIV-1-neutralizing activity. Cell. 2016;165(7):1621–31. https://doi.org/10.1016/j.cell.2016.05.024.
62. Li Z, Shen D, Hu S, et al. Pretargeting and bioorthogonal click chemistry-mediated endogenous stem cell homing for heart repair. ACS Nano. 2018;12(12):12193–200. https://doi.org/10.1021/ acsnano.8b05892.

63. Ziegler M, Wang X, Lim B, et al. Platelet-targeted delivery of peripheral blood mononuclear cells to the ischemic heart restores cardiac function after ischemia-reperfusion injury. Theranostics. 2017;7(13):3192–206. https://doi.org/10.7150/thno.19698.

64. Moores SL, Chiu ML, Bushey BS, et al. A novel bispecific antibody targeting EGFR and cMet is effective against EGFR inhibitor-resistant lung tumors. Cancer Res. 2016;76(13):3942–53. https://doi.org/10.1158/0008-5472.CAN-15-2833.

65. Geuijen CAW, De Nardis C, Maussang D, et al. Unbiased combinatorial screening identifies a bispecific IgG1 that potently inhibits HER3 signaling via HER2-guided ligand blockade. Cancer Cell. 2018;33(5):922–36.e10. https://doi.org/10.1016/j.ccell.2018.04.003

66. Wu AL, Kolumam G, Stawicki S, et al. Metabolic disease: amelioration of type 2 diabetes by antibody-mediated activation of fibroblast growth factor receptor 1. Sci Transl Med. 2011;3(113). https://doi.org/10.1126/scitranslmed.3002669

67. Kolumam G, Chen MZ, Tong R, et al. Sustained Brown fat stimulation and insulin sensitization by a humanized bispecific antibody agonist for fibroblast growth factor receptor 1/βKlotho complex. EBioMedicine. 2015;2(7):730–43. https://doi.org/10.1016/j.ebiom.2015.05.028.

68. Li Y, Hickson JA, Ambrosi DJ, et al. Abt-165, a dual variable domain immunoglobulin (dvd-ig) targeting dll4 and vegf, demonstrates superior efficacy and favorable safety profiles in preclinical models. Mol Cancer Ther. 2018;17(5):1039–50. https://doi.org/10.1158/1535-7163.MCT-17-0800.

69. Foxton RH, Uhles S, Grüner S, Revelant F, Ullmer C. Efficacy of simultaneous VEGF -A/ ANG −2 neutralization in suppressing spontaneous choroidal neovascularization. EMBO Mol Med. 2019;11(5). https://doi.org/10.15252/emmm.201810204

70. Regula JT, Lundh von Leithner P, Foxton R, et al. Targeting key angiogenic pathways with a bispecific Cross MA b optimized for neovascular eye diseases. EMBO Mol Med. 2017;9(7):985. https://doi.org/10.15252/emmm.201707895

71. Pernas S, Tolaney SM. HER2-positive breast cancer: new therapeutic frontiers and overcoming resistance. Ther Adv Med Oncol. 2019;11 https://doi.org/10.1177/1758835919833519.

72. Sampei Z, Igawa T, Soeda T, et al. Identification and multidimensional optimization of an asymmetric bispecific IgG antibody mimicking the function of factor VIII cofactor activity. PLoS One. 2013;8(2). https://doi.org/10.1371/journal.pone.0057479

73. Raso V, Griffin T. Hybrid antibodies with dual specificity for the delivery of ricin to immunoglobulin-bearing target cells. Cancer Res. 1981;41(6):2073–8.

74. Wolf P, Elsässer-Beile U. Pseudomonas exotoxin a: from virulence factor to anti-cancer agent. Int J Med Microbiol. 2009;299(3):161–76. https://doi.org/10.1016/j.ijmm.2008.08.003.

75. Yu YJ, Atwal JK, Zhang Y, et al. Therapeutic bispecific antibodies cross the blood-brain barrier in nonhuman primates. Sci Transl Med. 2014;6(261). https://doi.org/10.1126/scitranslmed.3009835

76. Wec AZ, Nyakatura EK, Herbert AS, et al. A "Trojan horse" bispecific-antibody strategy for broad protection against ebolaviruses. Science. 2016;354(6310):350–4. https://doi.org/10.1126/science.aag3267

77. De Goeij BECG, Vink T, Ten Napel H, et al. Efficient payload delivery by a bispecific antibody-drug conjugate targeting HER2 and CD63. Mol Cancer Ther. 2016;15(11):2688–97. https://doi.org/10.1158/1535-7163.MCT-16-0364.

78. Van Roy M, Ververken C, Beirnaert E, et al. The preclinical pharmacology of the high affinity anti-IL-6R Nanobody® ALX-0061 supports its clinical development in rheumatoid arthritis. Arthritis Res Ther. 2015;17(1). https://doi.org/10.1186/s13075-015-0651-0

79. Steeland S, Puimège L, Vandenbroucke RE, et al. Generation and characterization of small single domain antibodies inhibiting human tumor necrosis factor receptor 1. J Biol Chem. 2015;290 (7):4022–37. https://doi.org/10.1074/jbc.M114.617787.

80. Huston JS, Levinson D, Mudgett-Hunter M, et al. Protein engineering of antibody binding sites: recovery of specific activity in an anti-digoxin single-chain Fv analogue produced in Escherichia coli. Proc Natl Acad Sci U S A. 1988;85(16):5879–83. https://doi.org/10.1073/pnas.85.16.5879.
81. Holliger P, Prospero T, Winter G. "Diabodies": small bivalent and bispecific antibody fragments. Proc Natl Acad Sci U S A. 1993;90(14):6444–8. https://doi.org/10.1073/pnas.90.14.6444.
82. Moore PA, Zhang W, Rainey GJ, et al. Application of dual affinity retargeting molecules to achieve optimal redirected T-cell killing of B-cell lymphoma. Blood. 2011;117(17):4542–51. https://doi.org/10.1182/blood-2010-09-306449.
83. Wu C, Ying H, Grinnell C, et al. Simultaneous targeting of multiple disease mediators by a dual-variable-domain immunoglobulin. Nat Biotechnol. 2007;25(11):1290–7. https://doi.org/10.1038/nbt1345.
84. Steinmetz A, Vallée F, Beil C, et al. CODV-Ig, a universal bispecific tetravalent and multifunctional immunoglobulin format for medical applications. MAbs. 2016;8(5):867–78. https://doi.org/10.1080/19420862.2016.1162932.
85. Bostrom J, Haber L, Koenig P, Kelley RF, Fuh G. High affinity antigen recognition of the dual specific variants of Herceptin is entropy-driven in spite of structural plasticity. PLoS One. 2011;6(4). https://doi.org/10.1371/journal.pone.0017887
86. Wang L, He Y, Zhang G, et al. Retargeting T cells for HER2-positive tumor killing by a bispecific Fv-Fc antibody. PLoS One. 2013;8(9). https://doi.org/10.1371/journal.pone.0075589
87. Everett KL, Kraman M, Wollerton FPG, et al. Generation of Fcabs targeting human and murine LAG-3 as building blocks for novel bispecific antibody therapeutics. Methods. 2018.
88. Carter P, Ridgway JBB, Presta LG. 'Knobs-into-holes' provides a rational design strategy for engineering antibody CH3 domains for heavy chain heterodimerization. Immunotechnology. 1996;2(1):73. https://doi.org/10.1016/1380-2933(96)80685-3
89. Davis JH, Aperlo C, Li Y, et al. SEEDbodies: fusion proteins based on strand-exchange engineered domain (SEED) CH3 heterodimers in an fc analogue platform for asymmetric binders or immunofusions and bispecific antibodies. Protein Eng Des Sel. 2010;23(4):195–202. https://doi.org/10.1093/protein/gzp094.
90. Harris KE, Aldred SF, Davison LM, et al. Sequence-based discovery demonstrates that fixed light chain human transgenic rats produce a diverse repertoire of antigen-specific antibodies. Front Immunol. 2018;9(4):889. https://doi.org/10.3389/fimmu.2018.00889.
91. Mazor Y, Oganesyan V, Yang C, et al. Improving target cell specificity using a novel monovalent bispecific IgG design. MAbs. 2015;7(2):377–89. https://doi.org/10.1080/19420862.2015.1007816.
92. Fischer N, Elson G, Magistrelli G, et al. Exploiting light chains for the scalable generation and platform purification of native human bispecific IgG. Nat Commun. 2015;6:6113. https://doi.org/10.1038/ncomms7113.
93. Könning D, Rhiel L, Empting M, et al. Semi-synthetic vNAR libraries screened against therapeutic antibodies primarily deliver anti-idiotypic binders. Sci Rep. 2017;7(1). https://doi.org/10.1038/s41598-017-10513-9
94. Wang C, Vemulapalli B, Cao M, et al. A systematic approach for analysis and characterization of mispairing in bispecific antibodies with asymmetric architecture. MAbs. 2018;10(8):1226–35. https://doi.org/10.1080/19420862.2018.1511198.

Antibody–Drug Conjugates

8

Stephan Dickgiesser, Marcel Rieker, and Nicolas Rasche

Contents

Keywords

Antibody-drug conjugate · ADC · Linker · Payload · Warhead · Antibody conjugation · Monoclonal antibody · Linker-drug · Linker-payload · Targeted therapy

S. Dickgiesser (✉) · N. Rasche
ADCs and Targeted NBE Therapeutics, Merck KGaA, Darmstadt, Germany
e-mail: stephan.dickgiesser@merckgroup.com

M. Rieker
ADCs and Targeted NBE Therapeutics, Merck KGaA, Darmstadt, Germany

Clemens-Schöpf-Institut für Organische Chemie und Biochemie, Technische Universität Darmstadt, Darmstadt, Germany

© Springer Nature Switzerland AG 2021
F. Rüker, G. Wozniak-Knopp (eds.), *Introduction to Antibody Engineering*,
Learning Materials in Biosciences,
https://doi.org/10.1007/978-3-030-54630-4_8

189

What You Will Learn in This Chapter
Antibody–drug conjugates (ADCs) are complex molecules that consist of a monoclonal antibody (mAb) equipped with a small molecule drug. The antibody specifically guides a cytotoxic agent to diseased cells while sparing healthy tissue,
ultimately resulting in a drastically improved therapeutic index. This chapter
describes the manifold aspects of ADCs, beginning with the history of ADCs from
the first idea to key developments and drawbacks to the current situation with several
molecules approved for therapeutic use. Afterward, we focus on the critical
components of ADCs and describe the special requirements of mAbs utilized for
ADCs, the diverse types of drugs that are applied as ADC warheads, and different
classes of linkers that connect the mAb with the drug. In addition, conjugation
technologies used to attach small molecules to mAbs are reviewed. Several examples
from the clinics as well as preclinical development and early research are included to
illustrate the different concepts and molecular mechanisms described in each section.

8.1   Introduction to ADCs

Although much progress has been made over the last decades, cancer therapy is still a
difficult task. Especially the immense heterogeneity of tumors and the development of
drugs that selectively target the cancerous cells while sparing the healthy tissue are a major
challenge. Traditional chemotherapy, for example, makes use of the fact that most cancer
cells are rapidly dividing and proliferating. Cytotoxic chemotherapeutic agents that interfere with cell division (mitosis) therefore often lead to tumor regression. However, several
normal cell types, e.g., in the bone marrow or the digestive tract, are rapidly dividing as
well and are thus sensitive to anti-mitotic drugs eventually leading to the common side-
effects observed during chemotherapy. One of the major challenges in developing new
cancer therapies is therefore to improve the therapeutic index (TI) of a drug. The TI reflects
the ratio of the drug dose that causes the beneficial therapeutic effect to the dose that causes
toxicity (Fig. 8.1). In other words, an insufficient TI of a drug is the inability of this
compound to eradicate a tumor without causing unacceptable toxicity.

A discipline that focusses on the development of the next generation of cancer therapies
centered on the improvement of the TI is the field of targeted cancer therapies. Targeted
cancer therapies are a cornerstone of precision medicine and aim at blocking the growth
and spread of cancer by interfering with specific molecular targets that are involved in
tumor progression. In this discipline, a diverse set of approaches is being followed to target
the diseased tissue more specifically. It comprises, e.g., signal transduction inhibitors that
interfere with inappropriate cell signaling, gene modulators that allow modification of
protein expression levels, immunotherapies, antibody–drug conjugates (ADCs), and many
more. In this chapter, we would like to focus on ADCs (Fig. 8.2), which combine the

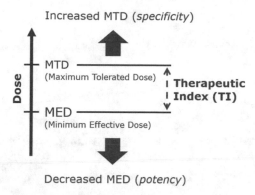

Fig. 8.1 The therapeutic index (TI) and its improvement by ADCs. The TI is characterized by the maximum dose above which toxicities occur (maximum tolerated dose, MTD) and the lowest dose that is required to reach a therapeutic effect (minimum effective dose, MED). ADCs increase the MTD and decrease the MED simultaneously by combining a target-specific mAb with a potent payload which ultimately results in an improved TI. This figure was adapted from [1]

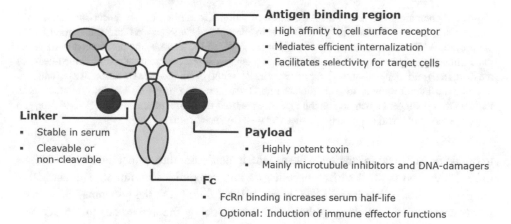

Fig. 8.2 Modular design of ADCs and critical parameters. An ADC comprises a monoclonal antibody, a payload, and a connecting linker. The antibody serves as a shuttle for the conjugated toxic payload and mediates specific antigen binding as well as a long serum half-life. The linker ensures high stability in serum but efficient payload release inside the cell

superior selectivity of monoclonal antibodies with the high potency of some cytotoxic agents.

The basic concept of ADCs is rather simple (Fig. 8.3). Prerequisite is an antigen that allows discrimination between cancer and healthy cells on a molecular basis. This can, for example, be a certain cell surface receptor, which is heavily upregulated in tumor cells. An antibody against such an antigen can serve as a targeting vehicle for a highly potent cytotoxic agent. To form the ADC, the cytotoxic agent needs to be covalently attached to the antibody via a linker that is stable in the circulation to avoid premature release of the

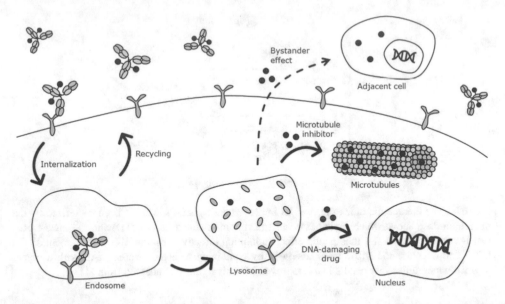

Fig. 8.3 Schematic overview of the ADC mode of action. Upon binding to a cancer-specific surface receptor (light blue), the resulting ADC–receptor complex is internalized and transported to the endosome. While receptor recycling can return the ADC to the surface, the anticipated route is the maturation of the endosome to a lysosome. Subsequent lysosomal degradation results in the release of the drug (red) and, depending on the applied toxin class, cell death induced by different mechanisms: microtubule inhibitors bind to microtubules and DNA-damaging drugs diffuse into the nucleus to induce DNA damage. Depending on the type of linker and drug, the intracellularly released drug can leave the target cell and act on adjacent cells as well (bystander effect)

toxin. After ADC administration in the patient, it distributes throughout the body and binds to its antigen on the cell surface once it gets into proximity to a tumor. The antibody–antigen complex is then internalized by the cell and directed to the lysosome via endogenous intracellular trafficking pathways. After reaching the lysosome, the ADC gets degraded and thereby releases its toxic cargo. The free toxin can then bind to its intracellular target and, thus, induce apoptosis and killing of the cancer cell. In some cases, the toxin can leave the cancer cell and act on the adjacent, ideally cancerous cells as well. This process is called the bystander effect and its extent depends on the applied linker and drug. Healthy cells, on the other hand, are mainly spared since the antibody should only bind and deliver the toxin to cancer cells that express the antigen.

8.2   Milestones of ADC Development

Historically, the concept of ADCs originates from the work of Paul Ehrlich carried out around 1900 (Fig. 8.4). He envisaged a drug, that behaves like a "magic bullet" and only harms diseased tissue while sparing healthy cells [2]. Ehrlich was able to prove his concept

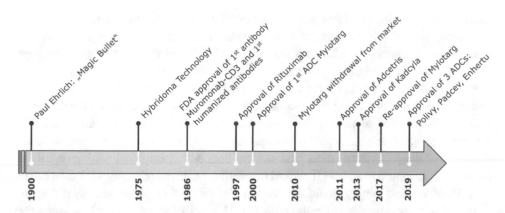

Fig. 8.4 Milestones of ADC development from first envision of the concept in 1900 to seven approved ADCs

after developing a dye that selectively killed trypanosomes, the parasites that cause sleeping sickness [3]. Besides, Paul Ehrlich was also one of the pioneers in the discovery of antibodies and therefore set some of the cornerstones for ADC development. However, after the introduction of the "magic bullet" concept in the early 1900s, ADC development was only advancing slowly until the 1980s. This can mainly be attributed to the fact that the production of selective antibodies was difficult and, thus, it was almost impossible to obtain large quantities of a specific antibody that could serve as the basis for an ADC. This changed after Georges Köhler and César Milstein reported their hybridoma technology in 1975 [4]. From this point on, the availability of monoclonal antibodies was no longer a problem and the way not only for therapeutic antibodies but also for ADCs had been paved. With Muromonab-CD3 (OKT3®), the first therapeutic monoclonal antibody was approved by the Food and Drug Administration (FDA) as an immunosuppressive drug in 1986. However, since Muromonab-CD3 is a murine antibody, its therapeutic application is associated with a rather high risk of immunogenicity. Patients therefore often develop anti-drug antibodies (ADAs) after a certain time of treatment, making them resistant toward this drug. However, recombinant DNA technology allowed the production of chimeric and, at a later stage, humanized antibodies whereby immunogenicity issues could drastically be reduced [5, 6]. These technological advancements finally led to the approval of the first mAb (rituximab) for the treatment of cancer in 1997. Rituximab is a chimeric antibody that binds to CD20, which is expressed on the surface of the majority of B cell lymphomas. Since antibodies are an integral component of ADCs, technological progress made in the field of antibodies also means progress in the field of ADCs, and the milestones mentioned above heavily contributed to the development of ADCs.

Besides the antibody, another important component of an ADC is obviously the cytotoxic payload attached to it. Some of the first investigated ADCs contained well-known chemotherapeutics with exceedingly high potencies like methotrexate, doxorubicin, or vinblastine. These were coupled to the antibody via different linker types and conjugation technologies that we will have a closer look at later in this chapter. However, in vivo evaluation of this first-generation conjugates in human tumor xenograft studies in mice showed a superior antitumor activity compared to that of the corresponding unconjugated drug [7, 8]. Encouraged by these data, two clinical phase I trials were initiated employing an ADC based on the murine KS1/4 antibody linked to methotrexate [9]. Unfortunately, no convincing therapeutic effect could be observed during these studies, which ultimately led to their discontinuation. One of the reasons for the failure was attributed to the development of ADAs against the murine mAb and the resulting fast clearance of the ADC. More successful was the antibody doxorubicin conjugate BR-96-Dox, which advanced to a phase II clinical trial for the treatment of metastatic breast cancer [10]. However, this conjugate did not show a significant therapeutic activity either and further development was therefore discontinued. This lack of success of the first generation of conjugates curbed the initial excitement about ADCs and, while seven mAbs were approved by the FDA in the 1990s, the first ADC approval was a long time coming. Finally, with gemtuzumab ozogamicin (Mylotarg®) the first ADC gained FDA approval for the treatment of acute myeloid leukemia (AML) in 2000 [11]. The basis for gemtuzumab ozogamicin is gemtuzumab, a humanized IgG4 type antibody directed against CD33, a transmembrane receptor expressed on cells of myeloid lineage [12]. As cytotoxic payload served calicheamicin, a highly potent DNA-damaging agent. Unfortunately, due to a lack of benefit for the patients, gemtuzumab ozogamicin was withdrawn from the market in 2010. Nevertheless, it was a milestone in the history of ADC development, and in the end, gemtuzumab ozogamicin was re-approved in 2017 since a lower recommended dose and a different dosing schedule led to an improved risk–benefit ratio.

The real breakthrough for the application of ADCs as therapeutics came with the approval of brentuximab vedotin (Adcetris®) in 2011 and trastuzumab emtansine (Kadcyla®) in 2013. These two ADCs gave a whole new momentum to the entire field, resulting in many development activities of different companies, the rise of new technologies and several human clinical trials. We will review some of these next-generation technologies and developments in the course of this chapter. An especially exciting year for ADCs was 2019 since the momentum gained with the approval of brentuximab vedotin and trastuzumab emtansine finally led to three late-stage clinical trials ending successfully with the approval of three ADCs. In total, seven ADCs are on the market today (listed here with international nonproprietary name (INN), followed by trade name, year of approval, and company): gemtuzumab ozogamicin (Mylotarg®, 2001/2017, Pfizer/Wyeth), brentuximab vedotin (Adcetris®, 2011, Seattle Genetics/Millennium), trastuzumab emtansine (Kadcyla®, 2013, Genentech/Roche), inotuzumab ozogamicin (Besponsa®, 2017, Pfizer/Wyeth), polatuzumab vedotin-piiq (Polivy®, 2019, Genentech/Roche), enfortumab vedotin-ejfv (Padcev®, 2019, Astellas/Seattle Genetics), and trastuzumab deruxtecan (Enhertu®, 2019, Daiichi Sankyo/AstraZeneca). The key parameters of these ADCs are depicted in Table 8.1 and several will be referred to in the following sections.

Table 8.1 Key parameters of ADCs approved for therapeutic use at the beginning of 2020

ADC INN	Trade name	Target	Toxin	Toxin MoA	Linker	Company	Indications	Approved	References
Gemtuzumab ozogamicin	Mylotarg	CD33	Calicheamicin	DNA damaging agent	Cleavable, hydrazone	Pfizer/Wyeth	Acute myeloid leukemia	2000 approval, 2010 withdrawal, 2017 re-approval	[13, 14]
Brentuximab vedotin	Adcetris	CD30	MMAE	Tubulin inhibitor	Cleavable, VCit	Seattle Genetics/Millennium	Hodgkin lymphoma, anaplastic large-cell lymphoma	2011	[13, 14]
Trastuzumab emtansine	Kadcyla	HER2	DM1	Microtubule disrupting agent	Non-cleavable, thioether	Genentech/Roche	HER2+ breast cancer	2013	[13, 14]
Inotuzumab ozogamicin	Besponsa	CD22	Calicheamicin	DNA damaging agent	Cleavable, hydrazone	Pfizer/Wyeth	Acute lymphocytic leukemia	2017	[13, 14]
Polatuzumab vedotin-piiq	Polivy	CD79b	MMAE	Tubulin inhibitor	Cleavable, VCit	Genentech/Roche	Diffuse large B cell lymphoma	2019	[13–15]
Enfortumab vedotin-ejfv	Padcev	Nectin-4	MMAE	Tubulin inhibitor	Cleavable, VCit	Astellas/Seattle Genetics	Urothelial cancers	2019	[13–15]
Trastuzumab deruxtecan	Enhertu	HER2	Camptothecin	DNA damaging agent	Cleavable, GGFG	Daiichi Sankyo/AstraZeneca	HER2+ breast cancer	2019	[13–15]

8.3   Targets and Antibodies Utilized for ADCs

The design of ADCs is a multidisciplinary endeavor since they are composed of large biotechnologically produced biomolecules and chemically synthesized, highly potent small molecule drugs. Both entities are produced separately and combined afterward to a highly complex hybrid molecule. Hence, the entire process of ADC development starting from the design of the individual components to the final production of the conjugate comes along with huge technical challenges.

The first hurdle is the identification of a target that is suitable for an ADC approach. This target should allow discrimination between the cancerous cells and the healthy tissue either by tumor-specific target overexpression or by alternative mechanisms such as exclusive presentation of a mutated protein variant on tumor cells. It should also be accessible to an antibody, which essentially limits the space to extracellular targets. In addition, fast internalization of the target upon binding of the ADC is important to reach the critical concentration of the cytotoxic agent inside the cell. An example of a target that fulfills these criteria is CD30. CD30 is a receptor presented by activated T and B cells and exhibits a high, uniform expression on cancer cells, low expression in normal tissue, and rapid internalization upon antibody binding [16]. A monoclonal antibody against CD30 is the basis for brentuximab vedotin (Adcetris®), an ADC used to treat relapsed or refractory Hodgkin lymphoma (HL) and systemic anaplastic large cell lymphoma (ALCL). On the contrary, not every ADC target necessarily needs to fulfill all these criteria. The target of trastuzumab emtansine (Kadcyla®), HER2, shows high expression levels in breast cancer tissue and internalizes rapidly upon antibody binding. However, it is also expressed in other tissues like the skin, heart, gastrointestinal (GI), respiratory, reproductive, and urinary tracts [17]. Nevertheless, the ADC is still well-tolerated due to a much lower HER2 expression on healthy cells than on cancer cells.

The antibody is the second crucial component of an ADC since it represents the main determinant for targeting and the pharmacokinetics (PK) of the whole molecule. The mAb should, therefore, bind its target with high affinity and selectivity to make sure the cytotoxic warhead is being delivered to the right cells. However, an antibody not only binds to its antigen, but it undergoes several additional interactions with a diverse set of Fc receptors. Of special importance is binding to the neonatal Fc receptor (FcRn) that subjects the antibody to a mechanism known as FcRn recycling. This process results in a drastically improved half-life of antibodies and is, aside from the high molecular size, one of the main reasons for the extraordinary long serum half-life of antibodies. Considering that a small molecular cytotoxic agent usually exhibits half-lives of only a few hours, preservation of FcRn binding is highly desirable for ADCs since attachment to an antibody can extend the toxin's half-live to days or even weeks. Special attention has therefore to be paid to not impair the mAb–FcRn interaction by, e.g., conjugation of the cytotoxic agent. Some engineering approaches even aim at boosting FcRn interaction to further improve the PK properties of antibodies [18–20]. Beside the FcRn receptor, many other Fc receptor classes exist which are responsible for Fc effector functions like antibody-dependent cell-mediated cytotoxicity (ADCC) and complement-dependent cytotoxicity (CDC). Although these mechanisms are often contributing to the therapeutic effect of mAbs, interactions with these receptors might be unfavorable for ADC approaches. Engagement of these secondary immune functions may have the unintended effect of reduced conjugate localization to the tumor or increased ADC uptake by healthy cells that express Fc receptors. For that reason, the mAb utilized for gemtuzumab ozogamicin (Mylotarg®) belongs to the IgG4 subclass that does not have ADCC or CDC activity whereas the immune-active IgG1 subclass is typically employed for therapeutic antibodies. It thus might be a valid strategy to use effector-silent antibody subclasses for ADCs or to conduct antibody engineering that aims at avoiding the issues mentioned above [21].

8.4   Cytotoxic ADC Payloads

The element of an ADC that ultimately causes its therapeutic effect is the conjugated cytotoxic agent, which is often referred to as "payload" or "warhead." Several molecule classes have proven successful in ADC context so far and even more with great potential are currently under investigation. Several of the toxins used as ADC payloads today, namely maytansinoids, auristatins, and calicheamicins, have been discovered because of their exceedingly high potency which made them very attractive as possible new anticancer drugs. However, these toxins had an unfavorable TI and the doses required to achieve an anticancer effect came along with serious toxicity, eventually ruling out their further development as standalone anticancer agents. With the advent of ADCs, however, the therapeutic application of these toxins experienced a renaissance. The vast majority of ADCs being developed today carry a cytotoxic payload, but it is worth mentioning that recent approaches also utilize alternative drugs such as immune-stimulatory payloads or antibiotics [22, 23]. In the following section, we will focus on different classes of cytotoxic ADC payloads and have a closer look at the payloads that have reached final success as components of currently approved ADCs.

Maytansine is a macrolide of the ansamycin type that was initially isolated from the bark of the African shrub *Maytenus* [24, 25]. It is a very potent antimitotic drug that inhibits tubulin polymerization and thereby arrests the cell cycle at the G2/M phase, which ultimately results in cell death by apoptosis [26, 27]. Maytansine and its derivatives, the maytansinoids, showed strong in vitro activity at concentrations in the low picomolar range on different cancer cell lines [27, 28]. The most common dose-limiting toxicities of these compounds are neurotoxicity, vomiting, and nausea. The originally isolated maytansine structure was, however, not very well suited for the covalent attachment to an antibody. Therefore, derivatives with modifications that allow conjugation to a mAb had to be developed [29]. One of the compounds resulting from these activities is DM1 (Fig. 8.5), the payload finally used for trastuzumab emtansine (Kadcyla®). Interestingly, using DM1 as an ADC component increased its maximum tolerated dose (MTD) by at least a factor of

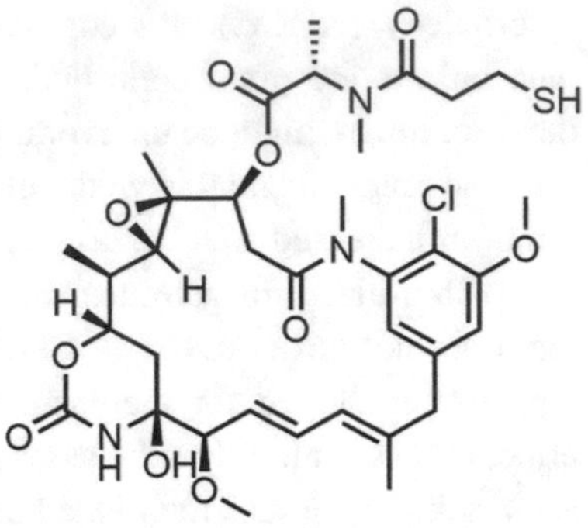

Fig. 8.5 Chemical structure of DM1, the payload utilized for trastuzumab emtansine

two, confirming the concept that incorporation of cytotoxic agents into ADCs improves the TI.

Auristatins are peptide-based molecules and dolastatin 10, the first described agent of this class, has been isolated from the sea slug *Dolabella auricularia*. Auristatins inhibit the polymerization of microtubules and suppress the tubulin-dependent guanosine-5′-triphosphate (GTP) hydrolysis. This leads to an interference with mitosis resulting in a cell cycle arrest in the G2/M phase, which finally results in cell death. The most common dose-limiting toxicities of auristatins are peripheral neuropathy (PN), fatigue, nausea, and diarrhea.

Today, mainly two dolastatin 10 analogs, namely monomethyl auristatin E (MMAE) (Fig. 8.6) and monomethyl auristatin F (MMAF) are being utilized as ADC payloads. However, so far only ADCs containing MMAE as active drugs have been approved for therapeutic use, namely brentuximab vedotin (Adcetris®), polatuzumab vedotin-piiq (Polivy®), and enfortumab vedotin-ejfv (Padcev®).

Calicheamicins are enediyne-containing anticancer antibiotics with special structural elements such as a methyl trisulfide group and a glycosylated hydroxyl-amino sugar. They were initially isolated from *Micromonospora echinospora calichensis* and show extraordinarily high antitumor activity. Calicheamicin $\gamma 1^1$, for example, showed antitumor activity in murine tumors at doses as low as 0.5–1.5 μg/kg [30, 31]. Calicheamicin makes up the payload in gemtuzumab ozogamicin (Mylotarg®) and inotuzumab ozogamicin (Besponsa®) (Fig. 8.7) where it is covalently linked to the mAb via an acid-labile linker. It is a very potent DNA damaging agent and binds to specific regions in the DNA minor grove where it induces double-strand breaks finally leading to apoptosis and cell death. The most common dose-limiting toxicities of calicheamicin include thrombocytopenia and neutropenia.

Camptothecin is a pentacyclic quinolone-based alkaloid that was originally isolated from the Asian tree *Camptotheca acuminate*. It inhibits DNA topoisomerase I, thus effectively stalling DNA replication in the S-phase ultimately resulting in the apoptotic cell death of tumor cells [32]. The initial compound, however, suffered from poor water solubility and the conversion of its lactone into an inactive carboxylate form under physiological conditions. Thus, several activities that aimed at improving the

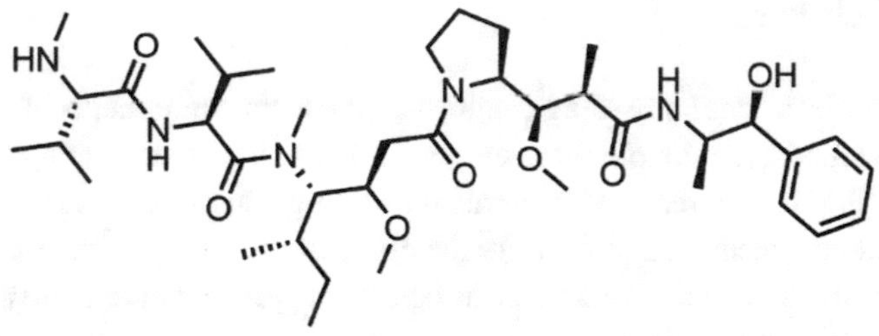

Fig. 8.6 Chemical structure of monomethyl auristatin E (MMAE), the cytotoxic component of brentuximab vedotin, polatuzumab vedotin, and enfortumab vedotin

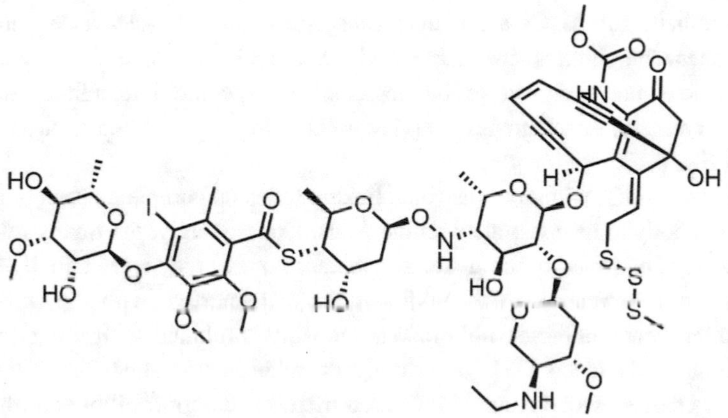

Fig. 8.7 Chemical structure of calicheamicin as found in the clinically approved ADCs gemtuzumab ozogamicin and inotuzumab ozogamicin

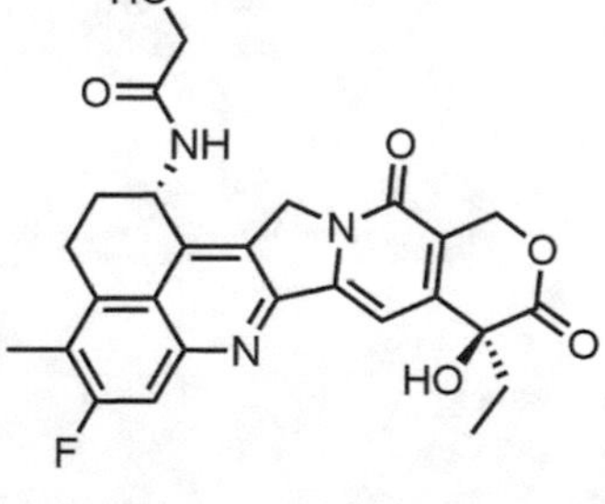

Fig. 8.8 Chemical structure of Dxd, which is part of the approved ADC trastuzumab deruxtecan

characteristics of camptothecin have been undertaken. These efforts resulted in the development of topotecan and irinotecan, which have been approved by the FDA as anticancer drugs. Another important derivative of camptothecin is the substituted exatecan Dxd (Fig. 8.8), which is the payload used for trastuzumab deruxtecan (Enhertu®). The most severe side-effects observed during camptothecin treatment include neutropenia and leukopenia.

8.5   ADC Linkers

According to the term "antibody–drug conjugate," the main components of an ADC are the drug and the antibody. To couple these entities, however, a linker that connects the mAb with the drug (Fig. 8.9) is required. Careful selection of this linker, taking both the mAb and the payload into account, is crucial for the efficacy and safety of the final ADC. In the bloodstream, the linker should be as stable as possible to prevent premature payload release which could otherwise cause systemic off-target toxicity. But once the ADC has reached the target cell, the payload has to be active without being hampered by an attached linker. In addition, the length and chemical nature of the linker can have strong effects on the pharmacokinetics and -dynamics of ADCs.

Linkers utilized for ADCs are mainly categorized into non-cleavable and cleavable ones. Non-cleavable linkers are stable both in the circulation and in cells, whereas cleavable linkers are designed to be degraded by specific intracellular mechanisms within the target cell. A schematic overview of the linkers described in the following is depicted in Fig. 8.10.

Release from ADCs with non-cleavable linkers relies on complete proteolytic degradation of the antibody to its constituent amino acids. Consequently, the finally released drug metabolites stay connected to the linker and the amino acid they were initially conjugated to (e.g., lysine or cysteine). Such modifications can interfere with the activity of the cytotoxic drug, hence non-cleavable linkers are only combined with drugs that are still active as amino acid adducts [33]. Moreover, these adducts are quite hydrophilic due to the amino acid's charges which results in reduced membrane permeability and, thereby, low

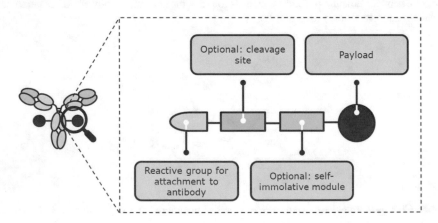

Fig. 8.9 Schematic illustration of ADC linkers. ADC linkers covalently connect the cytotoxic payload with the antibody and are typically synthesized together with the payload to a combined linker–payload structure. All linkers carry a reactive group that enables conjugation to the antibody. Many linkers also comprise a cleavage site, that allows improved payload release inside the target cell. Some cleavage mechanisms require an additional self-immolative module ensuring traceless release of the payload

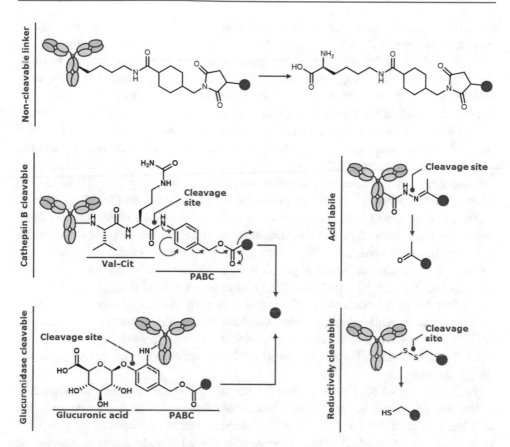

Fig. 8.10 Overview of commonly applied ADC linkers. ADC linkers can be divided into two classes: non-cleavable (top) and cleavable (bottom) linkers. Payloads released from ADCs with cleavable linkers stay connected to the amino acid used for linker–payload attachment. Cleavable linkers are degraded by enzymes or the chemical environment. Cathepsin B linkers contain peptide motifs (e.g., valine-citrulline) whereas glucuronidase linkers contain glucuronic acid. Both are combined with a self-immolative PABC module, which enables traceless drug release. Acid-labile linkers and disulfide linkers are degraded by acidic and reductive conditions, respectively

bystander activity [1]. On the other hand, non-cleavable linkers are highly stable in the circulation and thereby reduce the risk of unintended drug release. One prominent example of an ADC with a non-cleavable linker is trastuzumab emtansine (Kadcyla®). Here, the warhead is connected to the mAb by a non-cleavable thioether linkage resulting in a DM1-lysine adduct as the finally released active drug [1, 34].

Cleavable linkers are designed to specifically release the drug at the tumor site, preferentially only inside the cancerous cells. Ideally, this release is traceless and leaves the drug without any residual substituent like parts of the linker or amino acids, which in many cases also helps to improve bystander effects. The underlying cleavage mechanism typically employs factors that differ between the circulation and intracellular compartments, mostly the chemical environment or specific intracellular enzymes [1].

A class of linkers that rely on cleavage triggered by a change in the chemical environment is hydrazone containing linker systems. These linkers are designed to be stable at the blood's neutral pH but hydrolyze in acidic environments. As already described earlier, the internalization of ADCs typically leads to transportation to endosomal and lysosomal compartments. These compartments exhibit an increasingly acidic environment (pH ≈ 5), which finally triggers the release of the drug. Although this type of linker has been associated with nonspecific drug release, successful applications demonstrate its functionality [35]. In fact, gemtuzumab ozogamicin (Mylotarg®) and inotuzumab ozogamicin (Besponsa®) both rely on pH-sensitive hydrazone linkers [36, 37].

Disulfide containing linkers are another type of chemically cleavable linkers. These linkers benefit from the fact, that the intracellular compartments are typically much more reducing than the bloodstream. The reducing intracellular environment is created by an elevated concentration of reduced glutathione (1–10 mM) and reducing enzymes as compared to the circulation (≈5 μM) [1]. Of note, as a response to the reduced oxygen levels found in solid tumors, these effects might be even more pronounced in cancerous cells [38]. Consequently, disulfide-containing linkers release their drug predominantly in the intracellular space of tumor cells. By introducing sterically hindering substituents next to the disulfides, the stability of these linkers can be fine-tuned to further prevent premature drug release [39].

Peptide linkers are designed to release the drug upon selective enzymatic cleavage. Due to the low proteolytic activity in the blood, these linkers can be highly stable in the circulation while they are intended to be cleaved after internalization by lysosomal proteases such as cathepsins [40]. A frequently applied peptide linker comprises the dipeptide motif valine-citrulline (Val-Cit), which is rapidly hydrolyzed in the lysosome. It was initially designed for selective cleavage by the cysteine protease cathepsin B, but it was shown that drug release can be triggered by other proteases as well [41, 42]. Further successfully applied peptide motifs are Val-Ala or Gly-Gly-Phe-Gly; the latter one is used for trastuzumab deruxtecan (Enhertu®) [43, 44]. A frequently applied linker–drug of this class is composed of the Val-Cit peptide connected to the drug MMAE via a self-immolative para-aminobenzylcarbamate (PABC) spacer. It had been shown that substituents can drastically decrease the potency of MMAE. Therefore, much effort has been made to develop this linker system that allows traceless release of MMAE: Proteolytic cleavage of Val-Cit creates a PABC–MMAE intermediate that subsequently undergoes self-immolation of the PABC moiety and thereby releases the unmodified drug. MMAE can then either be active inside the same cell or diffuse to neighboring cells and elicit bystander effects [45]. This linker–drug is part of brentuximab vedotin (Adcetris®), polatuzumab vedotin-piiq (Polivy®), and enfortumab vedotin-ejfv (Padcev®) [46, 47].

β-glucuronidase cleavable linkers represent an additional class of enzyme cleavable linkers. β-glucuronidase is abundantly present in lysosomes but also overexpressed in some tumors. It selectively hydrolyzes the glycosidic bond of linkers equipped with β-glucuronic acid as a core unit and thereby triggers a self-immolating mechanism that ensures release of the unmodified payload. Interestingly, with the β-glucuronide acid as an integral part, these linkers are more hydrophilic and allow to deploy unusually hydrophobic drugs that would otherwise negatively impact the biophysical properties of the ADC [48].

Beside the cleavage site that ensures drug release, additional linker components have been developed to yield ADCs with improved properties. Especially the concept of hydrophilic linkers that, similar to β-glucuronide linkers, serve as solubility enhancers masking the payload's hydrophobicity has been pursued. For this purpose, non-charged polymers (e.g., based on polyacetal or polyethylene glycol) or negatively charged groups (e.g., sulfonates) have been integrated into linker structures [49–51]. Interestingly, these modifications positively influence ADC stability, allow higher cytotoxin loading, and resulted in ADCs with improved pharmacokinetics and an improved therapeutic index [49–51].

8.6   Conjugation Technologies for ADC Generation

While all three parts of an ADC—the antibody, the linker, and the cytotoxic payload—determine the key properties of the final conjugate, a similarly important parameter is the way these components are assembled. The linker and the payload are produced by chemical synthesis either as a combined linker–payload structure that is directly conjugated to the mAb or as individual components that are successively assembled during ADC generation. In both cases, a small molecule needs to be conjugated to a mAb without impairing its favorable properties, which is a major technical challenge. The main parameters that need to be controlled during ADC generation are the number of linker–drugs conjugated to each antibody, termed drug-to-antibody ratio (DAR), and the positions on the antibody surface the structures are attached to (conjugation sites). Both parameters can decisively influence several properties of an ADC including its stability and pharmacokinetic behavior and ultimately also its toxicity and efficacy profile [52–55]. On the one hand, warheads used for ADCs are mostly hydrophobic and an increasing DAR can significantly alter the overall hydrophobicity and severely disturb protein stability of the final conjugate. On the other hand, a certain amount of drug, depending on its potency, is required to reach a sufficiently active ADC. However, not only the DAR but also the conjugation site and chemistry heavily impact these parameters. For instance, several studies have shown, that certain sites show superior tolerance toward challenging payloads and result in more stable conjugates than others by providing a favorable microenvironment and steric shielding on the antibody surface [52–55]. Hence, finding a favorable combination of the individual components linker, drug and mAb as well as a suitable DAR, conjugation strategy and conjugation sites is key for the development of efficient and safe therapeutics. In the following, we will review the most established conjugation strategies for ADC generation and describe some techniques that are currently under development. Figure 8.11 provides a graphical overview of these technologies.

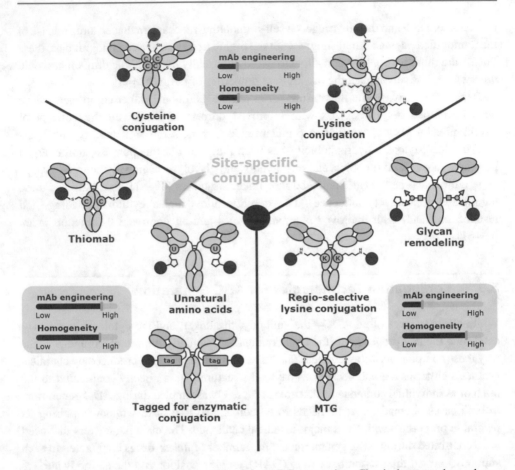

Fig. 8.11 Conjugation technologies utilized for ADC generation. Classical approaches rely on modification of surface lysines or hinge cysteines of non-engineered antibodies but usually result in rather heterogeneous products (except for conjugates with exhaustive cysteine conjugation). In order to increase ADC homogeneity, several site-specific conjugation strategies have been developed. Recombinant antibody engineering allows to introduce cysteines ("thiomab") or unnatural amino acids that serve as linker–payload attachment sites. Similarly, genetically introduced peptide tags enable modification by different enzymes, e.g., sortase A, formylglycine-generating enzyme or microbial transglutaminase (MTG). In contrast, recent approaches aim for site-specific conjugation of non-engineered, native mAbs. Regio-selective lysine conjugation can be achieved by pre-orienting lysine-reactive groups. In addition, removal of mAb glycans allows MTG-mediated conjugation whereas remodeled glycans can serve as attachment sites themselves

Antibodies are large macromolecules that carry a vast number of chemical groups on their surface making it difficult to conjugate linker–drugs in a targeted manner. Depending on the anticipated conjugation chemistry, linker–drugs have to be equipped with the required chemical groups and structures already during chemical synthesis. Traditionally, conjugation is performed using established methods from classical protein chemistry that target the side chains of lysine or cysteine residues—amino acids that naturally reside in every mAb. Lysine residues are abundantly present on the mAb surface and at least 40 of them are accessible for conjugation [56]. For lysine conjugation, linkers are usually equipped with N-hydroxysuccinimide esters that react with the lysine's ε-amino group to form stable amide bonds. Several ADCs that have reached the market (e.g., gemtuzumab ozogamicin (Mylotarg®)) are produced following this approach, demonstrating that it is a viable strategy to yield effective molecules [36]. However, due to the high number of accessible lysine residues that are distributed all over the mAb surface, this method results in heterogeneous mixtures containing a vast number of ADC species with different DARs and attachment sites [56]. Each of these species has different characteristics in several regards, e.g., efficacy, toxicity, aggregation, and stability. Also, accessible lysines can theoretically even reside in functional regions of the mAb where conjugation might influence key properties such as affinity to the target or mAb specific receptors [57].

Instead of targeting lysine residues, linker–payloads are often conjugated to the mAb cysteine residues. Cysteines are much less abundant than lysines and often not susceptible to conjugation strongly reducing the theoretical number of conjugation sites. In fact, mAbs of the IgG isotype comprise 32 conserved cysteines that form the characteristic disulfide bonds within each of the 12 domains as well as 4 bonds that connect the mAb chains in the hinge region. Classical cysteine conjugation such as maleimide-based Michael addition, however, targets the cysteine side chain's thiol that is not reactive when the thiols are trapped in disulfide bonds. Hence, conjugation to mAb cysteines requires disulfide reduction in order to generate reactive thiols for conjugation with maleimide-functionalized linkers. Interestingly, mild reduction conditions allow to exclusively reduce the hinge interchain disulfides of IgGs and thereby limit the number of conjugation sites to 8. Although this procedure partially removes the covalent interchain linkage, the final ADCs are sufficiently stabilized by non-covalent interactions [49]. Approval of several cysteine conjugated ADCs such as brentuximab vedotin (Adcetris®) and several ongoing clinical studies confirm the integrity and efficacy of ADCs derived from this process [58, 59].

Although the reduced number of conjugation sites of cysteine-targeted conjugation results in less heterogeneous ADCs than lysine conjugation, products are usually still a stochastic mixture of species since the aspired DAR is in the majority of all cases below 8. Trastuzumab deruxtecan (Enhertu®), which is produced by full loading of all interchain cysteines to yield a homogeneous conjugate with a DAR close to 8, is an exception so far [59]. In addition, there is no choice of conjugation site because the native cysteines are obviously at a fixed position within the mAb. However, several publications have shown that the conjugation site can have a tremendous effect on ADC stability and in vivo properties and that the interchain cysteines are not necessarily the perfect choice [54, 60, 61]. For that reason, several next-generation conjugation technologies have been developed allowing site-specific linker–drug attachment precisely to selected positions and thereby generate tailored, homogeneous ADCs [57, 62].

One frequently applied site-specific technology is based on cysteines as well, but in this case the cysteines are placed into the mAb primary structure by recombinant engineering.

After expression, such mAbs (often termed "thiomabs") carry the engineered cysteines as part of the mAb sequence at the anticipated positions. Unfortunately, thiol bearing compounds from the expression medium form disulfides with these engineered cysteines and thereby prevent direct conjugation [63]. To remove these thio-caps and enable conjugation, the mAbs have to be reduced which not only removes the caps but also opens the interchain disulfide bonds. Hence, in a subsequent step the interchain disulfides are re-oxidized leaving the engineered cysteines as sole free thiols. Such pretreated mAbs are then conjugated with, e.g., maleimide linker–payloads which exclusively react with the engineered cysteine residues [64, 65].

Conjugation to interchain as well as engineered cysteines is typically conducted using maleimide-functionalized linkers. Several studies have, however, demonstrated that the thio-succinimide linkage formed during maleimide conjugation can suffer from instability by undergoing retro-Michael addition or other thiol exchange reactions leading to premature release of linker–payload [66]. The extend of this instability strongly depends on the microenvironment of the conjugation site. Hence, a lot of effort was put into seeking the perfect position for engineered cysteines. Several identified positions not only allow to produce homogeneous ADCs but also come along with additional favorable properties in terms of conjugate stability and pharmacokinetic behavior [54, 64, 67]. As an alternative, several strategies for stabilization of maleimide-derived conjugates as well as alternative thiol chemistries that allow the formation of more stable products have been described [68–70].

Beside conjugation to engineered cysteines, multiple additional technologies for site-specific ADC generation exist and several rely on bond-forming enzymes. One prominent example is the bacterial enzyme microbial transglutaminase (MTG) that naturally catalyzes the formation of isopeptide bonds between glutamine and lysine residues. Although IgG antibodies have several glutamine and lysine residues in their primary structure, none of these is usually modified by MTG. Nevertheless, strategies that allow efficient and site-specific MTG-driven ADC generation have been developed. The most frequently applied strategies rely on either of two principles: first, certain peptide tags that contain a glutamine residue (e.g., LLQG) can be inserted into the antibody sequence by recombinant technologies. These tags can afterward be addressed by MTG that specifically conjugates linker–drugs to the peptide tag glutamine [71, 72]. Second, removal of the conserved IgG N-glycan from position 297 (EU numbering) (either by enzymatic cleavage or genetic engineering) exposes the adjacent glutamine Q295. This residue serves as an exquisite MTG substrate once the glycan has been removed [73]. In both cases, the linker has to be equipped with a primary amino group, marking it as a substrate for the enzyme. Both strategies have been successfully applied to produce homogenous ADCs with promising in vivo properties [61, 74].

As opposed to removing the conserved N-glycans for MTG conjugation, they can serve as linker–payload attachment sites themselves. For this, the existing glycans can be modified by chemical oxidation to allow subsequent linker–drug conjugation [75]. Alternatively, glycans can be remodeled either already during recombinant mAb expression by metabolic engineering or post-translationally by using glycan-modifying enzymes. Both approaches eventually result in mAbs with glycan-derived structures that are equipped with reactive groups. These groups serve as anchors for specific linker–payload conjugation in a subsequent step [43, 76, 77]. The utilized reactive groups are usually selected from the constantly growing toolbox of bio-orthogonal chemistry. A common choice for ADCs is azido groups that readily react with strained alkynes in copper-free click chemistry [78]. Hence, linker–drugs that are functionalized with such alkynes specifically react with azido-anchors introduced to the mAb glycan.

Other technologies for site-specific ADC generation are based on the introduction of unnatural amino acids (UAA). Such amino acids are not part of the naturally encoded proteinogenic amino acids and are therefore not found in proteins. However, expanding the genetic code allows the incorporation of UAAs precisely at selected positions during recombinant mAb expression. This is usually accomplished by using a specific set of engineered tRNA and tRNA synthetase in combination with a suitable orthogonal expression system. Similar to the aforementioned remodeled glycans, the UAAs can be equipped with biorthogonal reactive groups (e.g., azido groups) that serve as attachment sites for the linker–drug [79–81].

Beside the described techniques, numerous additional strategies for site-specific ADC generation have been developed. For instance, several approaches rely on enzymes that modify mAbs either co- or posttranslationally. Similar to MTG, these enzymes (e.g., sortase A, formylglycine-generating enzyme or phosphopantetheinyl transferases) recognize specific amino acid sequences that have to be engineered into the mAb sequence prior to expression [82–84]. Just like the other described methods, these processes can generate ADCs with excellent characteristics both in vitro and in vivo. However, most technologies for site-specific generation require manipulation of the mAb sequence or glycans. Although this is not necessarily a major issue, it bears certain risks (e.g., immunogenicity) and requires engineering of the mAb sequence and thereby excludes, e.g., already approved antibodies with established GMP production processes. Hence, several groups are seeking for approaches that allow site-specific conjugation to native mAbs without the need for sequence or glycan modification. As an example, one technology utilizes affinity peptides to pre-orientate amine-reactive groups to specific lysine residues and thereby limits conjugation to these positions [85]. Other groups re-bridge the hinge cysteine residues during cysteine conjugation or engineer MTG variants that allow conjugation of native mAbs without the need for peptide tags or removal of the mAb glycans [69, 86].

In summary, a plethora of different conjugation strategies for ADC production is available, and even more techniques are being developed. While currently approved ADCs are manufactured by classical cysteine or lysine conjugation, multiple preclinical studies have shown superior behavior of tailored ADCs from site-specific conjugation approaches and several of these ADCs are currently being investigated in clinical trials. However, the translation of preclinical data of site-specific ADCs into beneficial therapeutic effects still remains to be shown [43]. The choice of conjugation strategy depends on several factors including the linker, the drug, and the anticipated DAR. Hence, each conjugation technology comes along with individual advantages and disadvantages and multiple factors have to be considered to select the suitable conjugation strategy for each ADC individually.

Take Home Messages
- ADCs are conjugates consisting of a monoclonal antibody, a linker, and a cytotoxic payload.
- The combination of antibody specificity and drug potency results in an improved therapeutic index.
- The target antigen needs to be specific for cancer cells and typically internalizes upon ADC binding.
- The antibody mediates high specificity for the target antigen favorable pharmacokinetics.
- Most ADC payloads are either microtubule inhibitors or DNA damagers.
- Linkers that connect the mAb with the drug are either non-cleavable or cleavable.
- Cleavable linkers are degraded inside the target cell by chemical or enzymatical mechanisms to release the drug.
- Currently approved ADCs are produced by classical conjugation methods that mostly result in heterogeneous products.
- Several recently developed site-specific conjugation technologies enable the generation of homogeneous ADCs.

References

1. Chari RVJ, Miller ML, Widdison WC. Antibody-drug conjugates: an emerging concept in cancer therapy. Angew Chem Int Ed. 2014;53(15):3796–827. https://doi.org/10.1002/anie.201307628.
2. Strebhardt K, Ullrich A. Paul Ehrlich's magic bullet concept: 100 years of progress. Nat Rev Cancer. 2008;8(6):473–80. Available from: http://www.nature.com/articles/nrc2394
3. Prüll C-R. Part of a scientific master plan? Paul Ehrlich and the origins of his receptor concept. Med Hist. 2003;47(3):332–56. Available from: http://www.ncbi.nlm.nih.gov/pubmed/12905918.
4. Köhler G, Milstein C. Continuous cultures of fused cells secreting antibody of predefined specificity. Nature. 1975;256(5517):495–7. Available from: http://www.nature.com/articles/256495a0
5. Morrison SL, Johnson MJ, Herzenberg LA, Oi VT. Chimeric human antibody molecules: mouse antigen-binding domains with human constant region domains. Proc Natl Acad Sci. 1984;81 (21):6851–5. https://doi.org/10.1073/pnas.81.21.6851.
6. Jones PT, Dear PH, Foote J, Neuberger MS, Winter G. Replacing the complementarity-determining regions in a human antibody with those from a mouse. Nature. 1986;321(6069):522–5. Available from: http://www.nature.com/articles/321522a0
7. Laguzza BC, Nichols CL, Briggs SL, Cullinan GJ, Johnson DA, Starling JJ, et al. New antitumor monoclonal antibody-vinca conjugates LY203725 and related compounds: design, preparation,

and representative in vivo activity. J Med Chem. 1989;32(3):548–55. https://doi.org/10.1021/jm00123a007.

8. Trail P, Willner D, Lasch S, Henderson A, Hofstead S, Casazza A, et al. Cure of xenografted human carcinomas by BR96-doxorubicin immunoconjugates. Science. 1993;261(5118):212–5. https://doi.org/10.1126/science.8327892.

9. Elias DJ, Kline LE, Robbins BA, Johnson HC, Pekny K, Benz M, et al. Monoclonal antibody KS1/4-methotrexate immunoconjugate studies in non-small cell lung carcinoma. Am J Respir Crit Care Med. 1994;150(4):1114–22. https://doi.org/10.1164/ajrccm.150.4.7921445.

10. Tolcher AW, Sugarman S, Gelmon KA, Cohen R, Saleh M, Isaacs C, et al. Randomized phase II study of BR96-doxorubicin conjugate in patients with metastatic breast cancer. J Clin Oncol. 1999;17(2):478. https://doi.org/10.1200/JCO.1999.17.2.478.

11. Ducry L, Stump B. Antibody–drug conjugates: linking cytotoxic payloads to monoclonal antibodies. Bioconjug Chem. 2010;21(1):5–13. https://doi.org/10.1021/bc9002019.

12. Garnache-Ottou F. Expression of the myeloid-associated marker CD33 is not an exclusive factor for leukemic plasmacytoid dendritic cells. Blood 2004;105(3):1256–1264. Available from: https://doi.org/10.1182/blood-2004-06-2416

13. Khongorzul P, Ling CJ, Khan FU, Ihsan AU, Zhang J. Antibody–drug conjugates: a comprehensive review. Mol Cancer Res. 2020;18(1):3–19. https://doi.org/10.1158/1541-7786.MCR-19-0582.

14. Shim H. Bispecific antibodies and antibody–drug conjugates for cancer therapy: technological considerations. Biomolecules. 2020;10(3):360. Available from: https://www.mdpi.com/2218-273X/10/3/360

15. Morrison C. Fresh from the biotech pipeline—2019. Nat Biotechnol. 2020;38(2):126–31. Available from: http://www.nature.com/articles/s41587-019-0405-7

16. Falini B, Pileri S, Pizzolo G, Dürkop H, Flenghi L, Stirpe F, et al. CD30 (Ki-1) molecule: a new cytokine receptor of the tumor necrosis factor receptor superfamily as a tool for diagnosis and immunotherapy. Blood. 1995;85(1):1–14. Available from: http://www.ncbi.nlm.nih.gov/pubmed/7803786.

17. Press MF, Cordon-Cardo C, Slamon DJ. Expression of the HER-2/neu proto-oncogene in normal human adult and fetal tissues. Oncogene. 1990;5(7):953–62. Available from: http://www.ncbi.nlm.nih.gov/pubmed/1973830.

18. Lee CH, Kang TH, Godon O, Watanabe M, Delidakis G, Gillis CM, et al. An engineered human Fc domain that behaves like a pH-toggle switch for ultra-long circulation persistence. Nat Commun. 2019;10(1):5031. https://doi.org/10.1038/s41467-019-13108-2.

19. Dall'Acqua WF, Kiener PA, Wu H. Properties of human IgG1s engineered for enhanced binding to the neonatal Fc receptor (FcRn). J Biol Chem. 2006;281(33):23514–24.

20. Zalevsky J, Chamberlain AK, Horton HM, Karki S, Leung IWL, Sproule TJ, et al. Enhanced antibody half-life improves in vivo activity. Nat Biotechnol. 2010;28(2):157–9. Available from: http://www.nature.com/articles/nbt.1601

21. Bross PF, Beitz J, Chen G, Chen XH, Duffy E, Kieffer L, et al. Approval summary: gemtuzumab ozogamicin in relapsed acute myeloid leukemia. Clin Cancer Res. 2001;7(6):1490–6. Available from: http://www.ncbi.nlm.nih.gov/pubmed/11410481.

22. Peck M, Rothenberg ME, Deng R, Lewin-Koh N, She G, Kamath AV, et al. A phase 1, randomized, single-ascending-dose study to investigate the safety, tolerability, and pharmacokinetics of DSTA4637S, an anti-*Staphylococcus aureus* thiomab antibody-antibiotic conjugate, in healthy volunteers. Antimicrob Agents Chemother. 2019;63(6). https://doi.org/10.1128/AAC.02588-18

23. Gadd AJR, Greco F, Cobb AJA, Edwards AD. Targeted activation of toll-like receptors: conjugation of a toll-like receptor 7 agonist to a monoclonal antibody maintains antigen binding

and specificity. Bioconjug Chem. 2015;26(8):1743–52. https://doi.org/10.1021/acs.bioconjchem. 5b00302.

24. Kupchan SM, Komoda Y, Court WA, Thomas GJ, Smith RM, Karim A, et al. Tumor inhibitors. LXXIII. Maytansine, a novel antileukemic ansa macrolide from Maytenus ovatus. J Am Chem Soc. 1972;94(4):1354–6. https://doi.org/10.1021/ja00759a054

25. Kupchan SM, Komoda Y, Branfman AR, Richard G, Dailey J, Zimmerly VA. Letter: novel maytansinoids. Structural interrelations and requirements for antileukemic activity. J Am Chem Soc. 1974;96(11):3706–8. Available from: http://www.ncbi.nlm.nih.gov/pubmed/4833726.

26. Remillard S, Rebhun L, Howie G, Kupchan S. Antimitotic activity of the potent tumor inhibitor maytansine. Science. 1975;189(4207):1002–5. https://doi.org/10.1126/science.1241159.

27. Wolpert-Defilippes MK, Adamson RH, Cysyk RL, Johns DG. Initial studies on the cytotoxic action of maytansine, a novel ansa macrolide. Biochem Pharmacol. 1975;24(6):751–4. Available from: https://linkinghub.elsevier.com/retrieve/pii/0006295275902579

28. Kupchan SM, Sneden AT, Branfman AR, Howie GA, Rebhun LI, McIvor WE, et al. Structural requirements for antileukemic activity among the naturally occurring and semisynthetic maytansinoids. J Med Chem. 1978;21(1):31–7. Available from: http://www.ncbi.nlm.nih.gov/pubmed/563462.

29. Widdison WC, Wilhelm SD, Cavanagh EE, Whiteman KR, Leece BA, Kovtun Y, et al. Semisynthetic maytansine analogues for the targeted treatment of Cancer. J Med Chem. 2006;49(14):4392–408. https://doi.org/10.1021/jm060319f.

30. Lee MD, Manning JK, Williams DR, Kuck NA, Testa RT, Borders DB. Calicheamicins, a novel family of antitumor antibiotics. 3. Isolation, purification and characterization of calicheamicins beta 1Br, gamma 1Br, alpha 2I, alpha 3I, beta 1I, gamma 1I and delta 1I. J Antibiot (Tokyo). 1989;42(7):1070–87. Available from: http://www.ncbi.nlm.nih.gov/pubmed/2753814.

31. Lee MD, Dunne TS, Chang CC, Siegel MM, Morton GO, Ellestad GA, et al. Calicheamicins, a novel family of antitumor antibiotics. 4. Structure elucidation of calicheamicins .beta.1Br, .gamma.1Br, .alpha.2I, .alpha.3I, .beta.1I, .gamma.1I, and .delta.1I. J Am Chem Soc. 1992;114 (3):985–97. https://doi.org/10.1021/ja00029a030

32. Hsiang YH, Hertzberg R, Hecht S, Liu LF. Camptothecin induces protein-linked DNA breaks via mammalian DNA topoisomerase I. J Biol Chem. 1985;260(27):14873–8. Available from: http://www.ncbi.nlm.nih.gov/pubmed/2997227.

33. Erickson HK, Park PU, Widdison WC, Kovtun YV, Garrett LM, Hoffman K, et al. Antibody-maytansinoid conjugates are activated in targeted cancer cells by lysosomal degradation and linker-dependent intracellular processing. Cancer Res. 2006;66(8):4426–33. https://doi.org/10.1158/0008-5472.CAN-05-4489.

34. Lewis Phillips GD, Li G, Dugger DL, Crocker LM, Parsons KL, Mai E, et al. Targeting HER2-positive breast cancer with Trastuzumab-DM1, an antibody-cytotoxic drug conjugate. Cancer Res. 2008;68(22):9280–90. https://doi.org/10.1158/0008-5472.CAN-08-1776.

35. Jain N, Smith SW, Ghone S, Tomczuk B. Current ADC linker chemistry. Pharm Res. 2015;3526–40. https://doi.org/10.1007/s11095-015-1657-7

36. Sievers EL, Larson RA, Stadtmauer EA, Estey E, Löwenberg B, Dombret H, et al. Efficacy and safety of gemtuzumab ozogamicin in patients with CD33-positive acute myeloid leukemia in first relapse. J Clin Oncol. 2001;19(13):3244–54. https://doi.org/10.1200/JCO.2001.19.13.3244.

37. Lamb YN. Inotuzumab Ozogamicin: first global approval. Drugs. 2017;77(14):1603–10. Available from: https://www.ncbi.nlm.nih.gov/pubmed/28819740.

38. Cairns RA, Harris IS, Mak TW. Regulation of cancer cell metabolism. Nat Rev Cancer. 2011;11 (2):85–95. https://doi.org/10.1038/nrc2981.

39. Kellogg BA, Garrett L, Kovtun Y, Lai KC, Leece B, Miller M, et al. Disulfide-linked antibody–maytansinoid conjugates: optimization of in vivo activity by varying the steric

hindrance at carbon atoms adjacent to the disulfide linkage. Bioconjug Chem. 2011;22(4):717–27. https://doi.org/10.1021/bc100480a.

40. Sanderson RJ, Hering MA, James SF, Sun MMC, Doronina SO, Siadak AW, et al. In vivo drug-linker stability of an anti-CD30 dipeptide-linked auristatin immunoconjugate. Clin Cancer Res. 2005;11(2 Pt 1):843–52. Available from: http://www.ncbi.nlm.nih.gov/pubmed/15701875.

41. Caculitan NG, dela Cruz Chuh J, Ma Y, Zhang D, Kozak KR, Liu Y, et al. Cathepsin B is dispensable for cellular processing of cathepsin B-cleavable antibody-drug conjugates. Cancer Res. 2017;9:canres.2391.2017. https://doi.org/10.1158/0008-5472.CAN-17-2391.

42. Dubowchik GM, Firestone RA, Padilla L, Willner D, Hofstead SJ, Mosure K, et al. Cathepsin B-labile dipeptide linkers for lysosomal release of doxorubicin from internalizing immunoconjugates: model studies of enzymatic drug release and antigen-specific in vitro anti-cancer activity. Bioconjug Chem. 2002;13(4):855–69. https://doi.org/10.1021/bc025536j.

43. Beck A, Goetsch L, Dumontet C, Corvaïa N. Strategies and challenges for the next generation of antibody–drug conjugates. Nat Rev Drug Discov. 2017;10(5):345–52. Available from: http://www.ncbi.nlm.nih.gov/pubmed/20414207.

44. Ogitani Y, Aida T, Hagihara K, Yamaguchi J, Ishii C, Harada N, et al. DS-8201a, a novel HER2-targeting ADC with a novel DNA topoisomerase I inhibitor, demonstrates a promising antitumor efficacy with differentiation from T-DM1. Clin Cancer Res. 2016;22(20):5097–108.

45. Doronina SO, Toki BE, Torgov MY, Mendelsohn BA, Cerveny CG, Chace DF, et al. Development of potent monoclonal antibody auristatin conjugates for cancer therapy. Nat Biotechnol. 2003;21(7):778–84. Available from: http://www.nature.com/articles/nbt832

46. Senter PD, Sievers EL. The discovery and development of brentuximab vedotin for use in relapsed Hodgkin lymphoma and systemic anaplastic large cell lymphoma. Nat Biotechnol. 2012;30(7):631–7. Available from: http://www.nature.com/articles/nbt.2289

47. Challita-Eid PM, Satpayev D, Yang P, An Z, Morrison K, Shostak Y, et al. Enfortumab vedotin antibody-drug conjugate targeting nectin-4 is a highly potent therapeutic agent in multiple preclinical cancer models. Cancer Res. 2016;76(10):3003–13. https://doi.org/10.1158/0008-5472.CAN-15-1313.

48. Jeffrey SC, Nguyen MT, Moser RF, Meyer DL, Miyamoto JB, Senter PD. Minor groove binder antibody conjugates employing a water soluble β-glucuronide linker. Bioorganic Med Chem Lett. 2007;17(8):2278–80.

49. Lyon RP, Bovee TD, Doronina SO, Burke PJ, Hunter JH, Neff-LaFord HD, et al. Reducing hydrophobicity of homogeneous antibody-drug conjugates improves pharmacokinetics and therapeutic index. Nat Biotechnol. 2015;33(7):733–6. Available from: http://www.ncbi.nlm.nih.gov/pubmed/26076429.

50. Zhao RY, Wilhelm SD, Audette C, Jones G, Leece BA, Lazar AC, et al. Synthesis and evaluation of hydrophilic linkers for antibody–maytansinoid conjugates. J Med Chem. 2011;54(10):3606–23. https://doi.org/10.1021/jm2002958.

51. Yurkovetskiy AV, Yin M, Bodyak N, Stevenson CA, Thomas JD, Hammond CE, et al. A polymer-based antibody-vinca drug conjugate platform: characterization and preclinical efficacy. Cancer Res. 2015;75(16):3365–72. https://doi.org/10.1158/0008-5472.CAN-15-0129.

52. Strop P, Delaria K, Foletti D, Witt JM, Hasa-Moreno A, Poulsen K, et al. Site-specific conjugation improves therapeutic index of antibody drug conjugates with high drug loading. Nat Biotechnol. 2015;33(7):694–6. https://doi.org/10.1038/nbt.3274.

53. Tumey LN, Li F, Rago B, Han X, Loganzo F, Musto S, et al. Site selection: a case study in the identification of optimal cysteine engineered antibody drug conjugates. AAPS J. 2017. https://doi.org/10.1208/s12248-017-0083-7

54. Su D, Kozak KR, Sadowsky J, Yu S-F, Fourie-O'Donohue A, Nelson C, et al. Modulating antibody–drug conjugate payload metabolism by conjugation site and linker modification. Bioconjug Chem. 2018;29(4):1155–67. https://doi.org/10.1021/acs.bioconjchem.7b00785.

55. Shen B, Xu K, Liu L, Raab H, Bhakta S, Kenrick M, et al. Conjugation site modulates the in vivo stability and therapeutic activity of antibody-drug conjugates. Nat Biotechnol. 2012;30(2):184–9. https://doi.org/10.1038/nbt.2108.

56. Wang L, Amphlett G, Bla WA, Lambert JM, Zhang WEI. Structural characterization of the maytansinoid – monoclonal antibody immunoconjugate, huN901 – DM1, by mass spectrometry. Protein Sci. 2005;14:2436–46.

57. Agarwal P, Bertozzi CR. Site-specific antibody-drug conjugates: the nexus of bioorthogonal chemistry, protein engineering, and drug development. Bioconjug Chem. 2015;26(2):176–92.

58. Younes A, Yasothan U, Kirkpatrick P. Brentuximab vedotin. Nat Rev Drug Discov. 2012;11 (1):19–20. Available from: http://www.nature.com/articles/nrd3629

59. Tamura K, Tsurutani J, Takahashi S, Iwata H, Krop IE, Redfern C, et al. Trastuzumab deruxtecan (DS-8201a) in patients with advanced HER2-positive breast cancer previously treated with trastuzumab emtansine: a dose-expansion, phase 1 study. Lancet Oncol. 2019;20(6):816–26. https://doi.org/10.1016/S1470-2045(19)30097-X.

60. Panowski S, Bhakta S, Raab H, Polakis P, Junutula JR. Site-specific antibody drug conjugates for cancer therapy. MAbs. 2014;6(1):34–45.

61. Strop P, Delaria K, Foletti D, Witt JM, Hasa-Moreno A, Poulsen K, et al. Site-specific conjugation improves therapeutic index of antibody drug conjugates with high drug loading. Nat Biotechnol. 2015;33(7):694–6. Available from: https://www.ncbi.nlm.nih.gov/pubmed/26154005.

62. Perez HL, Cardarelli PM, Deshpande S, Gangwar S, Schroeder GM, Vite GD, et al. Antibody–drug conjugates: current status and future directions. Drug Discov Today. 2014;19(7):869–81. Available from: http://linkinghub.elsevier.com/retrieve/pii/S135964461300398X

63. Zhong X, He T, Prashad AS, Wang W, Cohen J, Ferguson D, et al. Mechanistic understanding of the cysteine capping modifications of antibodies enables selective chemical engineering in live mammalian cells. J Biotechnol. 2017;248:48–58. Available from: http://linkinghub.elsevier.com/retrieve/pii/S0168165617301049

64. Junutula JR, Raab H, Clark S, Bhakta S, Leipold DD, Weir S, et al. Site-specific conjugation of a cytotoxic drug to an antibody improves the therapeutic index. Nat Biotechnol. 2008;26(8):925–32.

65. Junutula JR, Bhakta S, Raab H, Ervin KE, Eigenbrot C, Vandlen R, et al. Rapid identification of reactive cysteine residues for site-specific labeling of antibody-Fabs. J Immunol Methods. 2008;332:41–52.

66. Ravasco JMJM, Faustino H, Trindade A, Gois PMP. Bioconjugation with maleimides: a useful tool for chemical biology. Chem A Eur J. 2019;25(1):43–59.

67. Ohri R, Bhakta S, Fourie-O'Donohue A, dela Cruz-Chuh J, Tsai SP, Cook R, et al. High-throughput cysteine scanning to identify stable antibody conjugation sites for maleimide- and disulfide-based linkers. Bioconjug Chem. 2018;acs.bioconjchem.7b00791. https://doi.org/10.1021/acs.bioconjchem.7b00791

68. Lyon RP, Setter JR, Bovee TD, Doronina SO, Hunter JH, Anderson ME, et al. Self-hydrolyzing maleimides improve the stability and pharmacological properties of antibody-drug conjugates. Nat Biotechnol. 2014;32(10):1059–62. Available from: http://www.ncbi.nlm.nih.gov/pubmed/25194818.

69. Badescu G, Bryant P, Bird M, Henseleit K, Swierkosz J, Parekh V, et al. Bridging disulfides for stable and defined antibody drug conjugates. Bioconjug Chem. 2014;25(6):1124–36. https://doi.org/10.1021/bc500148x.

70. Tumey LN, Charati M, He T, Sousa E, Ma D, Han X, et al. Mild method for succinimide hydrolysis on ADCs: impact on ADC potency, stability, exposure, and efficacy. Bioconjug Chem. 2014;25(10):1871–80. https://doi.org/10.1021/bc500357n.

71. Strop P, Liu S-H, Dorywalska M, Delaria K, Dushin RG, Tran T-T, et al. Location matters: site of conjugation modulates stability and pharmacokinetics of antibody drug conjugates. Chem Biol. 2013;20(2):161–7.   Available   from: http://linkinghub.elsevier.com/retrieve/pii/S1074552113000331

72. Siegmund V, Schmelz S, Dickgiesser S, Beck J, Ebenig A, Fittler H, et al. Locked by design: a conformationally constrained transglutaminase tag enables efficient site-specific conjugation. Angew Chem Int Ed. 2015;54(45):13420–4.

73. Jeger S, Zimmermann K, Blanc A, Grünberg J, Honer M, Hunziker P, et al. Site-specific and stoichiometric modification of antibodies by bacterial transglutaminase. Angew Chem Int Ed. 2010;49(51):9995–7. https://doi.org/10.1002/anie.201004243.

74. King GT, Eaton KD, Beagle BR, Zopf CJ, Wong GY, Krupka HI, et al. A phase 1, dose-escalation study of PF-06664178, an anti-Trop-2/Aur0101 antibody-drug conjugate in patients with advanced or metastatic solid tumors. Invest New Drugs. 2018;36(5):836–47. https://doi.org/10.1007/s10637-018-0560-6.

75. Zuberbühler K, Casi G, Bernardes GJL, Neri D. Fucose-specific conjugation of hydrazide derivatives to a vascular-targeting monoclonal antibody in IgG format. Chem Commun. 2012;48(56):7100. Available from: http://xlink.rsc.org/?DOI=c2cc32412a

76. van Geel R, Wijdeven MA, Heesbeen R, Verkade JMM, Wasiel AA, van Berkel SS, et al. Chemoenzymatic conjugation of toxic payloads to the globally conserved N-glycan of native mAbs provides homogeneous and highly efficacious antibody–drug conjugates. Bioconjug Chem. 2015;26(11):2233–42. https://doi.org/10.1021/acs.bioconjchem.5b00224.

77. Okeley NM, Toki BE, Zhang X, Jeffrey SC, Burke PJ, Alley SC, et al. Metabolic engineering of monoclonal antibody carbohydrates for antibody–drug conjugation. Bioconjug Chem. 2013;24 (10):1650–5. https://doi.org/10.1021/bc4002695.

78. Gordon CG, MacKey JL, Jewett JC, Sletten EM, Houk KN, Bertozzi CR. Reactivity of biarylazacyclooctynones in copper-free click chemistry. J Am Chem Soc. 2012;134(22):9199–208.

79. Zimmerman ES, Heibeck TH, Gill A, Li X, Murray CJ, Madlansacay MR, et al. Production of site-specific antibody–drug conjugates using optimized non-natural amino acids in a cell-free expression system. Bioconjug Chem. 2014;25(2):351–61. https://doi.org/10.1021/bc400490z.

80. Tian F, Lu Y, Manibusan A, Sellers A, Tran H, Sun Y, et al. A general approach to site-specific antibody drug conjugates. Proc Natl Acad Sci. 2014;111(5):1766–71. https://doi.org/10.1073/pnas.1321237111.

81. VanBrunt MP, Shanebeck K, Caldwell Z, Johnson J, Thompson P, Martin T, et al. Genetically encoded Azide containing amino acid in mammalian cells enables site-specific antibody–drug conjugates using click cycloaddition chemistry. Bioconjug Chem. 2015;26(11):2249–60. https://doi.org/10.1021/acs.bioconjchem.5b00359.

82. Beerli RR, Hell T, Merkel AS, Grawunder U. Sortase enzyme-mediated generation of site-specifically conjugated antibody drug conjugates with high in vitro and in vivo potency. PLoS One. 2015;10(7):e0131177. https://doi.org/10.1371/journal.pone.0131177

83. Huang BCB, Kim YC, Bañas S, Barfield RM, Drake PM, Rupniewski I, et al. Antibody-drug conjugate library prepared by scanning insertion of the aldehyde tag into IgG1 constant regions. MAbs. 2018;00(00):1–8. Available from: http://www.ncbi.nlm.nih.gov/pubmed/30252630

84. Grünewald J, Klock HE, Cellitti SE, Bursulaya B, McMullan D, Jones DH, et al. Efficient preparation of site-specific antibody-drug conjugates using phosphopantetheinyl transferases. Bioconjug Chem. 2015;26(12):2554–62.

85. Yamada K, Shikida N, Shimbo K, Ito Y, Khedri Z, Matsuda Y, et al. AJICAP: affinity peptide mediated regiodivergent functionalization of native antibodies. Angew Chem Int Ed. 2019;58 (17):5592–7. https://doi.org/10.1002/anie.201814215.
86. Dickgiesser S, Rieker M, Mueller-pompalla D, Schro C, Tonillo J, Warszawski S, et al. Site-specific conjugation of native antibodies using engineered microbial transglutaminases. Bioconjug Chem. 2020;31(4):1070-1076. https://doi.org/10.1021/acs.bioconjchem.0c00061.

Alternative Binding Scaffolds: Multipurpose Binders for Applications in Basic Research and Therapy

9

Doreen Koenning and Jonas V. Schaefer

Contents

D. Koenning (✉)
Department of Antibody-Drug Conjugates and Targeted NBE Therapeutics, Merck KGaA,
Darmstadt, Germany
e-mail: doreen.koenning@merckgroup.com

J. V. Schaefer (✉)
Chemical Biology and Therapeutics (CBT), Novartis Institutes for BioMedical Research, Basel,
Switzerland
e-mail: jonas.schaefer1@novartis.com

© Springer Nature Switzerland AG 2021
F. Rüker, G. Wozniak-Knopp (eds.), *Introduction to Antibody Engineering*,
Learning Materials in Biosciences,
https://doi.org/10.1007/978-3-030-54630-4_9

Keywords

Adnectin · Affibody · Affitin · Alternative binding scaffold · Anticalin · Avimer · Designed Ankyrin Repeat Proteins (DARPins) · Fibronectin · Fynomer · Kunitz domain · Knottin

What You Will Learn in This Chapter
Next to monoclonal antibodies (mAbs), alternative binding scaffolds have proven to be powerful tool reagents and therapeutic entities over the past decades. In contrast to their hetero-tetrameric, macromolecular counterparts, these affinity reagents are based on non-antibody structures and provide several advantages, like a smaller size, improved biophysical properties as well as the possibility to engage difficult-to-address target proteins through surface-exposed binding sites. With three alternative binding scaffolds being approved by the US Food and Drug Administration (FDA) to date and many more currently investigated in clinical trials, scaffold proteins are on a rise to become promising next-generation therapeutics. Additionally, these binders have successfully been employed (pre)clinically in a variety of applications, e.g., as imaging reagents for the detection and monitoring of various cancers, given that their small format enables an efficient distribution in cancerous tissue while at the same time ensuring a rapid elimination from the system by the kidney. In addition, alternative scaffolds are often more robust than mAbs and have also proven to be versatile and efficient affinity reagents for applications in basic research, such as affinity purifications and structural biology. Advances in high-throughput screening using different display technologies and protein engineering have facilitated the generation of novel binding scaffolds with high affinities and specificities and have paved the way for the expansion of the alternative binding scaffold toolbox.

9.1   What Are Alternative Binding Scaffolds, What Are Their Features, and What Is Their Potential?

Ever since the discovery of monoclonal antibodies (mAbs) by Köhler and Milstein in 1975 [1] and the approval of the first therapeutic mAb by the US Food and Drug Administration (FDA) in 1986, these binding proteins have rapidly taken the stage with so far over 80 mAbs having received marketing approval and a similar number being in late-stage clinical trials [2]. Besides being very potent therapeutics, antibodies have also proven to be efficient and versatile affinity reagents in a variety of applications ranging from clinical diagnostics [3] over basic research [4] to biotechnological tasks [5]. Immunoglobulins (IgGs) structurally are Y-shaped macromolecules containing multiple domains encoded by two copies each of two polypeptides—termed heavy and light chain—that functionally form two distinct regions in the antibody's structure. The conserved part—termed Fc region for "fragment, crystallizable"—at the bottom of the Y-structure is comprised of the two C-terminal domains of the heavy chains and mediates the interactions with certain cell surface receptors, thereby, e.g., leveraging immune effector functions or contributing to the long half-life of these molecules through FcRn (neonatal Fc receptor)-mediated recycling [6]. In contrast, the antibody's antigen binding is executed by the complementarity determining regions (CDRs) of the variable domains of both its heavy and light chain that together form the so-called Fab fragments (abbreviation for "Fragment antigen-binding") in the Y's arms. Due to this hetero-tetrameric structure that also contains various inter- and intrachain disulfide bonds, mAbs generally are too complex to be produced in microorganisms, but instead require eukaryotic expression systems with suitable folding and quality control mechanisms. These cellular machineries ensure, e.g., proper heavy and light chain pairing, the formation of the correct disulfide bonds as well as post-translational modifications like the specific glycosylation present on the Fc fragment that is required for FcRn interactions. As a consequence of their complex architectures and their sizes of approximately 150 kDa, antibodies are often too big for certain applications and further have problems to effectively penetrate tissues and tumors in medical applications [7, 8]. Also, their expression often suffers from low yields, causing rather high production costs which unfortunately make mAb-based therapeutics very costly and out-of-reach for a large part of the world's population, as such therapies require relatively high doses. Next to these aspects, the applicability of antibodies often is limited by their biophysical properties. While some of these binders do not exhibit a high thermal stability and/or only a limited protease resistance, long-term storage often also requires sophisticated buffer compositions to counterbalance their aggregation propensity [9]. Finally, due to the large number of disulfide bonds, proteins that can be targeted by antibodies are limited to extracellular molecules or the extracellular domains of membrane-associated receptors.

As a consequence of these challenging properties that partly limit the broad applicability of IgGs, alternative scaffold proteins with desirable functional and biophysical properties have gained more and more attention throughout the past two decades as they not only can complement antibody-based agents but in part even compensate for their pitfalls. The advancements in available display technologies, such as ribosomal, phage, and yeast display [10] (see also other chapters in this book) have enabled researchers to leverage a number of naturally occurring and stable non-antibody proteins and utilize them as scaffolds for the introduction of new and artificial affinity functionalities. Through extensive protein engineering efforts, many different proteins originating from either bacterial or plant proteins (e.g., the later discussed affibodies) and non-antibody human proteins (e.g., DARPins or Anticalins), but also non-human antibodies themselves (e.g., nanobodies, discussed in another chapter of this book) could be turned into promising alternative binding scaffolds with interesting features and properties. All of these are single-domain, monovalent proteins with small sizes ranging between 4 and 20 kDa and thus only have a fraction of the size of full-length IgGs. This allows scalable low-cost, high-yield

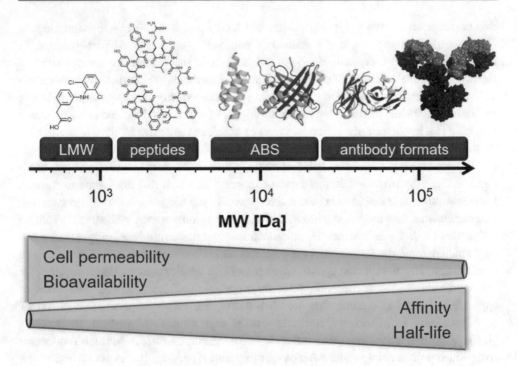

Fig. 9.1 Schematic depiction of molecular weights of low-molecular-weight (LMW) compounds, peptides, alternative binding scaffolds (ABS) as well as antibodies. Different molecular weights are accompanied by different biophysical properties

manufacturing in established bacterial expression systems or even through chemical synthesis—something impossible for antibodies. Also compared to other artificial ligands, such as low-molecular-weight (LMW) compounds or peptidic binders, these affinity reagents populate a size-range niche not yet covered by those or the different antibody formats available, also offering interesting characteristics regarding certain key criteria like bioavailability or affinity (see Fig. 9.1).

Moreover, alternative binding scaffolds generally feature rather beneficial biophysical properties, having high folding efficiencies, good solubilities, low aggregation propensities, and further increased thermal and proteolytic stabilities—all of these resulting in prolonged storage shelf lives. As an example, some scaffold proteins can be subjected to temperatures close to 100 °C without denaturation or a loss-of-function [11, 12], while others have even shown to be resistant toward very low pH values and proteases [13, 14]. The small size of these affinity reagents also enables the extension of the available target space as some of them can interact with cleft-like regions (e.g., areas close to the membrane of membrane-associated targets) or even catalytic sites of enzymes. The latter is enabled by the fact that the majority of these binders employ exposed loops for interacting with their targets (rather than the predominantly planar binding sites of IgGs), but at the same time do not have any disulfide bonds and thus even allow intracellular applications.

Opposite to their antibody-based counterparts, these alternative scaffolds further permit straightforward protein engineering approaches. These, for example, allow to perform covalent conjugations to fusion proteins (like albumin), site-directed modifications (e.g., the introduction of non-natural amino acids for enabling conjugations to small molecules likes toxins, radionucleotides, or near-infrared dyes) and also the generation of multi-specific formats through the multimerization of individual binding units [15]. Another important aspect with regard to the development of novel scaffolds is the opportunity to generate novel intellectual property (IP). Since antibodies have been around for several decades, their IP landscape has become more and more crowded, making it challenging to introduce new patentable antibody-based modalities [16]. This is easier to achieve if the novel scaffold is based on an unrelated protein being sufficiently different from existing affinity reagents.

For clinical applications, alternative binding scaffolds have already proven to possess the most important key characteristics (examples are given in Sect. 9.3). Since most of these reagents are derived from human proteins, they generally cause a low risk for immunogenicity. However, also those presented binders of non-human origin often exhibit a low immunogenic potential, most likely due to inefficient peptide-MHC presentation to the immune system [17]. Interestingly, as a side note, also binders that might potentially be immunogenic (e.g., due to a bacterial origin) could still be of value in short-term medical applications like imaging, among others.

The small size of alternative binding scaffolds does not only deliver attractive pharma-cokinetic properties and rapid tumor/tissue penetration in vivo [15] but also causes rapid renal clearance as their respective molecular weight is below the glomerular filtration limit (being approximately 70 kDa). As an example, the later discussed Adnectins are cleared from the bloodstream by the kidney's mechanical filtration, thus exhibiting an in vivo half-life of around 1–2 h in primates and approximately 20–30 min in rodents [18]. In compari-son, the half-life of a monoclonal antibody adds up to several days in primates and humans [19]. While this rapid clearance provokes some challenges for alternative binding scaffolds in long-term systemic therapeutic applications for chronic diseases, it also minimizes potentially long-lasting side effects in acute and local treatments as well as deleterious bystander cell interactions in liver and gut (e.g., when employing radiolabeled tracers for diagnostics).

Still, for some therapeutic applications, it is essential to prevent alternative binding scaffolds from being rapidly cleared from circulation. Therefore, different approaches have successfully been developed to increase their plasma half-lives from minutes to hours or even days [20, 21]. One popular tactic is to modify them by chemical conjugations to polyethylene glycol (PEG) chains. This process, termed PEGylation, causes an increase of both the molecular mass and the hydrodynamic radius of the resulting conjugates, conse-quently decreasing the rate of their glomerular filtration by the kidney [22]. As another advantage, PEGylation often further increases the water solubility, which is especially beneficial for hydrophobic molecules. Other half-life prolongation approaches are for example conjugations to polypeptide sequences containing a proline–alanine–serine

(PAS) amino acid motif [23] or genetic fusions to the IgG's Fc domain [24], which—as mentioned before—is effectively recycled through FcRn interactions. Finally, a promising approach is taking advantage of the long half-life of Human Serum Albumin (HSA), the most abundant protein in human blood plasma which stays in the system for an average of 3 weeks due to FcRn recycling [25]. Thus, fusing binders either to HSA directly or creating a bispecific molecule of the original affinity reagent and an albumin-binding domain has proven to be an effective way of dramatically prolonging circulatory times of associated binders [26].

As the discussed alternative scaffolds do no possess any Fc regions, they do not exhibit any immune effector functionalities. While the lack of immune system activation thus cannot support their therapeutic power, these affinity reagents as a consequence also favorably prevent adverse effects associated with complement-dependent cytotoxicity (CDC) or antibody-dependent cellular cytotoxicity (ADCC). Finally, the discussed bio-physical properties partly allow oral administration of scaffold-based therapeutics as well as improved formulations at a higher dose per injection volume compared to classical mAbs.

All of the aforementioned advantages and limitations of alternative binding scaffolds are summarized in Table 9.1 (next to those of mAbs). In addition, the individual scaffolds are discussed in detail in the next paragraph. Generally, these synthetic binding proteins can be divided into two different categories based on the structural location of their ligand-binding amino acids. While for the majority of the presented scaffolds these residues are inserted as a linear and continuous sequence in exposed loops (like it is the case for Adnectins, Anticalins, Avimers, Fynomers, Kunitz domains, and Knottins), there are also architectures which mediate target binding through modifications of their secondary structural elements, such as ß-sheets, partly even in a discontinuous fashion [27] (as it is the case for affibodies or DARPins). Irrespective of these differences, the artificial molecular recognition interfaces of all of the discussed binders usually are comprised of regions in which multiple randomized mutations have been introduced in a structure-guided design (numbers of randomized residues can be found in the table shown in Fig. 9.3). The resulting highly diverse libraries can be screened by different display technologies to identify those members that interact with the target of interest—depending on the experimental design—with high affinities and in the intended mode of action. The choice of the most suitable display platform depends on various aspects, like the library size to be screened. Some in vitro systems (like ribosome display) are not limited by the transformation efficiencies of the employed cellular systems and thus allow the screening of very big libraries with up to 10^{14} members. This does not restrict the diversity of the initial libraries but also allows the natural error rate of the used polymerases to be exploited or error-prone PCR steps to be integrated into such selection procedures, reducing the need for subsequent labor-intensive "affinity maturation". As covered in another chapter in this book, the various technologies also allow to accommodate different screening conditions (e.g., harsh screening buffers or the presence of additional ligands during the screen) and also must comply with IP considerations [28].

Table 9.1 High-level comparison of advantages and limitations of both alternative scaffold proteins and monoclonal antibodies

	Benefits/advantages	Limitations/challenges	"Mixed" features
Alternative binding scaffolds	• Single-domain proteins with small size, allowing low-cost, high-yield manufacturing in bacterial expression systems or through chemical synthesis • Good biophysical properties: high folding efficiencies and solubilities, low aggregation propensities and increased thermal and proteolytic stability, resulting in prolonged storage shelf-life • Ease of protein engineering (e.g., conjugation to fusion proteins; site-directed modifications, multimerizations) • Improved tumor/tissue penetration in vivo • Extended target space: interactions with cleft-like regions (e.g., close to membrane) and catalytic sites of enzymes, potentially intracellular applications (lack of disulfide bonds) • Patentability (when sufficiently different from existing affinity reagents)	• Limited half-life in vivo, potentially requiring half-life extension for therapeutic applications • Potential immunogenicity	• Rapid in vivo renal clearance: potentially problematic at systemic therapeutic applications for chronic or recurring diseases, but beneficial to prevent long-lasting side effects at treatments of acute conditions and at local applications of, e.g., radiolabeled tracers for diagnostics • No immune effector functions: no activation of immune system to support therapeutic power (i.e., engagement of complement system or opsonization of targeted cells), but also reduced probability of adverse CDC- or ADCC-associated effects
Monoclonal antibodies	• Long in vivo half-life due to large size and Fc-mediated recycling • Potential to leverage immune effector functions for therapeutic applications • Reduced risk of immunogenicity	• Multi-domain proteins, requiring sophisticated eukaryotic expression systems and causing high production cost • Potentially lower biophysical stability compared to abovementioned	• Bivalent IP situation: Freedom to operate (no IP-restrictions on scaffold), but also little chance for patentability

(continued)

Table 9.1 (continued)

	Benefits/advantages	Limitations/challenges	"Mixed" features
		scaffolds • Large size can impair tissue/tumor penetration • Target space limited to extracellular proteins and extracellular domains of membrane-associated targets • Limited options for protein conjugations and modifications • Predominantly planar binding site, causing potential cross-reactivities due to limited specificities	

In a nutshell, affinity reagents based on alternative binding scaffolds are in many ways equally suited or even superior to mAbs and thus are more and more employed for a large variety of applications, including therapeutic, but also non-therapeutic approaches as discussed in Sects. 9.3 and 9.4, respectively. However, first, the most advanced alternative binding scaffolds will be discussed in an alphabetical order in the following paragraphs (for a comparative timeline of their development and FDA approval see Fig. 9.2).

9.2 Details on the Most Advanced Alternative Binding Scaffolds

9.2.1 Tenth Type III Domain of Fibronectin (Monobodies, Adnectins)

One of the most established and well-characterized alternative scaffold proteins is the tenth type III domain of fibronectin (Fn3). Fn3 domains belong to the human immunoglobulin superfamily, but in contrast to antibodies do not possess any disulfide bonds and have a monomeric architecture with a size of approximately 10 kDa. More precisely, Fn3 domains comprise seven ß-strands, thereby resembling conventional antibody variable domains, also having three solvent-exposed, non-contiguous loops that can be employed for diversification (Fig. 9.3). This rather simple structure paired with the absence of disulfide bonds enables a straightforward, cost-efficient expression of these domains in bacterial systems. The immunoglobulin fold of Fn3 domains also features the benefit of a well-packed hydrophobic core that provides a stable protein framework, resulting in a high thermostability with a melting temperature of around 88 °C. An important aspect is the abundance of

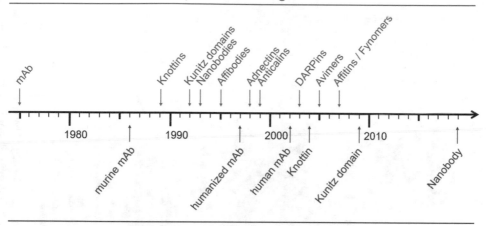

Fig. 9.2 Timeline highlighting the discovery years of the most prominent alternative binding scaffolds discussed within the scope of this chapter (shown in blue) as well as the years of the FDA's approval of the respective first member of these scaffold formats and mAbs (marked below the timeline in red)

fibronectin proteins in human blood, emphasizing their potential as a suitable scaffold for therapeutic applications with limited risk of potential immunogenicity. The tenth domain of fibronectin was first leveraged for protein engineering in 1998 and through random mutagenesis of two out of three surface-exposed loops, Fn3 domains targeting ubiquitin could be isolated from phage display libraries [29]. This study also elegantly demonstrated that the identified binders tolerated up to 12 mutations in their 94-amino acid sequence, highlighting the potential of Fn3 domains as flexible alternative binding scaffolds [30]. Up to now, several engineered Fn3 domains, commonly referred to as either Monobodies or Adnectins, have progressed into the clinic for a variety of different applications which will be reviewed in more detail in paragraph 3. Monobodies have further been fused to the green fluorescent protein (GFP) to be functionally expressed intracellularly as live-cell intrabody probes for, e.g., labeling specific states of Src activation [31] and both excitatory (anti-PSD-95) and inhibitory (anti-Gephyrin) synapses in neurons [32].

9.2.2   Affibodies

The affibody protein framework has been derived from the Z domain of the staphylococcal protein A. Naturally, this protein is able to bind the Fc-region of mammalian IgGs though its Z domain and thus represents both a virulence factor and a survival mechanism, facilitating the infection of the host organism. Affibodies consist of 58 amino acids

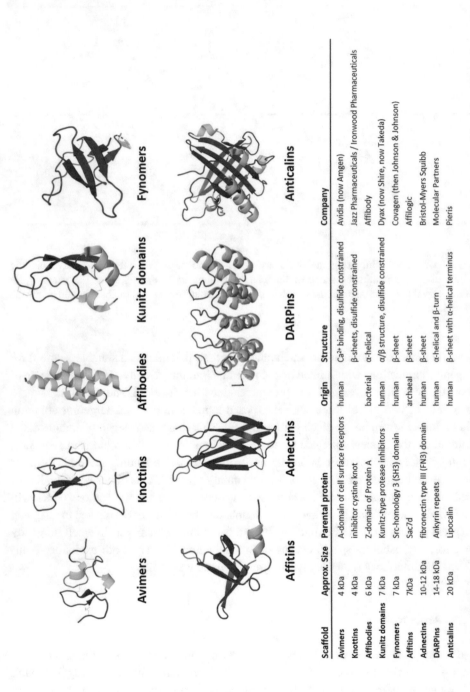

Scaffold	Approx. Size	Parental protein	Origin	Structure	Company
Avimers	4 kDa	A-domain of cell surface receptors	human	Ca²⁺ binding, disulfide constrained	Avidia (now Amgen)
Knottins	4 kDa	inhibitor cystine knot		β-sheets, disulfide constrained	Jazz Pharmaceuticals / Ironwood Pharmaceuticals
Affibodies	6 kDa	Z-domain of Protein A	bacterial	α-helical	Affibody
Kunitz domains	7 kDa	Kunitz-type protease inhibitors	human	α/β structure, disulfide constrained	Dyax (now Shire, now Takeda)
Fynomers	7 kDa	Src-homology 3 (SH3) domain	human	β-sheet	Covagen (then Johnson & Johnson)
Affitins	7kDa	Sac7d	archaeal	β-sheet	Affilogic
Adnectins	10-12 kDa	fibronectin type III (FN3) domain	human	β-sheet	Bristol-Myers Squibb
DARPins	14-18 kDa	Ankyrin repeats	human	α-helical and β-turn	Molecular Partners
Anticalins	20 kDa	Lipocalin	human	β-sheet with α-helical terminus	Pieris

Fig. 9.3 Structural depiction of the alternative binding scaffolds discussed in this chapter (not drawn to scale, arranged based on increasing molecular sizes). Employed coloring is based on secondary structure (helices in blue, β-sheets in red, disulfide bonds in yellow). Shown are representative structures

(corresponding to a size of about 6 kDa) and possess a cysteine-free structure comprising a bundle of three helices (see Fig. 9.3). Their inherent ability to bind proteins through a defined set of surface-exposed residues renders them ideal as an alternative scaffold. Affibodies were employed for engineering approaches as early as in 1997 when Nord and coworkers successfully randomized 13 amino acids in two helices of the Z domain [33]. Using phage display, they were able to isolate affibodies with micromolar affinities toward human insulin, a variant of the human apolipoprotein A1 as well as a DNA polymerase. In order to improve the affinity of the selected binders, another round of re-randomization was performed. In this affinity maturation, only a selected number of residues in either one of the two targeted helices was subjected to genetic randomization, yielding clones with affinities in the double-digit nanomolar range [27, 34]. This is not the highest possible affinity of this scaffold, as even higher ones in the picomolar range have been reported to date with an affibody variant targeting the cancer-related protein HER2 exhibiting an affinity of 22 pM [35].

Since affibodies also exhibit a high stability, they are popular binders to be employed as affinity ligands for a number of different chromatography resins, thereby facilitating the purification of various proteins and also enabling an efficient regeneration of the respective columns by alkali treatment [36]. In addition, this scaffold's high stability was also leveraged by fusing a representative to a DNA polymerase. Upon binding the polymerase with high affinity, a particular affibody blocks the interaction between the enzyme and its DNA template at ambient temperature but releases the polymerase at elevated temperatures. As a consequence, the resulting DNA polymerase/affibody fusion protein shows increased fidelity and an improved performance during polymerase chain reactions (PCRs) in a "hot start" approach since unspecific binding of the enzyme to DNA at lower temperatures can be avoided [37].

Affibody scaffolds have also been developed as tracers for imaging applications, facilitating the detection, monitoring, and characterization of tumorous tissue [37]. In that manner, preclinical imaging studies in mice have been conducted using the aforementioned high-affinity anti-HER2 affibody variant and confirmed the potential of affibodies for in vivo tumor imaging [35].

Fig. 9.3 (continued) for Avimers (PDB: 1AJJ), Knottins (PDB: 1HYK), Affibodies (PDB: 2KZI), Kunitz domains (PDB: 1KTH), Fynomers (PDB: 4AFS), Affitins (PDB: 1SSO), Adnectins (PDB: 1TTG), DARPins (PDB: 1SVX), and Anticalins (PDB: 4GH7), respectively. Below the structures, the tabular overview summarizes the respective properties of the individual scaffolds, including the name of the company who first/primarily commercializes these binders

9.2.3   Affitins/Nanofitins

Scaffold proteins derived from hyperthermophilic archaea have likewise been developed as alternative binding proteins. These binders are referred to as Nanofitins or Affitins and are based on the DNA-binding proteins Sac7d and Sso7d. Nanofitins are attractive scaffolds due to their high thermo-, chemical, and pH stability as well as the absence of any glycosylation sites and disulfide bonds [11, 38, 39]. Whereas Sac7d is derived from *Sulfolobus acidocaldarius*, Sso7d originally was found in the archaeon *Sulfolobus solfataricus*. Structurally, these proteins form a β-barrel with a C-terminal α-helix and their ligand-binding site is located on the top of a rigid β-sheet. Both Sac7d and Sso7d exhibit a molecular weight of around 7 kDa and have been selected for binding to various disease-relevant targets, such as the epidermal growth factor receptor (EGFR), K-Ras and Notch1 using yeast surface display [40, 41]. The EGFR-targeting Sac7d scaffold has subsequently been conjugated to an imaging tracer and successfully employed for in vivo positron emission tomography (PET) imaging in tumor-bearing mice [42]. Other non-therapeutic applications for Nanofitins/Affitins include the design of affinity resins for the purification of proteins such as human IgG or their utilization for protein chip arrays [43, 44].

A recent addition to this scaffold family is the protein Aho7c which is naturally found in *Acidianus hospitalis*. Aho7c is slightly smaller than Sac7d, having a length of only 60 instead of 66 amino acids [43, 45]. Aho7c variants engaging the epithelial cell adhesion molecule (EpCAM) have been identified upon screening of a randomized Aho7c library with the best-performing variant binding to this target with picomolar affinity.

9.2.4   Avimers

The Avimer scaffold is based on the A-domain found in various cell surface receptors, such as the human low-density-related protein (LRP) or the very-low density-related lipoprotein receptor (VLDLR) [17]. Avimers exhibit a length of around 35 amino acids, corresponding to a molecular weight of approximately 4 kDa, and contain three pairs of disulfide bonds that stabilize its structure as do some complexed calcium ions. Several A-domains engaging different sites on the same biological ligand can be combined and Avimers consisting of up to eight engineered A-domains have been produced in soluble form in bacterial cells [46]. This scaffold is likewise characterized by its extraordinary thermostability and has been shown to be stable at temperatures between 50 and 80 °C for 2 weeks [47]. Moreover, the structure of Avimers is conserved by only 12 residues which need to be protected during protein engineering, offering a high degree of flexibility [46]. One Avimer variant targeting interleukin-6 (termed 'C326') with a combination of three A-domains has been investigated in a phase I clinical trial for the treatment of Crohn's disease; however, its development has not been pursued.

9.2.5   Cysteine-Knot Miniproteins (Knottins)

Naturally occurring cysteine-knot miniproteins exhibit a defined three-dimensional and rigid core structure characterized by an interlocked arrangement of several disulfide bonds (Fig. 9.3). As such, they are commonly referred to as "knottins" and have been isolated from a plethora of different species, including fungi, arthropoda, and vertebrata, likely representing a case of convergent evolution [48, 49]. Naturally, these miniproteins have been described to act as protease inhibitors and ion channel blockers or to comprise antimicrobial activity. Owing to their constrained structure, cysteine-knot miniproteins are highly stable toward proteolytic as well as chemical denaturation and have also been shown to withstand temperatures of 100 °C [48, 50]. Knottins can also be exposed to acids, bases, and serum without exhibiting a loss of their structural integrity. This robustness opens up new avenues for the development of these binders for oral administration, as they can withstand the harsh conditions of the human gut [50]. In addition, their interwoven disulfide bond pattern—being characterized by one disulfide bond threading a macrocycle formed by the remaining disulfides—gives rise to several loop structures that can be modified for molecular recognition. Throughout several rational and combinatorial protein engineering approaches, knottins proofed to be amendable to a myriad of structural and sequence modifications, culminating in the removal of selected binding loops and subsequent replacement by another loop structure from a completely different protein [47, 49]. This "grafting" of a novel, foreign loop structure onto the knottin scaffold altered the binding properties of the resulting chimeric miniproteins. With a length of 30–50 amino acids, cystine-knot miniproteins can be produced recombinantly in the periplasm of bacterial cells or even through chemical peptide synthesis.

Cystine-rich miniproteins have been engineered for a wide array of applications ranging from diagnostics to therapy. Indeed, one of the first FDA-approved alternative binding scaffolds was the knottin peptide Ziconotide which was approved for the treatment of severe chronic pain in 2004 (also see paragraph 3 for more information). Another knottin called Linaclotide was also approved by the FDA in 2012. Moreover, knottins targeting integrin receptors have been employed for tumor imaging as their small size enables an efficient tumor uptake, but at the same time ensures a rapid clearance from non-targeted tissue through the kidney [51].

9.2.6   Designed Ankyrin Repeat Proteins

Designed Ankyrin Repeat Proteins—abbreviated DARPins—are binders of high affinity and specificity with a scaffold derived from natural ankyrin proteins which are intracellular adaptor molecules that mediate diverse protein–protein interactions [47, 52]. As the name implies, these affinity reagents consist of smaller repeat units that are fused to N- and C-terminal capping motifs to shield their hydrophobic core, resulting in high solubilities and expression yields in bacterial hosts [52, 53]. DARPins are typically comprised of three ankyrin repeats and possess a molecular weight in the range of 14–17 kDa [44]. Each of these repeats consists of 33 amino acids and a β-sheet followed by two anti-parallel α-helices (Fig. 9.3). The individual repeats are connected through a loop from one repeat to the β-turn of the subsequent one. For randomization, six of the 33 residues per ankyrin unit can be randomized, with the potential diversity being further increased as the two or three repeats within one single DARPin binder are of different sequences. To even further enhance their binding properties, new libraries have recently been designed that include additional randomized residues in both capping repeats. This broadened binding surface proved to be of advantage in covering a broad range of epitopes in selections as well as, in some cases, helped to obtain binders with picomolar affinity in only a single round of ribosome display [54]. Inspired by the antibody heavy chain's hypervariable CDR3 being the most important loop for mediating antibody–antigen interactions, the flexibility and binding interface of the rather rigid DARPin backbone was further modified in another library set up by the introduction of an elongated 19-amino acid loop into the central repeat [54]. The resulting "LoopDARPins" thus possess a rather convex binding surface with a carefully designed, restricted amino acid composition and retain the favorable thermostability of the original DARPin scaffold.

As many of the other scaffolds, DARPins are highly thermostable and have shown to be resistant toward chemical and proteolytic degradation. One rather unique feature is their concave shape [55] which allows DARPins to recognize three-dimensional rather than linear epitopes. Consequently, their targets do not have to possess a cavity, and their large surface area results in a potentially larger binding interface when compared to globular proteins [47]. This feature, together with the ability to subtract certain epitopes during the selection process, allows the generation of DARPins that specifically discriminate between different conformational [56, 57] and post-translationally modified states of target proteins, and to also enhance their selectivity among different states of targets present at different subcellular localizations. DARPins addressing post-translationally different proteins can, for example, differentiate between either the active or inactive form of the kinase ERK2 (extracellular signal-regulated kinase 2). These selected binders allow the detection of the respective kinase forms in cellular lysates [58], but also to identify the activation status of ERK2 directly inside cells as real-time, real-space sensors [59]. DARPins have been selected successfully using phage display, yeast display, and ribosomal display against a wide variety of intracellular and extracellular protein targets typically with sub-nanomolar affinity. In addition to being very stable, mainly monomeric and resistant to aggregation, DARPins have the advantage of stable expression in the cytoplasm of mammalian cells, due to the absence of any disulfide bonds and glycosylation patterns [53]. This allows them to be fully functional inside mammalian cells and to bind to intracellular targets, thus enabling their use as intrabodies [59–61]. Furthermore, as it is the case for other scaffolds, it is relatively straightforward to create genetic fusions of DARPins with GFP or other proteins for the design of biosensors [58] and use them in a variety of cells and organisms (e.g., mammalian cells, Drosophila larvae, and zebrafish) [62]. DARPins have also proven to possess functionalities important for therapeutical applications. DARPins targeting

EpCAM has been generated and fused to Exotoxin A, a toxin derived from the bacterium *Pseudomonas aeruginosa*. The resulting conjugate exhibited potent cell killing on EpCAM-positive tumor cells in vitro and demonstrated an efficient localization to tumor tissue in in vivo experiments in mice [63]. Other DARPins are under clinical investigation, as highlighted in Sect. 9.3. Moreover, such binders have been conjugated to diverse cargo, among them radiolabels, in order to generate efficient tracers for tumor imaging. A DARPin binding the cancer-related receptor HER2 with high affinity has been subjected to preclinical imaging studies in mice, demonstrating this scaffold's potential as a versatile imaging tool [64].

9.2.7 Fynomers

Fynomers are based on the human Src-homology 3 (SH3) protein domain of Fyn tyrosine kinases and engineered fynomer variants exhibit both high thermostability and a lack of aggregation. In addition, fynomers are also devoid of cysteine residues, making them suitable for recombinant expression in bacterial cells and potential intracellular applications [17, 65]. These scaffolds comprise a size of around 7 kDa and their structure is characterized by a pair of anti-parallel β-sheets which is joined by two loops, creating the binding site (Fig. 9.3). A fusion protein of a fynomer targeting the pro-inflammatory cytokine interleukin-17 A (IL-17A) with the anti-tumor necrosis factor-alpha antibody Adalimumab was evaluated in a phase I/II clinical trial for the treatment of rheumatoid arthritis. However, the study of this fusion protein ("COVA322") was terminated based on the observed safety profile.

9.2.8 Kunitz Domain Inhibitors

Kunitz domains represent one of the earliest examples of alternative binding scaffolds that have been engineered for molecular recognition. Kunitz domains are peptides comprising approximately 60 amino acids and have a molecular mass of around 7 kDa [17]. They are naturally derived from the active motif of reversible serine and trypsin protease inhibitors and exhibit a hydrophobic core as well as a rather irregular secondary structure (see Fig. 9.3). These domains are also referred to as "Kunitz domain inhibitors" and are structurally stabilized by three pairs of disulfide bonds. Their secondary structure gives rise to three surface-exposed loops which have been subjected to modifications without destabilizing the scaffold [66]. Some of the earliest attempts aimed at the engineering of bovine [67, 68] and human [69] pancreatic trypsin inhibitor scaffolds toward an altered enzyme specificity. To do so, the protruding loop structures were diversified by random mutagenesis and the resulting phage-displayed libraries screened for binding toward the enzymes neutrophil elastase and α-chymotrypsin, abrogating the scaffold's ability to bind trypsin.

9.2.9 Lipocalins (Anticalins)

Lipocalins are single polypeptide chains with sizes of 150–190 amino acids and thus are among the largest alternative scaffolds with a molecular weight of around 20 kDa [70]. They are defined by a structurally conserved β-barrel comprising eight anti-parallel β-strands that are aligned around a central axis in a conical manner (Fig. 9.3). Lipocalins are secreted proteins that are naturally responsible for the transport and sequestration of hydrophobic compounds from cells [70]. To fulfill this function, these affinity reagents exhibit an intrinsic ligand-binding site which predestines them for the use as alternative binding scaffolds. In addition, lipocalins exhibit high thermal stability and are devoid of complex glycosylation patterns which permits the recombinant production in bacterial expression systems. With regard to protein engineering, the β-barrel structure gives rise to a ligand-binding pocket that is formed by four surface-exposed loops. A distinctive characteristic of lipocalins is the possibility to generate high-affinity binders toward low-molecular-weight proteins or hapten-like ligands [71]. Generally, engineered lipocalins with altered ligand-binding properties are referred to as "Anticalins". Anticalins are typically generated by randomizing 20 amino acids of a human lipocalin scaffold on average [44]. It has been demonstrated that anticalins with engineered residues located directly in the ligand-binding cavity were able to bind the small molecules fluorescein, digoxigenin, or colchicine with (sub-)nanomolar affinities, respectively [72–74]. Such toxin-targeting anticalin scaffolds could be useful as therapeutic antidotes for the treatment of acute toxification. In a more recent approach, an anticalin targeting petrobactin, a substance secreted by *Bacillus anthracis*, with picomolar affinity was developed [75]. This anticalin has been shown to suppress bacterial growth and may be useful in the treatment of the anthrax disease.

A recent approach describes the incorporation of a non-natural amino acid at a sterically permissive position in the ligand-binding pocket of a human lipocalin scaffold [76]. The resulting anticalin structure was able to reversibly complex pyranose monosaccharides which is usually difficult to achieve since most proteins—including monoclonal antibodies—exhibit a suboptimal affinity for oligosaccharides in aqueous solution. Highly specific, carbohydrate-binding anticalins could be useful in a therapeutic setting as they may enable the recognition of cancer-related cell surface oligosaccharides [44].

9.3 Alternative Binding Scaffolds in the Clinic

Over the past years, several representatives of the aforementioned alternative binding scaffolds have made it to the clinic, either as therapeutic entities or for the purpose of tumor imaging (for a summary including the corresponding clinical trial numbers, see Table 9.2. An explanation of the different clinical trial categories can be found in reference [77]). The first scaffold to be approved by the FDA in 2004 was the knottin Ziconotide, which is marketed under the brand name "Prialt®" first by the company Jazz

Table 9.2 Tabular overview of selected alternative scaffold proteins (in alphabetical order) being either FDA-approved or in clinical development as of January 2020 (clinical trial numbers refer to entries at clinicaltrials.gov)

Scaffold	Name	Target	Indication	Originator	Clinical trials
Affibodies	ABY-025	HER2	Tumor imaging	Affibody	Phase I/II studies completed (NCT01858116 and NCT01216033), Phase II/III study active (NCT03655353)
	ABY-029	EGFR	Tumor imaging	Affibody	Three early phase I studies suspended (NCT02901925, NCT03154411, and NCT03282461)
	ABY-035	IL17A	Plaque Psoriasis	Affibody	Phase I study active (NCT03580278), phase II study active (NCT03591887), phase I/II study completed (NCT02690142)
Anticalins	PRS-080	Hepcidin	Anemia of chronic kidney disease	Pieris	Phase I study completed (NCT02340572), phase II study completed (NCT03325621)
	PRS-050 ("Angiocal®")	VEGF	Solid tumors	Pieris	Phase I study completed (NCT01141257)
	PRS-060	IL4-Rα	Moderate to severe asthma	Pieris	Two-phase I studies completed (NCT03384290 and NCT03921268), one phase I study active (NCT03574805)
	PRS-343 (Antibody-Anticalin fusion protein)	Antibody: HER2 Anticalin: CD137	Solid tumors	Pieris	Two-phase I studies active (NCT03330561 and NCT03650348)
DARPins	Abicipar pegol®	VEGF	Macular degeneration	Molecular Partners	Phase III study completed (NCT02462486)
	MP0250	VEGF, HGF	Solid tumors	Molecular Partners	Phase I/II study completed (NCT02194426), Two-phase I/II studies active (NCT03418532 and NCT03136653)
	MP0274	HER2	Solid tumors	Molecular Partners	Phase I study active (NCT03084926)
	MP0310	4-1BB, FAP	Advanced solid tumors	Molecular Partners	Phase I study active (NCT04049903)

(continued)

Table 9.2 (continued)

Scaffold	Name	Target	Indication	Originator	Clinical trials
Knottins	Ziconotide ("Prialt®")	Calcium ion-channel	Severe chronic pain	Jazz Pharmaceuticals	Approved in 2004
	Linaclotide ("Linzess®")	Guanylate cyclase C receptor	Irritable bowel disease and chronic constipation	Ironwood Pharmaceuticals	Approved in 2012
Kunitz domain inhibitors	DX-88 ("Kalbitor® Ecallantide")	Kallikrein	Hereditary angioedema	Dyax (now Shire)	Approved in 2009
	DX-890 ("Depelestat")	Neutrophil elastase	Pulmonary fibrosis	Dyax (now Shire)	Phase IIa study completed (NCT00455767)

Table adapted from reference [17]. For Abicipar pegol®, only the most advanced clinical study is listed

Pharmaceuticals and now by TerSera Therapeutics [17]. Ziconotide is derived from a component of the venom of the fish-eating marine cone snail which leverages its venom to immobilize its prey. This knottin is a non-opioid analgesic antagonist that interrupts the signaling of pain receptors by blocking calcium ion channels expressed in those nerves responsible for pain signal transduction. Therefore, it has proven to help patients who suffer from severe chronic pain when administered intrathecally (i.e., into the spinal canal), this way reaching the cerebrospinal fluid without the need to actively crossing the blood–brain barrier [17].

Linaclotide was the second knottin approved by the FDA in 2012 and is now marketed as "Linzess®" by Allergan (originally by Ironwood Pharmaceuticals) for the treatment of irritable bowel disease and chronic constipation. This miniprotein functions as an agonist for the guanylate cyclase C receptor in the intestine and can be administered orally due to its exceptional stability. Activation of the guanylate cyclase C receptor results in an increase of cyclic guanosine monophosphate which in turn enhances the secretion of bicarbonate and chloride ions into the intestinal lumen. Consequently, the fluid secretion into the intestine is increased, ultimately promoting intestinal transit [78].

Next to Ziconotide and Linaclotide, the Kunitz domain inhibitor DX-88 was the third alternative scaffold to receive marketing approval by the FDA in 2009 for the treatment of a rare, inherited blood disorder referred to as hereditary angioedema (HAE) [17, 55]. HAE manifests in intermittent and acute attacks of edemas of both the mucosa and the skin and is caused by a malfunctioning inhibitor protein found in plasma that targets the serine protease kallikrein [79]. DX-88 is marketed under the name "Kalbitor® Ecallantide" by the company Dyax (which was acquired by Shire in 2015) and competes with this mutated natural inhibitor for binding to kallikrein, thereby neutralizing its effect. A second kunitz domain termed DX-890, or "Depelestat", was also developed by Dyax and has been evaluated in a phase II study for the treatment of pulmonary fibrosis [17]. Depelestat inhibits the protease neutrophil elastase which is released by activated neutrophils, a cell type predominantly present in the lungs of fibrosis patients. As this serine protease can degrade components of the extracellular matrix, its inhibition can stop the lung injuries appearing in the absence of this kunitz domain [80, 81].

Besides the three abovementioned approved alternative binders, several affinity reagents of other scaffold classes are currently under clinical investigations. The most advanced binder is Abicipar pegol®, a PEGylated and thus half-life extended DARPin that targets the vascular endothelial growth factor (VEGF). This binder has recently successfully completed a phase III trial for the treatment of patients suffering from neovascular age-related macular degeneration (AMD), the leading cause of permanent vision loss in elderly people in the developed world. Untreated, newly formed immature blood vessels grow into the space beneath the retinal pigment epithelium and cause tissue disruption and, ultimately, the degeneration of the neurosensory retina. This process can be blocked by VEGF inhibitors which thus are the standard of care for AMD. While current antibody-based anti-VEGF treatments require frequent intravitreal injections, less frequent dosing might be required for the DARPin-based therapeutic due to its high stability and efficacy.

However, Abicipar pegol[®] is not the only DARPin currently being in clinical studies. MP0250 is an anti-angiogenic molecule which differentiates itself from Abicipar pegol[®] not just in its oncology indication, but also in its architectural design. It is constructed in a multimeric format consisting of three different individual DARPins. One DARPin targets VEGF while a second binder engages with the hepatocyte growth factor (HGF) and two copies of the same DARPin interact with serum albumin [82]. Next to exhibiting an extended half-life through the HSA interactions, this multi-DARPin has the potential to overcome the development of drug resistance, the major drawback of conventional VEGF antagonists (such as the monoclonal antibody Bevacizumab (Avastin[®])), as it targets two different cancer-related pathways simultaneously. MP0250 has successfully completed a phase I/II study and is currently being investigated in a phase II trial for the treatment of multiple myeloma and in a phase Ib/II study for the treatment of non-small cell lung cancer. It has recently received Orphan Drug Designation for multiple myeloma by the FDA, a designation to create financial incentives for supporting the development of new drugs and biologics for rare diseases [83]. Another half-life-extended DARPin, MP0274, comprises again an HSA-binding unit as well as a second DARPin binding HER2 with high affinity. MP0274 is intended for the treatment of solid tumors and has entered a first-in-human phase I clinical trial in 2017. The most recent addition to the clinical development landscape of the DARPin scaffold is MP0310 which just entered a phase I trial for the treatment of advanced solid tumors. This bispecific DARPin construct simultaneously targets 4-1BB (also known as CD137), a member of the tumor necrosis factor family that acts as a co-stimulatory immune checkpoint molecule, as well as the fibroblast activation protein [84]. It is designed to activate immune T cells specifically in the tumor and not systemically in the rest of the body, thus potentially delivering greater efficacy with a lower risk of systemic side effects and toxicities.

Another scaffold prominently present in clinical investigations is that of Anticalins with several of them currently being in early and later stage clinical trials. PRS-050 ("Angiocal") is a PEGylated and half-life extended Anticalin that binds VEGF, thereby inhibiting tumor angiogenesis, and is intended for the treatment of solid tumors. PRS-050 has successfully been evaluated in a phase I study in patients with advanced solid tumors with no observable signs of toxicity or immunogenicity [85]. Another PEGylated Anticalin, PRS-080, targets hepcidin (hepatic bactericidal protein) which plays an important role in the regulation of iron metabolism [86]. Hepcidin interacts with the iron transporter ferroportin, which is expressed on the surface of many different cells, and subsequently causes its internalization and degradation. As such, the export of iron from the body's depositories by ferroportin is blocked and its availability reduced. Antagonizing hepcidin with PRS-080 has the potential to improve the availability of iron in patients suffering from functional iron deficiency (FID). The safety and tolerability of PRS-080 has been demonstrated in a phase I study and paved the way for an ongoing phase II2b study that investigates the safety in anemic chronic-kidney disease patients suffering from FID [87]. The Anticalin variant PRS-060, on the other hand, antagonizes the interleukin 4 receptor IL-4Rα and has so far been evaluated in two completed phase I studies. It is currently investigated for the treatment of moderate to severe asthma in healthy volunteers in a third phase I study. In contrast to systemically administered antibody-based therapeutics which have been developed for asthma, PRS-060 can be administered locally to the lung through inhalation which potentially requires lower doses [86]. Finally, a bispecific antibody-Anticalin fusion protein termed PRS-343 has recently entered phase I studies. PRS-343 consists of an antibody targeting the HER2 receptor which is fused to an Anticalin that engages with CD137, a costimulatory immune receptor expressed on activated T cells, B cells, and natural killer cells [44, 86]. This fusion protein was designed to specifically activate T cells in the tumor microenvironment which ultimately could lead to tumor eradication.

Next to human-derived alternative binding scaffolds, the abovementioned affibodies are investigated in clinical studies as well. However, due to potential immunogenicity, these binders of bacterial origin were first considered rather for short-term diagnostic or local therapeutic than long-term systemic indications. The HER2-binding affibody ABY-025 was therefore conjugated to different imaging tracers and has successfully been investigated in two-phase I/II clinical studies for the imaging of breast cancer. Since ABY-025 engages a region on the HER2 receptor that is different from the epitopes addressed by the therapeutic anti-HER2 antibodies Trastuzumab (Herceptin®) and Pertuzumab (Perjeta®), it potentially enables tumor imaging in patients undergoing antibody therapy [44]. ABY-025 is currently investigated in a phase II/III clinical trials. A phase 0 first-in-human study encompassed the administration of another affibody termed ABY-029 to a single patient, targeting EGFR while being conjugated to a near-infrared fluorophore [88]. Such labeled reagents could be employed in fluorescence-guided surgery, enabling an efficient removal of the tumor upon binding to a receptor specifically overexpressed on tumor cells, thereby sparing normal tissue. In this proof-of-concept study, a microdose of ABY-029 was administered to the patient approximately 4 h prior to the surgery and the removed tumor tissue was subsequently analyzed for its fluorescence signal. The study demonstrates that ABY-029 fluorescence was readily detectable ex vivo and positively correlated with EGFR expression.

While ABY-025 and ABY-029 were employed rather for diagnostic applications, interestingly another affibody variant termed ABY-035 has recently completed its phase I/II study and is actively investigated in a phase I as well as a phase II clinical trial. ABY-035 is a bispecific molecule, targeting both subunits of IL-17A as well as HSA and is intended for the treatment of moderate-to-severe psoriasis. The phase I/II study demonstrated favorable safety and tolerability in subcutaneous, but also intravenous administration, paving the way for further development.

While the abovementioned examples highlight some promising success stories of alternative binding scaffolds in clinical applications, unfortunately, some of them have also not reached the hope set in them. Two Adnectins have been investigated in clinical trials for the treatment of glioblastoma and hypercholesteromia, respectively. While both scaffolds were well-tolerated, both programs were subsequently discontinued due to either a failure to meet clinical significance or for undisclosed reasons [89, 90]. A third Adnectin, BMS-986089 ("talditercept alfa"), had already been evaluated in several phase I clinical trials for the treatment of Duchenne Muscular Dystrophy; however, the results of a phase I/II study indicated that the primary endpoint would most likely not be met. As such, the clinical studies encompassing talditercept alfa were recently discontinued, which currently leaves no Adnectin scaffold in clinical development. A similar fate also struck the fynomer scaffold as mentioned above.

9.4   Alternative Binding Scaffolds as Tools in Basic Research and Beyond

Besides their development for therapeutic purposes, several alternative scaffolds have proven to be valuable tools for applications in laboratory research (see Fig. 9.4). A prime example is the aforementioned affibody variant fused to a DNA polymerase, increasing the polymerase's fidelity and now being widely used for PCRs by researchers worldwide ([37]; see Sect. 9.2.2 for details). Affibodies have also extensively been employed for affinity purification, encompassing functionalized resins which for example leverage the intrinsic ability of affibodies to interact with immunoglobulin Fc domains (marketed under the name "MabSelect SuRe" by GE Healthcare) [36]. In addition, new specificities have been introduced by protein engineering, generating affibody-based capturing ligands that effectively engage proteins such as human factor VIII [91]. Beyond being useful for chromatographic applications, affibodies targeting transferrin or human immunoglobulin A have been employed successfully for protein depletion from complex biological samples such as cerebrospinal fluid or human serum [92, 93]. Moreover, these affinity reagents have been fused with reporter enzymes, such as horseradish peroxidase, in order to function as detection reagents in enzyme-linked immunosorbent assays (ELISA), dot blot and immunohistochemistry assays, providing a colorimetric read-out [94, 95]. Alternatively,

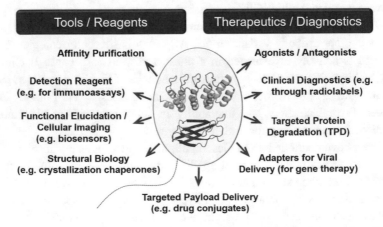

Fig. 9.4 Overview of different applications of alternative binding scaffolds, either falling in the "therapeutics/diagnostics" or "tool/reagent" category covered in paragraphs 3 and 4, respectively

fusion proteins of affibodies with the Fc portion of a human IgG have been utilized for the visualization of target proteins in western blot analyses [96]. Further, affibodies have successfully been employed as capturing ligands on protein microarrays, demonstrating specific binding to their respective target protein and no observable cross-reactivity with other unrelated proteins [97]. When coupled to fluorophores, members of this scaffold class have been shown to be suitable for the detection and quantification of cancer-relevant receptors on tumor cell lines by flow cytometry [95, 98], and also as biosensors for fluorescence-based energy transfer (FRET)-based applications [99].

Cysteine-knot miniproteins, on the other hand, have been primarily developed for their application in therapy and, specifically, for tumor imaging. However, knottins that interact with integrins have also efficiently been employed as tool reagents for the characterization of novel integrin-binding peptides [100]. In a competition ELISA setup, different integrins were immobilized and subsequently incubated with a mixture of both a knottin recognizing several different integrin variants as well as different concentrations of integrin-binding peptides. Peptides comprising higher affinities for the immobilized integrins than this tool knottin could be identified, as those were influencing the detected knottin binding in a concentration-dependent manner. The use of chemically synthesizable knottins instead of recombinant anti-integrin antibodies could potentially reduce the costs of these screenings and facilitate standardization, especially in a high-throughput assay format.

Adnectins, or monobodies, have proven to be excellent tools for studying protein–protein interactions and further provided great insights into structural analyses of different RAS (rat sarcoma) proteins. The human RAS GTPases N-, H-, and K-RAS play an important role in oncogenesis and a mutation in one of those GTPases is commonly present in a variety of different tumor types [101]. Structural and biochemical analyses encompassing a monobody that targets the H- and K-RAS GTPases have aided in defining a previously unknown interface in the RAS proteins responsible for their dimerization [102]. In addition, this binder also efficiently inhibited H- and K-RAS mediated signaling upon binding. Another monobody has been identified as a specific inhibitor for a member of the sarcoma homology 2 (SH2) protein family, an important representative of the so-called protein interaction domains (PID) [103]. PIDs are recognition modules that can interact with a plethora of different biological ligands and understanding their individual functions and contributions is crucial for deciphering cellular protein networks. However, PIDs such as the SH2 family are present in a myriad of different proteins found in the human body and share high structural homology, making it difficult to identify highly specific affinity tools which allow the engagement of only one desired family member while sparing all others. A highly specific monobody variant exclusively interacting with the SH2 domain of the protein kinase Abl, however, could be identified by phage display [103]. The solved co-crystal structure of the kinase–monobody complex delivered valuable insights into the inhibition mechanism and thus might provide guidance for the design of novel specific SH2 inhibitors for therapeutic purposes.

Alternative scaffold proteins based on the lipocalin scaffold have likewise been employed as tools in molecular biology. An anticalin binding the cancer-related protein

cytotoxic T-lymphocyte-associated antigen 4 (CTL-A4) has been used in immunohistochemistry applications [74]. Conjugated to the dye fluorescein, this anticalin enabled
the staining of several non-malignant as well as breast cancer tissue sections with no
unspecific cross-reactivity, demonstrating its potential in clinical diagnostics. Since
anticalins are further able to accommodate not only protein-based but also small-molecule
ligands, such a binder targeting digoxigenin could be well-suited for the quantitative
detection of this steroid in human serum. Dioxigenin is the aglycon of digoxin, a
cardioactive compound used in patients with heart insufficiency. Since digoxin unfortunately exhibits a low therapeutic window, its concentration in patients needs to be monitored closely during therapy. In this context, anticalins detecting both digoxin and
digoxigenin could be well-suited for the development of digoxin quantification
assays [104].

In the same way, alternative binding scaffolds based on ankyrin repeat proteins have
been employed for a wide range of applications beyond the clinic and tumor imaging. A
DARPin variant targeting HER2 with picomolar affinity, for instance, has successfully
been utilized as a tool reagent for immunohistochemistry staining, identifying the amplification status of HER2 in paraffin-embedded human breast cancer tissue sections [105]. A
comparison with an FDA-approved detection monoclonal antibody demonstrated that the
DARPin was indeed a valuable alternative regarding the detection of HER2 expression in
breast cancer tissue. DARPins have also been successfully employed as "co-crystallization
chaperones", i.e., helper proteins that aid in the crystallization of other proteins by forming
a stabilized, non-covalent complex [106]. Many proteins have shown to be difficult to
crystallize, mainly because their potential instabilities and inherent flexibility can hamper
the formation of well-ordered crystals. In that manner, protein binder might help to stabilize
the protein of interest, even in a particular structural state important for understanding the
target's functionality. In contrast to antibody domains, which have also been employed as
crystallization chaperones for decades, DARPins and other alternative binding scaffolds
might be obtained easier and in a more cost-efficient way [107]. Thus, it is not surprising
that up to date several dozen targets have been crystallized in complex with DARPins, e.g.,
the polo-like kinase 1 [108], caspase-2 [109], or the multidrug exporter AcrB [110], among
many others. Interestingly, not only DARPins interacting directly with the protein of
interest have been beneficial for such structural studies: A binder interacting with the
Maltose-binding protein (MBP) has successfully been used for the crystallization of the
MBP-fusion to the human dual-specificity phosphatase 1 [111]. More recently, an elegant
strategy to employ rigid helix fusions has opened new opportunities and expanded the
crystallization space. For this purpose, the C- and N-terminal DARPin capping repeats
were re-designed to create a shared helix with β-lactamase in a rigid way [112]. While not
interfering with the DARPin's binding to its target, such rigid fusions to larger and easier to
crystallize proteins have proven to yield high-resolution structures, partly also through the
rigid fusion of several DARPins to each other in different geometries, thereby generating a
range of alternative crystal packings [113]. Obviously, such rigid DARPin fusions are also
of great interest for cryogenic electron microscopy (cryo-EM) applications for which an

increased size could enable structural investigations of complexes currently too small. Finally, an even further expansion of this shared helix approach was recently achieved, employing protein engineering to construct a well-crystallizing fusion-construct that shields the target-binding surface from crystal contacts and thus might be a more generalizable route for fast high-throughput structural studies [114].

Another interesting application describes DARPin binders which target and thereby activate the dye malachite green through the formation of a homodimeric structure [115]. This effect on their target allows the usage of these binders as so-called fluorogen activators (FAPs). Unlike traditional fluorescent proteins (like the aforementioned GFP), FAPs do not fluoresce unless being bound to specific small-molecule fluorogens and thus enable real-time detection of proteins in living cells in a noninvasive way with high signal-to-noise ratios [116]. While commonly used FAPs are based on antibody single-chain variable fragments (and thus are partly restricted by their limitations highlighted in another chapter of this book), the published DARPin-based FAP nicely illustrates the potential of alternative binding scaffolds for these applications. This FAP can be utilized for a wide range of applications supporting the elucidation of protein functionalities, ranging from in situ detection to cell-surface and cytosolic imaging of desired target proteins.

Affitins have likewise been employed for applications in basic research, such as increasing the fidelity of a DNA polymerase [117], for affinity chromatography [118], cellular immunostainings [41], and tumor imaging [42].

For Avimers and Kunitz domains, however, suitable applications as research tools have yet to be publicly reported.

All of the abovementioned applications clearly highlight the great potential different alternative binding scaffolds have in supporting not only medical doctors in the clinic but also researchers as tools in a broad range of applications, allowing to gather improved insights into many different biological questions in an improved and more affordable way.

9.5   Future Directions

Over the past two decades, the field of alternative binding scaffolds has continuously evolved, with both novel affinity classes and innovative applications constantly being developed. While many of the new binder classes conceptually resemble those binder classes described in detail above, a few newer scaffolds include orthogonal design approaches or underlying rationales and thus will be briefly mentioned in a selected manner.

The quest to push the size limits of new alternative scaffold binders toward even smaller affinity reagents, and thus aiming to further improve solid tumor penetration and biodistribution, led to the identification of the T7 phage gene 2 (Gp2) protein [119]. Opposite to the aforementioned scaffolds, Gp2 was identified not directly from natural examples of protein–protein interactors, but rather by a systemic mining of the Protein Data Bank and exploring known protein topologies. The search identified Gp2 as a promising alternative scaffold with a lack of disulfide bonds, a large available binding surface as well as a structural tolerance toward mutations. The 45-residue Gp2 scaffold consists of an α-helix located opposite of a β-sheet and features two solvent-exposed loops that can be targeted for diversification. A highly thermostable Gp2 protein that binds EGFR has been developed and recently been employed for PET studies in tumor-bearing mice with high specificity [120].

Another alternative binding scaffold was developed entirely by computational design [121], aiming at identifying a novel IgG-binding protein that could be used for affinity chromatography and the purification of human antibodies. The initial computational approach considered the knowledge that the majority of available human IgG-binding proteins address the same consensus region on the IgG's Fc region (such as the natural affibody scaffold). In addition, histidine residues present in the Fc interface were taken into account, intending to design a scaffold protein exhibiting a pH-sensitivity. By using hotspot-guided protein interface design, scaffold molecules able to host the identified hotspot positions and exhibiting beneficial properties were identified. Several of these redesigned scaffolds were expressed on the surface of yeast cells and analyzed for their binding to a human antibody. The best-performing scaffold variant was based on the bacterial enzyme pyrazinamidase from *Pyrococcus horikoshii*. As this scaffold naturally does not bind to human Fc regions, the computationally redesigned binding interface apparently was responsible for this altered binding specificity. The pyrazinamidase scaffold was subsequently subjected to error-prone PCR and screened by yeast display to identify variants with a higher affinity at neutral pH, but reduced binding in an acidic environment. This resulting pH-sensitive pyrazinamidase scaffold could be expressed well in bacterial cells and showed remarkable thermostability. It, therefore, was immobilized on solid support and applied for the purification of a mixture of human IgGs. Elution of the bound antibodies was carried out upon performing a mild elution step at a slightly acidic pH, demonstrating the feasibility of this computationally redesigned, pH-sensitive alternative binder.

Another approach for the design of a novel alternative binding scaffold is based on the aspiration of not having to perform individual selections every time a new protein of interest should be targeted. Having a toolbox of pre-selected "binding modules" which could be assembled into a single protein on demand to create new affinity reagents with the intended specificities would be of great value [122]. The armadillo repeat proteins, being abundant eukaryotic proteins involved in several cellular processes, including signaling, transport, and cytoskeletal regulation [123], were identified as a suitable scaffold for this purpose. As these repeat proteins recognize linear, stretched-out peptides in an extended conformation, they offer a conserved binding mode in which each repeat is complementary to a piece of the target peptide and thus satisfy the requirements for the generation of a synthetic modular peptide-binding scaffold [124]. Combining directed evolution, computational analyses, and insights from structural biology, this endeavor brought together all methods of protein engineering and has already resulted in the identification of modular sequence-specific peptide binders [125].

Next to the conceptualization of novel protein scaffolds, new and improved therapeutic applications of the already existing affinity reagent classes are constantly surfacing these days (see Fig. 9.4). One example is the involvement of alternative binding scaffolds in targeted payload delivery as conjugates to small cytotoxic drugs. While this approach has been around for a while and the first results with antibody-drug conjugates (ADC) were promising [126], it soon became clear that this procedure was not as straightforward as anticipated, requiring thoughtful combinations of antibody, linker, and drugs in the context of a suitable target and a defined cancer indication [127]. One of the potential issues was the hard to control drug-to-antibody-ration (DAR), indicating the average number of drugs conjugated to the chosen antibody. This is challenging to reliably control as antibodies generally are difficult to be manipulated for payload conjugation via the conventional conjugation and linker chemistries, being too big to be synthesized chemically or not having unique side chains or chemical groups available for site-directed conjugations. Here, alternative binding scaffolds might offer various opportunities for the site-specific incorporation of suitable residues, allowing a flexible and stoichiometrically defined payloading by for instance the modification of inserted non-natural amino acids through strain-promoted alkyne-azide cycloaddition click chemistry [128] or the enzymatic modification of specifically inserted protein tags [129].

Another promising application of alternative binding scaffolds is their usage as adapters mediating cell-specific viral delivery. Following a series of setbacks in the 1990s, gene therapy has re-gained momentum these days with the recent regulatory approval of two therapeutics employing adeno-associated virus (AAV) vectors [130]. Key to the success of such therapies is the cell-specific delivery of the genetic material for gene replacement, silencing, or editing, respectively. As the engineering of the viral capsids themselves are challenging, one interesting approach employs affinity reagents to act as adapters for the viral delivery to the cell types of choice. Such approaches have been successfully performed for Adenoviruses [131] as well as for AAVs and lentiviral vectors [132] and might help to further bring this important therapeutic field forward.

Finally, alternative binding scaffolds might also demonstrate their potential in targeted protein degradation (TPD). This popular approach aims at knocking down target proteins within cells by causing them to become polyubiquitinated and subsequently recognized and degraded by the proteasome [133], thereby allowing to potentially tackle so far "undruggable targets". While the majority of endeavors in this field are employing small molecule "degraders" (e.g., the so-called Proteolysis Targeting Chimeras (PROTACs)) to achieve proteasomal degradation [134], protein binders might also be used in the design of complexes between the required E3 ubiquitin ligase and the targets to be degraded. First reports already demonstrated the feasibility of employing alternative binding scaffolds for TPD [135] and thus it can be foreseen that additional results will follow shortly.

Taken together, current reports clearly show that the field of alternative binding scaffolds continues to evolve and novel scaffolds are being added to the collection on a regular basis. Since these molecular entities exhibit various advantages over conventional monoclonal antibodies, they will play a sustainable role in the development of next-generation therapeutics and affinity reagents for innovative applications.

Take Home Messages
- Alternative binding scaffolds are non-antibody molecules with favorable properties and functionalities which are able to interact with a variety of complex target proteins
- Alternative scaffolds generally are designed by mutating amino acid residues in surface-exposed areas of naturally occurring binding proteins
- Their small size, biophysical properties, and other biochemical features are advantageous for many therapeutic and biotechnological applications
- Alternative binding scaffolds have already been approved for therapeutic purposes or are currently undergoing clinical trials.
- The field of alternative binding scaffolds is continuously growing, with new scaffolds emerging on a regular basis, allowing more and more innovative applications.

References

1. Köhler G, Milstein C. Continuous cultures of fused cells secreting antibody of predefined specificity. Nature. 1975;256:495–7. https://doi.org/10.1038/256495a0.
2. Kaplon H, Muralidharan M, Schneider Z, Reichert JM. Antibodies to watch in 2020. MAbs. 2020;12:1703531. https://doi.org/10.1080/19420862.2019.1703531.
3. Ansar W, Ghosh S. Monoclonal antibodies: a tool in clinical research. Indian J Clin Med. 2013;4:S11968. https://doi.org/10.4137/IJCM.S11968.
4. Gao Y, Huang X, Zhu Y, Lv Z. A brief review of monoclonal antibody technology and its representative applications in immunoassays. J Immunoass Immunochem. 2018;39:351–64. https://doi.org/10.1080/15321819.2018.1515775.
5. Moser AC, Hage DS. Immunoaffinity chromatography: an introduction to applications and recent developments. Bioanalysis. 2010;2:769–90. https://doi.org/10.4155/bio.10.31.
6. Pyzik M, Rath T, Lencer WI, et al. FcRn: the architect behind the immune and nonimmune functions of IgG and albumin. J Immunol. 2015;194:4595–603. https://doi.org/10.4049/jimmunol.1403014.
7. Li Z, Krippendorff B-F, Sharma S, et al. Influence of molecular size on tissue distribution of antibody fragments. MAbs. 2016;8:113–9. https://doi.org/10.1080/19420862.2015.1111497.
8. Shah DK, Betts AM. Antibody biodistribution coefficients: inferring tissue concentrations of monoclonal antibodies based on the plasma concentrations in several preclinical species and human. MAbs. 5:297–305. https://doi.org/10.4161/mabs.23684.
9. Schaefer JV, Sedlák E, Kast F, et al. Modification of the kinetic stability of immunoglobulin G by solvent additives. MAbs. 2018;10:607–23. https://doi.org/10.1080/19420862.2018.1450126.
10. Ministro J, Manuel AM, Goncalves J. Therapeutic antibody engineering and selection strategies. Adv Biochem Eng Biotechnol. 2019;171:55–86.

11. Traxlmayr MW, Kiefer JD, Srinivas RR, et al. Strong enrichment of aromatic residues in binding sites from a charge-neutralized hyperthermostable Sso7d scaffold library. J Biol Chem. 2016;291:22496–508. https://doi.org/10.1074/jbc.M116.741314.

12. McCrary BS, Edmondson SP, Shriver JW. Hyperthermophile protein folding thermodynamics: differential scanning calorimetry and chemical denaturation of Sac7d. J Mol Biol. 1996;264:784–805. https://doi.org/10.1006/jmbi.1996.0677.

13. Colgrave ML, Craik DJ. Thermal, chemical, and enzymatic stability of the cyclotide kalata B1: the importance of the cyclic cystine knot †. Biochemistry. 2004;43:5965–75. https://doi.org/10.1021/bi049711q.

14. Kintzing JR, Cochran JR. Engineered knottin peptides as diagnostics, therapeutics, and drug delivery vehicles. Curr Opin Chem Biol. 2016;34:143–50. https://doi.org/10.1016/j.cbpa.2016.08.022.

15. Vazquez-Lombardi R, Phan TG, Zimmermann C, et al. Challenges and opportunities for non-antibody scaffold drugs. Drug Discov Today. 2015;20:1271–83. https://doi.org/10.1016/j.drudis.2015.09.004.

16. Skerra A. Alternative non-antibody scaffolds for molecular recognition. Curr Opin Biotechnol. 2007;18:295–304. https://doi.org/10.1016/j.copbio.2007.04.010.

17. Simeon R, Chen Z. In vitro-engineered non-antibody protein therapeutics. Protein Cell. 2018;9:3–14. https://doi.org/10.1007/s13238-017-0386-6.

18. Lipovšek D, Carvajal I, Allentoff AJ, et al. Adnectin–drug conjugates for Glypican-3-specific delivery of a cytotoxic payload to tumors. Protein Eng Des Sel. 2018;31:159–71. https://doi.org/10.1093/protein/gzy013.

19. Wang W, Lu P, Fang Y, et al. Monoclonal antibodies with identical fc sequences can bind to FcRn differentially with pharmacokinetic consequences. Drug Metab Dispos. 2011;39:1469–77. https://doi.org/10.1124/dmd.111.039453.

20. Strohl WR. Fusion proteins for half-life extension of biologics as a strategy to make biobetters. BioDrugs. 2015;29:215–39. https://doi.org/10.1007/s40259-015-0133-6.

21. Kontermann RE. Half-life extended biotherapeutics. Expert Opin Biol Ther. 2016;16:903–15. https://doi.org/10.1517/14712598.2016.1165661.

22. JevsÌŒevar S, Kunstelj M, Porekar VG. PEGylation of therapeutic proteins. Biotechnol J. 2010;5:113–28. https://doi.org/10.1002/biot.200900218.

23. Schlapschy M, Binder U, Borger C, et al. PASylation: a biological alternative to PEGylation for extending the plasma half-life of pharmaceutically active proteins. Protein Eng Des Sel. 2013;26:489–501. https://doi.org/10.1093/protein/gzt023.

24. Capon DJ, Chamow SM, Mordenti J, et al. Designing CD4 immunoadhesins for AIDS therapy. Nature. 1989;337:525–31. https://doi.org/10.1038/337525a0.

25. Kim J, Bronson CL, Hayton WL, et al. Albumin turnover: FcRn-mediated recycling saves as much albumin from degradation as the liver produces. Am J Physiol Liver Physiol. 2006;290: G352–60. https://doi.org/10.1152/ajpgi.00286.2005.

26. Sleep D, Cameron J, Evans LR. Albumin as a versatile platform for drug half-life extension. Biochim Biophys Acta – Gen Subj. 2013;1830:5526–34. https://doi.org/10.1016/j.bbagen.2013.04.023.

27. Nygren P-Å, Skerra A. Binding proteins from alternative scaffolds. J Immunol Methods. 2004;290:3–28. https://doi.org/10.1016/j.jim.2004.04.006.

28. Storz U. Intellectual property protection. MAbs. 2011;3:310–7. https://doi.org/10.4161/mabs.3.3.15530.

29. Koide A, Bailey CW, Huang X, Koide S. The fibronectin type III domain as a scaffold for novel binding proteins. J Mol Biol. 1998;284:1141–51. https://doi.org/10.1006/jmbi.1998.2238.

30. Bloom L, Calabro V. FN3: a new protein scaffold reaches the clinic. Drug Discov Today. 2009;14:949–55. https://doi.org/10.1016/j.drudis.2009.06.007.
31. Gulyani A, Vitriol E, Allen R, et al. A biosensor generated via high-throughput screening quantifies cell edge Src dynamics. Nat Chem Biol. 2011;7:437–44. https://doi.org/10.1038/nchembio.585.
32. Gross GG, Junge JA, Mora RJ, et al. Recombinant probes for visualizing endogenous synaptic proteins in living neurons. Neuron. 2013;78:971–85. https://doi.org/10.1016/j.neuron.2013.04.017.
33. Nord K, Gunneriusson E, Ringdahl J, et al. Binding proteins selected from combinatorial libraries of an α-helical bacterial receptor domain. Nat Biotechnol. 1997;15:772–7. https://doi.org/10.1038/nbt0897-772.
34. Gunneriusson E, Nord K, Uhlén M, Nygren P-Å. Affinity maturation of a Taq DNA polymerase specific affibody by helix shuffling. Protein Eng Des Sel. 1999;12:873–8. https://doi.org/10.1093/protein/12.10.873.
35. Orlova A, Magnusson M, Eriksson TLJ, et al. Tumor imaging using a picomolar affinity HER2 binding affibody molecule. Cancer Res. 2006;66:4339–48. https://doi.org/10.1158/0008-5472.CAN-05-3521.
36. Löfblom J, Feldwisch J, Tolmachev V, et al. Affibody molecules: engineered proteins for therapeutic, diagnostic and biotechnological applications. FEBS Lett. 2010;584:2670–80. https://doi.org/10.1016/j.febslet.2010.04.014.
37. Frejd FY, Kim K-T. Affibody molecules as engineered protein drugs. Exp Mol Med. 2017;49:e306. https://doi.org/10.1038/emm.2017.35.
38. Gera N, Hussain M, Wright RC, Rao BM. Highly stable binding proteins derived from the hyperthermophilic Sso7d scaffold. J Mol Biol. 2011;409:601–16. https://doi.org/10.1016/j.jmb.2011.04.020.
39. Mouratou B, Schaeffer F, Guilvout I, et al. Remodeling a DNA-binding protein as a specific in vivo inhibitor of bacterial secretin PulD. Proc Natl Acad Sci. 2007;104:17983–8. https://doi.org/10.1073/pnas.0702963104.
40. Kauke MJ, Traxlmayr MW, Parker JA, et al. An engineered protein antagonist of K-Ras/B-Raf interaction. Sci Rep. 2017;7:5831. https://doi.org/10.1038/s41598-017-05889-7.
41. Gocha T, Rao BM, DasGupta R. Identification and characterization of a novel Sso7d scaffold-based binder against Notch1. Sci Rep. 2017;7:12021. https://doi.org/10.1038/s41598-017-12246-1.
42. Goux M, Becker G, Gorré H, et al. Nanofitin as a new molecular-imaging agent for the diagnosis of epidermal growth factor receptor over-expressing tumors. Bioconjug Chem. 2017;28:2361–71. https://doi.org/10.1021/acs.bioconjchem.7b00374.
43. Kalichuk V, Renodon-Cornière A, Béhar G, et al. A novel, smaller scaffold for Affitins: showcase with binders specific for EpCAM. Biotechnol Bioeng. 2018;115:290–9. https://doi.org/10.1002/bit.26463.
44. Gebauer M, Skerra A. Engineering of binding functions into proteins. Curr Opin Biotechnol. 2019;60:230–41. https://doi.org/10.1016/j.copbio.2019.05.007.
45. Kalichuk V, Kambarev S, Béhar G, et al. Affitins: ribosome display for selection of Aho7c-based affinity proteins. Methods Mol Biol. 2020;2070:19–41.
46. Silverman J, Lu Q, Bakker A, et al. Multivalent avimer proteins evolved by exon shuffling of a family of human receptor domains. Nat Biotechnol. 2005;23:1556–61. https://doi.org/10.1038/nbt1166.
47. Weidle UH, Auer J, Brinkmann U, et al. The emerging role of new protein scaffold-based agents for treatment of cancer. Cancer Genomics Proteomics. 2013;10:155–68.

48. Moore SJ, Cochran JR. Engineering knottins as novel binding agents. Methods Enzymol. 2012;503:223–51.
49. Avrutina O. Synthetic cystine-knot miniproteins – valuable scaffolds for polypeptide engineering. Adv Exp Med Biol. 2016;917:121–44.
50. Werle M, Schmitz T, Huang H-L, et al. The potential of cystine-knot microproteins as novel pharmacophoric scaffolds in oral peptide drug delivery. J Drug Target. 2006;14:137–46. https://doi.org/10.1080/10611860600648254.
51. Miao Z, Ren G, Liu H, et al. An engineered knottin peptide labeled with 18 F for PET imaging of integrin expression. Bioconjug Chem. 2009;20:2342–7. https://doi.org/10.1021/bc900361g.
52. Plückthun A. Designed Ankyrin repeat proteins (DARPins): binding proteins for research, diagnostics, and therapy. Annu Rev Pharmacol Toxicol. 2015;55:489–511. https://doi.org/10.1146/annurev-pharmtox-010611-134654.
53. Jost C, Plückthun A. Engineered proteins with desired specificity: DARPins, other alternative scaffolds and bispecific IgGs. Curr Opin Struct Biol. 2014;27:102–12. https://doi.org/10.1016/j.sbi.2014.05.011.
54. Schilling J, Schöppe J, Plückthun A. From DARPins to LoopDARPins: novel LoopDARPin design allows the selection of low picomolar binders in a single round of ribosome display. J Mol Biol. 2014;426:691–721. https://doi.org/10.1016/j.jmb.2013.10.026.
55. Wurch T, Pierré A, Depil S. Novel protein scaffolds as emerging therapeutic proteins: from discovery to clinical proof-of-concept. Trends Biotechnol. 2012;30:575–82. https://doi.org/10.1016/j.tibtech.2012.07.006.
56. Mittal A, Böhm S, Grütter MG, et al. Asymmetry in the homodimeric ABC transporter MsbA recognized by a DARPin. J Biol Chem. 2012;287:20395–406. https://doi.org/10.1074/jbc.M112.359794.
57. Pecqueur L, Duellberg C, Dreier B, et al. A designed ankyrin repeat protein selected to bind to tubulin caps the microtubule plus end. Proc Natl Acad Sci. 2012;109:12011–6. https://doi.org/10.1073/pnas.1204129109.
58. Kummer L, Parizek P, Rube P, et al. Structural and functional analysis of phosphorylation-specific binders of the kinase ERK from designed ankyrin repeat protein libraries. Proc Natl Acad Sci. 2012;109:E2248–57. https://doi.org/10.1073/pnas.1205399109.
59. Kummer L, Hsu C-W, Dagliyan O, et al. Knowledge-based design of a biosensor to quantify localized ERK activation in living cells. Chem Biol. 2013;20:847–56. https://doi.org/10.1016/j.chembiol.2013.04.016.
60. Amstutz P, Binz HK, Parizek P, et al. Intracellular kinase inhibitors selected from combinatorial libraries of designed Ankyrin repeat proteins. J Biol Chem. 2005;280:24715–22. https://doi.org/10.1074/jbc.M501746200.
61. Parizek P, Kummer L, Rube P, et al. Designed Ankyrin repeat proteins (DARPins) as novel isoform-specific intracellular inhibitors of c-Jun N-terminal kinases. ACS Chem Biol. 2012;7:1356–66. https://doi.org/10.1021/cb3001167.
62. Brauchle M, Hansen S, Caussinus E, et al. Protein interference applications in cellular and developmental biology using DARPins that recognize GFP and mCherry. Biol Open. 2014;3:1252–61. https://doi.org/10.1242/bio.201410041.
63. Martin-Killias P, Stefan N, Rothschild S, et al. A novel fusion toxin derived from an EpCAM-specific designed Ankyrin repeat protein has potent antitumor activity. Clin Cancer Res. 2011;17:100–10. https://doi.org/10.1158/1078-0432.CCR-10-1303.
64. Goldstein R, Sosabowski J, Livanos M, et al. Development of the designed ankyrin repeat protein (DARPin) G3 for HER2 molecular imaging. Eur J Nucl Med Mol Imaging. 2015;42:288–301. https://doi.org/10.1007/s00259-014-2940-2.

65. Schlatter D, Brack S, Banner DW, et al. Generation, characterization and structural data of chymase binding proteins based on the human Fyn kinase SH3 domain. MAbs. 2012;4:497–508. https://doi.org/10.4161/mabs.20452.

66. Dennis MS, Lazarus RA. Kunitz domain inhibitors of tissue factor-factor VIIa. I. Potent inhibitors selected from libraries by phage display. J Biol Chem. 1994;269:22129–36.

67. Roberts BL, Markland W, Ley AC, et al. Directed evolution of a protein: selection of potent neutrophil elastase inhibitors displayed on M13 fusion phage. Proc Natl Acad Sci. 1992;89:2429–33. https://doi.org/10.1073/pnas.89.6.2429.

68. Roberts BL, Markland W, Siranosian K, et al. Protease inhibitor display M13 phage: selection of high-affinity neutrophil elastase inhibitors. Gene. 1992;121:9–15. https://doi.org/10.1016/0378-1119(92)90156-J.

69. Röttgen P, Collins J. A human pancreatic secretory trypsin inhibitor presenting a hypervariable highly constrained epitope via monovalent phagemid display. Gene. 1995;164:243–50. https://doi.org/10.1016/0378-1119(95)00441-8.

70. Richter A, Eggenstein E, Skerra A. Anticalins: exploiting a non-Ig scaffold with hypervariable loops for the engineering of binding proteins. FEBS Lett. 2014;588:213–8. https://doi.org/10.1016/j.febslet.2013.11.006.

71. Hosse RJ. A new generation of protein display scaffolds for molecular recognition. Protein Sci. 2006;15:14–27. https://doi.org/10.1110/ps.051817606.

72. Barkovskiy M, Ilyukhina E, Dauner M, et al. An engineered lipocalin that tightly complexes the plant poison colchicine for use as antidote and in bioanalytical applications. Biol Chem. 2019;400:351–66. https://doi.org/10.1515/hsz-2018-0342.

73. Schlehuber S, Beste G, Skerra A. A novel type of receptor protein, based on the lipocalin scaffold, with specificity for digoxigenin. J Mol Biol. 2000;297:1105–20. https://doi.org/10.1006/jmbi.2000.3646.

74. Schonfeld D, Matschiner G, Chatwell L, et al. An engineered lipocalin specific for CTLA-4 reveals a combining site with structural and conformational features similar to antibodies. Proc Natl Acad Sci. 2009;106:8198–203. https://doi.org/10.1073/pnas.0813399106.

75. Dauner M, Eichinger A, Lücking G, et al. Reprogramming human siderocalin to neutralize petrobactin, the essential iron scavenger of anthrax bacillus. Angew Chem Int Ed. 2018;57:14619–23. https://doi.org/10.1002/anie.201807442.

76. Edwardraja S, Eichinger A, Theobald I, et al. Rational design of an anticalin-type sugar-binding protein using a genetically encoded boronate side chain. ACS Synth Biol. 2017;6:2241–7. https://doi.org/10.1021/acssynbio.7b00199.

77. Umscheid CA, Margolis DJ, Grossman CE. Key concepts of clinical trials: a narrative review. Postgrad Med. 2011;123:194–204. https://doi.org/10.3810/pgm.2011.09.2475.

78. Linaclotide. LiverTox: cinical and research information on drug-induced liver injury. National Institute of Diabetes and Digestive and Kidney Diseases, Bethesda (MD) [Updated 2019 May 13]. https://www.ncbi.nlm.nih.gov/books/NBK548021/.

79. Cicardi M, Levy RJ, McNeil DL, et al. Ecallantide for the treatment of acute attacks in hereditary angioedema. N Engl J Med. 2010;363:523–31. https://doi.org/10.1056/NEJMoa0905079.

80. Chua F. Neutrophil elastase: mediator of extracellular matrix destruction and accumulation. Proc Am Thorac Soc. 2006;3:424–7. https://doi.org/10.1513/pats.200603-078AW.

81. Dunlevy FK, Martin SL, de Courcey F, et al. Anti-inflammatory effects of DX-890, a human neutrophil elastase inhibitor. J Cyst Fibros. 2012;11:300–4. https://doi.org/10.1016/j.jcf.2012.02.003.

82. Binz HK, Bakker TR, Phillips DJ, et al. Design and characterization of MP0250, a tri-specific anti-HGF/anti-VEGF DARPin® drug candidate. MAbs. 2017;9:1262–9. https://doi.org/10.1080/19420862.2017.1305529.

83. Miller KL. Do investors value the FDA orphan drug designation? Orphanet J Rare Dis. 2017;12:114. https://doi.org/10.1186/s13023-017-0665-6.
84. Link A, Hepp J, Reichen C, et al. Abstract 3752: preclinical pharmacology of MP0310: a 4-1BB/FAP bispecific DARPin drug candidate promoting tumor-restricted T-cell costimulation. In: Immunology. American Association for Cancer Research, 2018. p 3752.
85. Mross K, Richly H, Fischer R, et al. First-in-human phase I study of PRS-050 (angiocal), an anticalin targeting and antagonizing VEGF-A, in patients with advanced solid tumors. PLoS One. 2013;8:e83232. https://doi.org/10.1371/journal.pone.0083232.
86. Rothe C, Skerra A. Anticalin® proteins as therapeutic agents in human diseases. BioDrugs. 2018;32:233–43. https://doi.org/10.1007/s40259-018-0278-1.
87. Renders L, Budde K, Rosenberger C, et al. First-in-human Phase I studies of PRS-080#22, a hepcidin antagonist, in healthy volunteers and patients with chronic kidney disease undergoing hemodialysis. PLoS One. 2019;14:e0212023. https://doi.org/10.1371/journal.pone.0212023.
88. Samkoe KS, Shahzad Sardar H, Gunn JR, et al. Measuring microdose ABY-029 fluorescence signal in a primary human soft-tissue sarcoma resection. In: Pogue BW, Gioux S, editors. Molecular-guided surgery: molecules, devices, and applications V: SPIE; 2019. p. 38.
89. Schiff D, Kesari S, de Groot J, et al. Phase 2 study of CT-322, a targeted biologic inhibitor of VEGFR-2 based on a domain of human fibronectin, in recurrent glioblastoma. Investig New Drugs. 2015;33:247–53. https://doi.org/10.1007/s10637-014-0186-2.
90. Mullard A. Nine paths to PCSK9 inhibition. Nat Rev Drug Discov. 2017;16:299–301. https://doi.org/10.1038/nrd.2017.83.
91. Nord K, Nord O, Uhlén M, et al. Recombinant human factor VIII-specific affinity ligands selected from phage-displayed combinatorial libraries of protein A. Eur J Biochem. 2001;268:4269–77. https://doi.org/10.1046/j.1432-1327.2001.02344.x.
92. Grönwall C, Sjöberg A, Ramström M, et al. Affibody-mediated transferrin depletion for proteomics applications. Biotechnol J. 2007;2:1389–98. https://doi.org/10.1002/biot.200700053.
93. Rönnmark J, Grönlund H, Uhlén M, Nygren P-Å. Human immunoglobulin A (IgA)-specific ligands from combinatorial engineering of protein A. Eur J Biochem. 2002;269:2647–55. https://doi.org/10.1046/j.1432-1033.2002.02926.x.
94. Rönnmark J, Kampf C, Asplund A, et al. Affibody-β-galactosidase immunoconjugates produced as soluble fusion proteins in the *Escherichia coli* cytosol. J Immunol Methods. 2003;281:149–60. https://doi.org/10.1016/j.jim.2003.06.001.
95. Lundberg E, Höidén-Guthenberg I, Larsson B, et al. Site-specifically conjugated anti-HER2 Affibody® molecules as one-step reagents for target expression analyses on cells and xenograft samples. J Immunol Methods. 2007;319:53–63. https://doi.org/10.1016/j.jim.2006.10.013.
96. Rönnmark J, Hansson M, Nguyen T, et al. Construction and characterization of affibody-Fc chimeras produced in *Escherichia coli*. J Immunol Methods. 2002;261:199–211. https://doi.org/10.1016/S0022-1759(01)00563-4.
97. Renberg B, Nordin J, Merca A, et al. Affibody molecules in protein capture microarrays: evaluation of multidomain ligands and different detection formats. J Proteome Res. 2007;6:171–9. https://doi.org/10.1021/pr060316r.
98. Lyakhov I, Zielinski R, Kuban M, et al. HER2- and EGFR-specific affiprobes: novel recombinant optical probes for cell imaging. Chembiochem. 2010;11:345–50. https://doi.org/10.1002/cbic.200900532.
99. Engfeldt T, Renberg B, Brumer H, et al. Chemical synthesis of triple-labelled three-helix bundle binding proteins for specific fluorescent detection of unlabelled protein. Chembiochem. 2005;6:1043–50. https://doi.org/10.1002/cbic.200400388.

100. Bernhagen D, De Laporte L, Timmerman P. High-affinity RGD-knottin peptide as a new tool for rapid evaluation of the binding strength of unlabeled RGD-peptides to α v β 3 , α v β 5 , and α 5 β 1 integrin receptors. Anal Chem. 2017;89:5991–7. https://doi.org/10.1021/acs.analchem. 7b00554.

101. Muñoz-Maldonado C, Zimmer Y, Medová M. A comparative analysis of individual RAS mutations in cancer biology. Front Oncol. 2019;9:1088. https://doi.org/10.3389/fonc.2019. 01088.

102. Spencer-Smith R, Koide A, Zhou Y, et al. Inhibition of RAS function through targeting an allosteric regulatory site. Nat Chem Biol. 2017;13:62–8. https://doi.org/10.1038/nchembio. 2231.

103. Wojcik J, Hantschel O, Grebien F, et al. A potent and highly specific FN3 monobody inhibitor of the Abl SH2 domain. Nat Struct Mol Biol. 2010;17:519–27. https://doi.org/10.1038/nsmb.1793.

104. Skerra A. Alternative binding proteins: anticalins – harnessing the structural plasticity of the lipocalin ligand pocket to engineer novel binding activities. FEBS J. 2008;275:2677–83. https://doi.org/10.1111/j.1742-4658.2008.06439.x.

105. Theurillat J-P, Dreier B, Nagy-Davidescu G, et al. Designed ankyrin repeat proteins: a novel tool for testing epidermal growth factor receptor 2 expression in breast cancer. Mod Pathol. 2010;23:1289–97. https://doi.org/10.1038/modpathol.2010.103.

106. Sennhauser G, Grütter MG. Chaperone-assisted crystallography with DARPins. Structure. 2008;16:1443–53. https://doi.org/10.1016/j.str.2008.08.010.

107. Mittl PR, Ernst P, Plückthun A. Chaperone-assisted structure elucidation with DARPins. Curr Opin Struct Biol. 2020;60:93–100. https://doi.org/10.1016/j.sbi.2019.12.009.

108. Bandeiras TM, Hillig RC, Matias PM, et al. Structure of wild-type Plk-1 kinase domain in complex with a selective DARPin. Acta Crystallogr Sect D Biol Crystallogr. 2008;64:339–53. https://doi.org/10.1107/S0907444907068217.

109. Schweizer A, Roschitzki-Voser H, Amstutz P, et al. Inhibition of caspase-2 by a designed Ankyrin repeat protein: specificity, structure, and inhibition mechanism. Structure. 2007;15:625–36. https://doi.org/10.1016/j.str.2007.03.014.

110. Sennhauser G, Amstutz P, Briand C, et al. Drug export pathway of multidrug exporter AcrB revealed by DARPin inhibitors. PLoS Biol. 2006;5:e7. https://doi.org/10.1371/journal.pbio. 0050007.

111. Gumpena R, Lountos GT, Waugh DS. MBP-binding DARPins facilitate the crystallization of an MBP fusion protein. Acta Crystallogr Sect F Struct Biol Commun. 2018;74:549–57. https://doi.org/10.1107/S2053230X18009901.

112. Batyuk A, Wu Y, Honegger A, et al. DARPin-based crystallization chaperones exploit molecular geometry as a screening dimension in protein crystallography. J Mol Biol. 2016;428:1574–88. https://doi.org/10.1016/j.jmb.2016.03.002.

113. Wu Y, Honegger A, Batyuk A, et al. Structural basis for the selective inhibition of c-Jun N-terminal kinase 1 determined by rigid DARPin–DARPin fusions. J Mol Biol. 2018;430:2128–38. https://doi.org/10.1016/j.jmb.2017.10.032.

114. Ernst P, Honegger A, van der Valk F, et al. Rigid fusions of designed helical repeat binding proteins efficiently protect a binding surface from crystal contacts. Sci Rep. 2019;9:16162. https://doi.org/10.1038/s41598-019-52121-9.

115. Schütz M, Batyuk A, Klenk C, et al. Generation of fluorogen-activating designed ankyrin repeat proteins (FADAs) as versatile sensor tools. J Mol Biol. 2016;428:1272–89. https://doi.org/10.1016/j.jmb.2016.01.017.

116. Xu S, Hu H-Y. Fluorogen-activating proteins: beyond classical fluorescent proteins. Acta Pharm Sin B. 2018;8:339–48. https://doi.org/10.1016/j.apsb.2018.02.001.

117. Wang Y, Prosen DE, Mei L, et al. A novel strategy to engineer DNA polymerases for enhanced processivity and improved performance in vitro. Nucleic Acids Res. 2004;32:1197–207. https://doi.org/10.1093/nar/gkh271.
118. Béhar G, Renodon-Cornière A, Mouratou B, Pecorari F. Affitins as robust tailored reagents for affinity chromatography purification of antibodies and non-immunoglobulin proteins. J Chromatogr A. 2016;1441:44–51. https://doi.org/10.1016/j.chroma.2016.02.068.
119. Kruziki MA, Bhatnagar S, Woldring DR, et al. A 45-amino-acid scaffold mined from the PDB for high-affinity ligand engineering. Chem Biol. 2015;22:946–56. https://doi.org/10.1016/j.chembiol.2015.06.012.
120. Kruziki MA, Case BA, Chan JY, et al. 64 cu-labeled Gp2 domain for PET imaging of epidermal growth factor receptor. Mol Pharm. 2016;13:3747–55. https://doi.org/10.1021/acs.molpharmaceut.6b00538.
121. Strauch E-M, Fleishman SJ, Baker D. Computational design of a pH-sensitive IgG binding protein. Proc Natl Acad Sci. 2014;111:675–80. https://doi.org/10.1073/pnas.1313605111.
122. Reichen C, Hansen S, Plückthun A. Modular peptide binding: from a comparison of natural binders to designed armadillo repeat proteins. J Struct Biol. 2014;185:147–62. https://doi.org/10.1016/j.jsb.2013.07.012.
123. Coates J. Armadillo repeat proteins: beyond the animal kingdom. Trends Cell Biol. 2003;13:463–71. https://doi.org/10.1016/S0962-8924(03)00167-3.
124. Parmeggiani F, Pellarin R, Larsen AP, et al. Designed Armadillo repeat proteins as general peptide-binding scaffolds: consensus design and computational optimization of the hydrophobic core. J Mol Biol. 2008;376:1282–304. https://doi.org/10.1016/j.jmb.2007.12.014.
125. Reichen C, Hansen S, Forzani C, et al. Computationally designed armadillo repeat proteins for modular peptide recognition. J Mol Biol. 2016;428:4467–89. https://doi.org/10.1016/j.jmb.2016.09.012.
126. Sassoon I, Blanc V. Antibody–drug conjugate (ADC) clinical pipeline: a review. Methods Mol Biol. 2013;1045:1–27.
127. Birrer MJ, Moore KN, Betella I, Bates RC. Antibody-drug conjugate-based therapeutics: state of the science. JNCI J Natl Cancer Inst. 2019;111:538–49. https://doi.org/10.1093/jnci/djz035.
128. Merten H, Schaefer JV, Brandl F, et al. Facile site-specific multiconjugation strategies in recombinant proteins produced in bacteria. Methods Mol Biol. 2019;2033:253–73.
129. Zhang Y, Auger S, Schaefer JV, et al. Site-selective enzymatic labeling of designed Ankyrin repeat proteins using protein farnesyltransferase. Methods Mol Biol. 2019;2033:207–19.
130. Wang D, Tai PWL, Gao G. Adeno-associated virus vector as a platform for gene therapy delivery. Nat Rev Drug Discov. 2019;18:358–78. https://doi.org/10.1038/s41573-019-0012-9.
131. Dreier B, Honegger A, Hess C, et al. Development of a generic adenovirus delivery system based on structure-guided design of bispecific trimeric DARPin adapters. Proc Natl Acad Sci. 2013;110:E869–77. https://doi.org/10.1073/pnas.1213653110.
132. Hartmann J, Münch RC, Freiling R-T, et al. A library-based screening strategy for the identification of DARPins as ligands for receptor-targeted AAV and lentiviral vectors. Mol Ther Methods Clin Dev. 2018;10:128–43. https://doi.org/10.1016/j.omtm.2018.07.001.
133. Hanzl A, Winter GE. Targeted protein degradation: current and future challenges. Curr Opin Chem Biol. 2020;56:35–41. https://doi.org/10.1016/j.cbpa.2019.11.012.
134. Chamberlain PP, Hamann LG. Development of targeted protein degradation therapeutics. Nat Chem Biol. 2019;15:937–44. https://doi.org/10.1038/s41589-019-0362-y.
135. Fulcher LJ, Hutchinson LD, Macartney TJ, et al. Targeting endogenous proteins for degradation through the affinity-directed protein missile system. Open Biol. 2017;7:170066. https://doi.org/10.1098/rsob.170066.

Chimeric Antigen Receptor (CAR) Redirected T Cells

10

Astrid Holzinger and Hinrich Abken

Contents

A. Holzinger · H. Abken (✉)
Regensburg Center for Interventional Immunology (RCI), Department Genetic Immunotherapy,
University Hospital Regensburg, Regensburg, Germany
e-mail: hinrich.abken@ukr.de

© Springer Nature Switzerland AG 2021
F. Rüker, G. Wozniak-Knopp (eds.), *Introduction to Antibody Engineering*,
Learning Materials in Biosciences,
https://doi.org/10.1007/978-3-030-54630-4_10

251

Keywords

Adoptive cell therapy · Chimeric antigen receptor · T cell · Clinical trial · Costimulation · Cancer

Abbreviations

ACT	Adoptive cell therapy
ADCC	Antibody-dependent cellular cytotoxicity
AICD	Activation-induced cell death
AML	Acute myeloid leukemia
APRIL	A proliferation induced ligand
B-ALL	B cell-acute lymphocytic leukemia
BCMA	B cell maturation antigen
BiTE	Bispecific T cell engager
BTK	Bruton tyrosine kinase
CAR	Chimeric antigen receptor
CEA	Carcinoembryonic antigen
CLL	Chronic lymphocytic leukemia
CRS	Cytokine release syndrome
CSC	Cancer stem cell
CTLA-4	Cytotoxic T lymphocyte-associated antigen-4
CXCR2	C-X-C motif chemokine receptor-2
DARPin	Designed ankyrin repeat protein
dcCAR	Dual chain CAR

DLBCL	Diffuse large B cell lymphoma
EBV	Epstein-Barr virus
EGFR	Epithelial growth factor receptor
EMA	European Medical Agency
FAP	Fibroblast activation protein
FcεRI	Fc ε receptor-I
FcγR	Fc γ receptor
FDA	Food and Drug Administration
FITC	Fluorescein isothiocyanate
GM-CSF	Granulocyte macrophage-colony stimulating factor
GMP	Good manufacturing practice
GvHD	Graft-versus-host disease
HLH	Hemophagocytic lymphohistiocytosis
ICANS	Immune effector cell-associated neurotoxicity syndrome
iCAR	Inhibitory CAR
iCasp9	Inducible Cas9
IDO	Indoleamine 2,3-dioxygenase
IFN	Interferon
Ig	Immunoglobulin
IL	Interleukin
ITAM	Immunoreceptor tyrosine activation motif
MAS	Macrophage activation syndrome
MDSCs	Myeloid-derived suppressor cells
MHC	Major histocompatibility complex
NK	Natural killer
PD-1	Programmed cell death-1
PD-L	Programmed cell death ligand
PKA	Protein kinase A
scFv	Single-chain fragment of variable region
synNotch	Synthetic Notch
TALEN	Transcription activator-like effector nuclease
TCR	T cell receptor
TIGIT	T cell immunoreceptor with Ig and ITIM domains
TIL	Tumor-infiltrating lymphocyte
TLR	Toll-like receptor
Treg	Regulatory T cell
TRUCK	T cells redirected for unrestricted cytokine-mediated killing
VEGF	Vascular endothelial growth factor
VLS	Vascular leakage syndrome
ZFN	Zinc finger nuclease

What You Will Learn in This Chapter

Adoptive immunotherapy with chimeric antigen receptor (CAR) redirected T cells is a new therapeutic paradigm with a new class of therapeutic "drugs" producing complete and lasting remissions in patients with certain subtypes of so far refractory B cell malignancies. CAR T cells are derived from the patient's own immune cells that are ex vivo genetically engineered with a CAR that binds a defined target and transmits an activation signal upon target engagement. The CAR activates cytolytic T cells to destruct and eliminate cognate cancer cells in an efficient, repetitive, and lasting fashion. However, various barriers limit the widespread use of CAR T cells in other types of cancer or in other diseases. Current challenges include designing the optimal CAR, actively penetrating the tumor lesion, and sustaining T cell antitumor activation without inducing toxicities to healthy tissues. The chapter reviews the generations of CARs, strategies to enhance CAR T cell efficacy and selectivity, and to establish efficient manufacturing processes for a growing number of patients.

10.1 The Prototype Chimeric Antigen Receptor (CAR, T-Body, Immunoreceptor)

The power of the patient's T cells to recognize and destroy cancer cells in a specific and efficient fashion became obvious in the mid-1980s when Steven A. Rosenberg (NIH) re-infused tumor-infiltrating lymphocytes (TILs), isolated from melanoma biopsies and extensively amplified ex vivo, to melanoma patients [1]. These and further developments have drawn the concept of adoptive cell therapy to treat tumor patients with autologous immune cells which have the capacity to induce tumor regressions and long-term remissions in a substantial number of patients [2]. Such amplified TILs are assumed to exhibit specificity for particular tumor antigens and to exhibit migratory capacities to enter the tumor lesion; the antigen specificity, however, mostly remains unknown.

In the following years, tremendous efforts were made to provide defined targeting specificity to patient's T cells; genetic engineering with T cell receptor (TCR) chains, or with a recombinant chimeric antigen receptor (CAR), as discussed herein, turned out to be one of the successful strategies. In contrast to the TCR, the CAR is a composite, one-polypeptide chain transmembrane receptor that provides both targeting specificity and T cell activation capacities upon target recognition. The prototype CAR is modularly composed of four major domains: at the extracellular terminus a single-chain fragment of variable region (scFv) antibody for target recognition, a spacer linking to the transmembrane domain, and intracellular signaling moieties mostly derived from the TCR/CD3ζ or the Fc ε receptor-I (FcεRI) with or without a linked costimulatory domain (Fig. 10.1). Engagement of cognate antigen on the surface of target cells initiates activation of the CAR engineered immune cell resulting in a lasting T cell response [3, 4]. Each CAR domain has

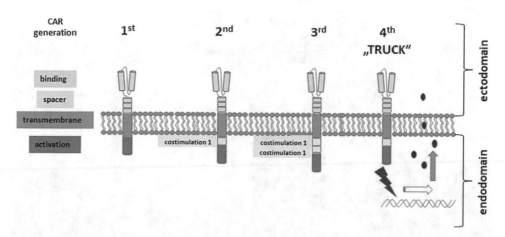

Fig. 10.1 The growing world of Chimeric Antigen Receptors (CARs). The first generation CAR harbors antigen-binding and spacer moieties in the extracellular domain (ectodomain), the transmembrane domain, and the activating moiety derived from the CD3ζ or FcεRI γ chain in the intracellular domain (endodomain). The CAR of second generation harbors in addition a costimulatory domain, the CAR of third generation two costimulatory domains in the endodomain. The fourth-generation CAR ("TRUCK", "T cells redirected for universal cytokine-mediated killing") consists of a second- or third-generation CAR that induces the expression of a transgenic protein upon CAR signaling. While depicted in the canonical design, a plethora of CAR variants was engineered, each optimized for different applications

a distinct function and impacts on the redirected T cell activation; for each specific application, the CAR design requires some adjustments in order to provide optimized targeting and T cell activation as described below in detail.

The basic design of the prototype CAR was firstly proposed by Zelig Eshhar and colleagues (Weizmann Institute of Science) by demonstrating that a modular composite receptor, at first named "T-body" or "immunoreceptor", with an antibody derived binding domain for binding and a TCR signaling domain for T cell activation provides both specific recognition and activation to the engineered T cell [5]. During the following years, a number of modifications were introduced allowing to target a broad variety of antigens and to signal through various intracellular domains in order to initiate and redirect defined immune cell functions (Table 10.1).

The canonical CAR is modularly composed of an antibody-derived binding domain, a spacer, a transmembrane domain, and intracellular primary and costimulatory signaling domains. Each domain can be built up by different modules that affect distinct variables and finally impact T cell activation and performance.

CAR engineered cytolytic T cells redirected toward cancer cells are the most studied application of the technology; adoptive transfer of autologous anti-CD19 CAR T cells became the first Food and Drug Administration (FDA) approved therapy with genetically engineered cells in the United States [6, 7]; subsequently, the European Medical Agency

Table 10.1 CAR building blocks and their alternative modules

	Alternative modules	Variable	Impact on
Binding domain	• Cytokine • Peptide (DARPin, adnectin) • Ligand or receptor • Nanobody • scFv • VH and VL antibody chains	• Affinity • Specificity • Solubility • Position of targeted epitope • Antigen density on target cells	• Target recognition • Activation threshold • Activation-induced cell death • Exhaustion • Terminal differentiation
Spacer/hinge domain	• CD4 • CD7 • CD8 • CD28 • IgD • IgG1 • IgG4	• Length (IgG1 CH1-CH2-CH3) • Dimerizing capacity • FcγR receptor binding capacity (IgG1, IgG4)	• Flexibility • Stability • Dimerization • Access to target antigen • Signal transmission • T cell-target cell distance
Transmembrane domain	• CD3ε, CD3ζ • CD4 • CD7 • CD8α • CD28 • FcεRIγ • NKG2D • OX40	• Incorporation into TCR/CD3	• Half-life time on cell surface • Signaling capacities
Signaling domain	• CD3ζ (3 ITAMs) • DAP10 • DAP12 • FcεRIγ (1 ITAM) • Fyn • Lck • Syk • Zap70	• Activation of downstream signaling proteins	• Primary T cell activation
Costimulatory domain	• CD2 • CD27 • CD28 • CD40-MyD88 • CD137 (4-1BB) • CD244	• PI3K/Akt/mTOR pathway activation (CD28) • IL-2 release (CD28) • Wnt/β-catenin pathway activation (4-1BB)	• Duration and quality of T cell activation • T cell survival and persistence • TGF-β resistance (CD28) • Glycolysis and effector memory cell polarization (CD28) • Oxidative metabolism and

(continued)

Table 10.1 (continued)

Alternative modules	Variable	Impact on
• ICOS • NKG2D • OX40		central memory cell polarization (4-1BB)

(EMA) approved. Other applications using other host cells and a different CAR design are increasingly also studied in early phase clinical trials.

10.2   Modular Composition of a CAR: The Extracellular CAR Domains

10.2.1 The Binding Domain

The antigen-binding domain is traditionally composed of a single-chain fragment of variable region (scFv) derived from a murine or human monoclonal antibody. The scFv is engineered by joining the variable (V) regions of the heavy (H) and light (L) antibody chains by a short flexible linker, mostly the $(Gly_4Ser)_3$ peptide, resulting in the order VH-linker-VL or VL-linker-VH. The glycine residues are providing the flexibility to allow proper folding of the VH and VL to form the antigen-binding pocket while serine is providing some solubility to the linker. The spatial position and interaction of the VH and VL chains can substantially affect affinity, specificity, and solubility. Sufficiently high binding affinity is essential for inducing T cell activation; however, there is an affinity ceiling above which CAR mediated T cell activation is not improved but results in activation-induced cell death (AICD) [8–14]. Some scFvs in the context of a CAR induce AICD, exhaustion, and terminal T cell differentiation by tonic signaling independently of antigen engagement [15, 16]. Beyond affinity and antigen density on the cell surface, the position of the targeted epitope within the antigen matters; in general targeting a membrane-proximal epitope induces stronger T cell activation than targeting a membrane distal epitope [17].

By using an antibody for targeting, the CAR mediated recognition is MHC-independent. This is of benefit when targeting tumors with deficiencies in the MHC-peptide processing and presentation machinery. By recognizing targets through an antibody, moreover, antigens like carbohydrates, lipids, or structural variants of an antigen can be recognized that are invisible to classical T cells. The canonical CAR recognizes antigens on the surface of a target cell; MHC-peptide recognizing CARs can also be engineered by using an antibody with specificity for a defined MHC–peptide complex, enabling sensing of intracellular antigens presented by the MHC [18–22]. As an example, a CAR with an scFv that recognizes NY-ESO-1 peptide in the context of HLA-A2 provides TCR-like recognition of a target like the corresponding TCR [23].

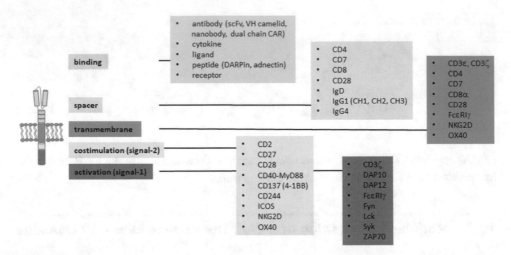

Fig. 10.2 Variants in the design of a second-generation CAR. Due to the modular composition, various alternative domains can be used to set up a CAR, some of them are exemplarily listed. Variations of the individual moieties allow to modulate CAR binding and signaling, finally to tune cellular functions, metabolism, and antitumor activities of the engineered T cells. Other variants were designed to improve binding affinity, accessibility to the target, signaling strength, and persistence on the cell membrane, among others. Finally, safety precautions resulted in further modifications in the CAR design

The prototype CAR harbors an scFv antibody for antigen recognition derived from a natural antibody; naturally occurring small single-chain antibodies, e.g., nanobodies derived from camelids heavy chain antibodies, can likewise be used as the binding domain of a CAR [24] (Fig. 10.2). If an scFv is not available, the conversion of a natural antibody into an scFv format, however, does not always conserve the binding affinity and specificity. To circumvent the limitation, an alternative CAR design was reported that uses the natural antibody format with the immunoglobulin (Ig) heavy chain linked to the transmembrane and intracellular CAR domain; the co-expressed Ig light chain spontaneously associates with the heavy chain forming a fully functional antibody anchored by a transmembrane domain to the T cell surface [25]. Such an immunoglobulin chain heterodimer, a so-called dual chain CAR (dcCAR), binds through the natural antibody V region of the heavy and light chain to its cognate antigen and allows targeting of antigens for which an scFv antibody is not available.

As an alternative to antibodies, recombinant binding domains with antibody-like specificity can be integrated into the CAR-like designed ankyrin repeat proteins (DARPins), which are composed of 33 amino acids ankyrin repeats and form a β-turn followed by two anti-parallel α-helices and a loop reaching the β-turn of the next repeat [26]. As a proof-of-concept, such DARPin CAR targets Her2 in a specific fashion [26, 27]. Adnectin CARs with a fibronectin derived binding domain were engineered to target epithelial growth factor receptor (EGFR) with high selectivity [28].

Any other binding domain is also suitable in the context of a CAR, e.g., the binding moiety of a cytokine receptor, a cytokine, or a natural ligand. Zetakine CARs harbor a cytokine linked to a transmembrane and intracellular signaling domain. For instance, a CAR with mutated IL-13 in the extracellular domain mediates binding to IL-13 receptor-α 2 with improved binding capacities thereby allowing preferential targeting of solid tumors compared to healthy tissues [29–32]. Examples for ligand-based CARs are the APRIL containing CAR binding to B cell maturation antigen (BCMA) [33]. A CAR with the FLT3 ligand recognizes FLT3$^+$ acute myeloid leukemia (AML) cells [34]; a CAR with granulocyte macrophage-colony stimulating factor (GM-CSF) as binding domain targets juvenile myelomonocytic leukemia by binding to CD116 [35]; a CAR with the NK cell receptor NKG2D targets cancer cells with the respective ligands [36].

10.2.2 The Spacer Domain

The prototype CAR frequently harbors a "spacer" between the antigen-binding and the transmembrane domain to provide access to the target antigen; the linked hinge region provides some flexibility. The type of spacer and the length is crucial for optimal target binding and signal transmission. The spacer is not mandatory since some CARs are active when the scFv is directly fused onto the transmembrane domain, other CARs are only active with a spacer of a certain length. The spacer is typically derived from the IgG1 or IgG4 constant domain or from the extracellular CD28, CD4, or CD8 [4, 37, 38] (Fig. 10.2). The general idea is that the spacer provides an optimal distance between the engineered T cell and the cognate target cell in order to allow the formation of a productive contact zone to the target cell [39]. Based on the knowledge of TCR-MHC interaction, it is generally believed that the CAR-target interaction needs to create a distance of about 15 nm between the T cell and the target cell for optimal formation of the contact zone [40]. Consequently, a longer spacer is required to target membrane-proximal epitopes while a smaller spacer is optimal for a more distal epitope; the distance can easily be adjusted by using a spacer of different lengths [41–44]. In this context, the IgG derived spacer has the advantage that the length can be adjusted by incorporating one, two, or three moieties of CH1-CH2-CH3 [45]. The IgG1 derived spacer domain moreover favors stabilization of the CAR molecule; however, it can potentially bind to Fc γ receptors (FcγR) (CD64) on myeloid cells, mediating "off-tumor" activation of both CAR T cells and myeloid cells. This can be prevented by deleting the IgG1 CH2 domain within the spacer or by deleting the Asn297 glycosylation site within the Fc receptor binding moieties of IgG1 [37, 46]. A similar modification can be introduced into the IgG4 spacer [47].

10.2.3 The Transmembrane Region

The membrane-spanning region of a typical CAR is derived from type-1 membrane-spanning proteins like CD3ζ, CD4, CD8α, CD28, or OX40 and consists of 20–23 hydrophobic amino acids, rich in leucines, isoleucines, and valines [4] (Fig. 10.2). CARs with CD3ζ transmembrane domain seem to incorporate into the TCR/CD3 complex more efficiently than others, thereby being more robust in expression and signaling [48]; CARs with CD28 transmembrane region seem to be likewise stable [49]. On the other hand, CARs with CD8α hinge and transmembrane domain are less prone to AICD; however, they mediate less T cell activation indicated by reduced release of pro-inflammatory cytokines [50].

10.2.4 Intracellular Signaling Domains

The "first generation" CARs signal entirely through the primary signaling domain (signal-1). The TCR/CD3ζ intracellular chain harbors three immunoreceptor tyrosine activation motifs (ITAMs), the FcεRI γ-chain one ITAM, which serve as specific adaptors for downstream signaling proteins upon phosphorylation. Thereby the CAR is utilizing the endogenous TCR signaling machinery for initiating the cascade of T cell activation events. Downstream kinases like Lck or Fyn can likewise be used as primary signals in CAR-mediated T cell activation, among others (Fig. 10.2). The CAR signaling domain alone is sufficient to associate with downstream kinases and to initiate a productive activation cascade since the CAR is also functional in TCR knock-out cells [51] and in non-T cells like NK cells [52]. However, CARs with only the primary signal are insufficiently activated to allow long-term persistence; such CAR T cells rapidly enter apoptosis.

According to the "two signal hypothesis", resting T cells require in addition a costimulatory signal for sustained activation. This is the rationale to combine the primary with a costimulatory signal (signal-2), like CD28 or 4-1BB (CD137), in a "second generation" CAR. CARs with alternative costimulatory domains, like OX40 [53], ICOS [54, 55], CD27 [56], CD40-MyD88 [57, 58], CD2 [59], and CD244 [60], are also preclinically and clinically explored. For reasons of accessibility for downstream kinases, CD28 is usually at the membrane proximal position followed by CD3ζ in the distal position; OX40 and 4-1BB are active in the membrane distal and proximal position. In contrast to first generation CARs, second generation CAR T cells show a robust and durable response with respect to cytokine release, amplification, and antitumor activity.

CAR-mediated costimulation impacts in a different fashion T cell effector functions, survival, persistence, and metabolism. For instance, 4-1BB CAR T cells persisted for more than 6 months in the blood of patients whereas CD28 CAR T cells were mostly undetectable beyond 3 months [61]. After repetitive stimulation CD28 CAR T cells reprogram toward CD45RO⁺ CCR7⁻ effector memory cells which then require OX40 to prolong persistence and to escape rapid exhaustion; 4-1BB CAR T cells predominantly convert to

CD45RO$^+$ CCR7$^+$ central memory cells and persist long-term in the peripheral blood [61–63]. CD28 CAR signaling activates the PI3K/Akt/mTOR pathway which stimulates aerobic glycolysis, increases glucose uptake and ATP generation [64] while 4-1BB CAR signaling stimulates the Wnt/β-catenin pathway and fatty acid oxidation which increases catabolism and mitochondrial respiratory chain capacities [65]. Canonical Wnt/β-catenin favors the formation of central memory cells and long-term survival of T cells while CD28-induced PI3K/Akt signaling sustains the effector cell immediate response [66–70]. In high doses both CD28 and 4-1BB CAR T cells eradicate large established tumors in preclinical models; at lower doses, CD28 CAR T cells show a larger degree of exhaustion than 4-1BB CAR T cells with the result that 4-1BB CAR T cells more efficiently eradicated a large tumor load in the long-term [71]. On the other hand, T cells with 4-1BB CAR are still sensitive to tumor-mediated inhibition through TGF-β, which is less the case for CD28 CAR T cells [72]. Therefore, the criteria for selecting a CAR design depends on multiple parameters including T cell persistence, resistance to repression, the pattern of costimulatory and coinhibitory ligands, the antigen density on tumor cells, the CAR density on T cells among others.

"Third generation" CARs combine two costimulatory domains along with the primary activation signal. The benefit of combined costimulation is documented only in a few cases, for instance, CD28-OX40 costimulation through a CAR provides benefit for T cells progressed in terminal maturation [73].

10.3   TRUCK: The "Fourth Generation" CAR T Cell

CAR T cells can be used to deliver transgenic proteins "on demand" to a predefined, targeted tissue. Therefore, CAR T cells were engineered with a CAR inducible expression cassette for an additional "payload". The "fourth generation" CAR T cells, so-called TRUCKs ("T cells redirected for universal cytokine-mediated killing"), are designed to deposit a secreted transgenic protein in the targeted tissue to achieve therapeutically effective concentrations (Fig. 10.3) [74]. The expression of the "payload" is under control of the NFAT$_6$-IL-2 minimal promoter which is activated upon CAR signaling; the payload can be any protein or peptide encoded by the transgene. The strategy is of particular interest in order to combine the redirected CAR T cell attack with the action of a locally deposited, biologically active protein while avoiding systemic toxicity. Such TRUCKs deposit the transgenic product as long as the CAR T cell remains activated [75, 76], for instance, IL-12 or IL-18 as a transgenic immune modifier to shape the targeted tumor environment in a specific fashion without causing systemic toxicity [75–81]. Accumulated, TRUCK produced IL-12 recruited innate immune cells, like NK cells and macrophages, in the CAR targeted tumor tissue where in turn a secondary antitumor immune response is initiated [77]. IL-12 TRUCKs resisted suppression by Treg cells [78] and showed an increased cytokine release and expansion [82]. The strategy is of particular interest when fast progressing tumors accumulate antigen loss cancer cells which give rise to tumor relapses

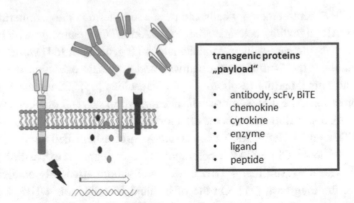

Fig. 10.3 The fourth generation of CAR T cells ("TRUCKs"). Upon antigen engagement, a second or third-generation CAR induces the expression of a transgenic protein that is acting intracellularly, integrated into the membrane or released by the CAR T cell. Released transgenic proteins are aimed at modifying the tumor environment, re-shaping the tumor tissue, or attracting a secondary immune response. The CAR T cells are induced to release the protein only when the CAR is engaging its target, thereby delivering the "payload" at a predefined site of delivery as a "TRUCK". Such CAR T cells are understood as "living factories" for the precise and continuous delivery of therapeutic cell products

invisible to CAR T cells. TRUCKs delivering IL-18 converted CAR T cells toward T-bet^{high} FoxO1^{low} effector cells resisting exhaustion and showing superior cytolytic activity against large established tumors [80]. IL-18 CAR T cell treatment changed the overall tumor immune cell landscape, in particular by increasing the numbers of CD206^{-} M1 macrophages and NKG2D^{+} NK cells and reducing suppressor cells like Tregs, suppressive CD103^{+} DCs and M2 macrophages. Other examples are TRUCKs that release catalase to protect the T cells from oxidative stress-mediated repression [83] and heparanase to improve T cell penetration through the tumor stroma [84].

So-called "armored CARs" producing 4-1BB ligand provide increased costimulation through stimulating the 4-1BB pathway [85]. T cells engineered to secrete Toll-like receptor (TLR) ligands can stimulate the TLR pathway on T cells and antigen-presenting cells which then can activate a panel of T cells for a broad antitumor attack [86, 87]. CAR T cells targeting Muc16ecto, which is the membrane-retained residue of mucin-16 (CA-125), and secreting IL-12 are evaluated in a clinical trial (NCT02498912) [88]. IL-15 TRUCK cells showed improved amplification and antitumor activity [89], however, demand a suicide gene in case of uncontrolled amplification [90] due to their leukemic potential [91]. "Armored" CARs were engineered by co-expressing the IL-7 receptor-α chain along with the CAR to restore responsiveness to IL-7 and to promote a Th1 response [92, 93]. Similarly, in prostate cancer with increased IL-4 levels, co-expressing the IL-4 binding/IL-7 signaling receptor improved the antitumor activity of anti-PSCA CAR T cells [94]. On the other hand, a co-expressed dominant-negative TGF-β DNRII receptor on CAR

T cells can compete with TGF-β to improve T cell antitumor activity in the presence of TGF-β in a melanoma model [95]. Applications beyond providing a cytokine or cytokine receptor can likewise be envisaged like delivering the soluble HVEM ectodomain which targets the tumor vasculature in order to sustain tumor penetration [96].

10.4   CAR T Cells with Dual Specificities, Combinatorial Antigen Recognition, and Conditional CARs

The prototype CAR redirects specificity toward one antigen by one scFv; several reasons demand CARs with multiple specificities: (1) tumor lesions may be heterogeneous with respect to targetable antigens or lose the antigen during tumor progression; (2) targeting of both cancer cells and tumor stroma or vasculature may be beneficial or required to increase the therapeutic efficacy. Applying two or more CAR T cell products to the cancer patient is basically feasible, although requires manufacturing two cell products under GMP conditions. For instance, CD19 and CD22 CAR T cells applied by sequential infusions are currently evaluated in a trial for the treatment of refractory B-ALL (NCT03620058). Alternatively, T cells are engineered with two CARs, so-called bicistronic CARs that express two full-length CARs, each targeting a different antigen (Fig. 10.4). A CAR with dual specificities is engineered by linking two scFvs to each other by a flexible linker. Such a bispecific CAR with scFvs arranged in tandem ("tandem-CAR", "Tan-CAR") or in a loop structure, targets two different antigens and is capable to provide T cell activation upon binding to either antigen ("OR" gating); target cells with one of the cognate antigens are still recognized [97]. Apart from two linked scFvs, diabodies, two-in-one antibodies, and dual variable domain antibodies can likewise be used to engineer a CAR with two specificities. Tri-specific CARs harbor a combination of monovalent and/or multi-specific CAR constructs to target three antigens, further broadening targeting specificities.

One example is a bispecific CAR with linked anti-CD20 and anti-CD19 scFv designed to mitigate relapse of a B cell leukemia/lymphoma that frequently occurs upon CD19 CAR T cell treatment [98]. Relapse by CD19$^-$ CD20$^+$ leukemic cells is aimed at being controlled by CD20-CD19 bispecific CAR T cells. The order of scFvs within the bispecific CAR depends on the spatial accessibility of the targeted epitope within the particular antigen. The CD20-CD19 CAR binds equally to both, CD19 and CD20, and redirects T cell activation while the CD19-CD20 CAR is far less efficient. Other examples of bispecific CARs are targeting CD19 and CD123 for the treatment of B-ALL [99] or CD19 co-targeting CD22 [42], ROR1 [100] and immunoglobulin kappa light chain (Igκ) [101].

Multi-specific CAR T cells are currently under clinical evaluation in early phase trials. T cells engineered with both a CD19 and CD22 CAR through a bicistronic expression cassette are in trial for the treatment of B-ALL of high risk or relapse after stem cell transplant (NCT03289455). Bivalent CARs with CD19 and CD22 specificities are evaluated for the treatment of B-ALL or B cell lymphoma (NCT03241940, NCT03233854, NCT03330691, NCT03448393, NCT03019055). A bispecific CAR

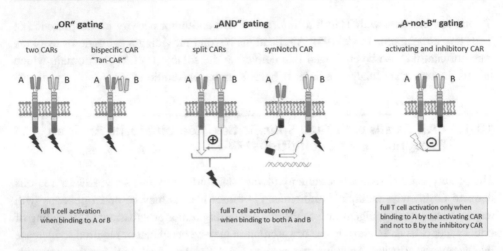

Fig. 10.4 CAR-mediated T cell activation based on logic gating in order to prove selectivity in tumor targeting. Two CARs are co-expressed, each CAR is capable of fully activating T cells upon engagement of the respective target antigen; the same applies for a bispecific CAR (tandem CAR, Tan-CAR). In this case, T cells are activated by engagement of cells that express either antigen A or B or both ("OR" gating). In order to activate CAR T cells only by cells with both A and B antigen, the CAR recognizing A provides only basic activation through CD3ζ, which is insufficient for full and persistent activation, while the CAR recognizing B provides costimulation; simultaneous signaling through both CARs induces full T cell activation ("AND" gating). The synNotch CAR recognizing antigen A drives transcription and expression of a second CAR that drives full T cell activation upon recognizing antigen B. An inhibitory CAR with an ITIM (immunoreceptor tyrosine-based inhibitory motif) prevents signaling through the activating CAR as long as engaging its cognate antigen B. When antigen B is absent, the CAR can activate the T cell when binding to antigen A ("A-not-B" gating)

targeting two epitopes on BCMA is currently evaluated in a phase II trial (NCT03758417) [102].

When engaging both antigens, bispecific CARs bind to target cells with higher avidity than mono-specific CARs which ideally stabilizes the synapses and improves the T cell response against target cells with low antigen levels. This situation can be used to redirect CAR T cells only against those cells that express both antigens but not one (Fig. 10.4). Such "AND" gating of antigen patterns is aimed at improving targeting selectivity for cancer cells while sparing healthy cells. Two co-expressed antigens are recognized by two co-expressed CARs, each recognizing a particular antigen and providing the primary or costimulatory signal, respectively. Thereby, only simultaneous engagement of both antigens provides both signals required for full T cell activation; engagement of one antigen is not sufficient. Examples of combinatorial antigen recognition are targeting ErbB2 by the CD3ζ CAR and Muc1 by the CD28 CAR [103], or targeting mesothelin by the CD3ζ CAR and folate receptor-α by the CD28 CAR [104]. Bispecific and split CAR T cells targeting CD13 and TIM3 eradicated patient-derived acute myeloid leukemic cells in preclinical models with reduced toxicity to bone marrow and peripheral myeloid cells [105].

The combination of primary and costimulatory signal on two CARs can be enforced by hetero-dimerizing the two polypeptide chains; one chain consists of the extracellular and transmembrane together with the primary signaling moiety, the second chain provides the costimulatory signaling moiety and is linked to the first chain by adding a "dimerizer". The CARs remain silent until the dimerizer allows hetero-dimerization and formation of a functional CAR [106–108]. The T cell activity can be fine-tuned by titrating the dimerizer dose; withdrawal of the dimerizer terminates CAR signaling. The clear "on-off" T cell activation by combinatorial antigen recognition is only achieved when each CAR signal alone is not sufficient for driving T cell activation. Since CD3ζ signaling in pre-activated T cells may re-induce a T cell response, de-tuning the signal or reducing binding affinities will be required to achieve a balanced activation threshold [109].

An alternative strategy of tunable CARs is based on synthetic Notch (synNotch) receptors [110, 111]. Notch is composed of an extracellular receptor, a transmembrane domain, and an intracellular transcription regulator that upon activation mediates proteolysis of the internal domain releasing the intracellular transcription regulator. These properties are used to control the transcription of an authentic CAR of different specificity than the Notch receptor. The advantage is that the antigen recognition by the Notch receptor is based on logic "AND" gating and controls CAR T cell function in a spatially defined fashion [111]. Upon CD19 binding, the CD19 specific synNotch receptor releases the transcriptional effector domains Gal4-VP64 or TetR-VP64 to induce the expression of an anti-mesothelin CAR. The T cell is only activated when both the synNotch ligand and the CAR ligand are recognized in the targeted tissue. Another example is targeting ROR1 by the CAR and EpCAM or B7-H3 by the Notch receptor [112]. While the "on-switch" for the CAR determines the selectivity in targeting cancer cells, the strategy may be limited by the time required for inducing sufficient levels of CAR expression and decaying the CAR when no Notch signaling furthermore occurs.

The TCR specificity of the engineered T cell can also be used to provide a more complex recognition profile. T cells with γδ TCR recognize phospho-antigens characteristic for tumor cells with deregulated metabolism; transgenic expression of a GD2 specific CAR with a costimulatory domain will only provide full T cell activation when both the γδ TCR and the CAR are engaging their cognate antigens; GD2$^+$ healthy cells are not attacked due to lack of γδ TCR activation [113].

While a number of targetable antigens are expressed by both healthy and cancer cells, some antigens are only expressed by healthy cells but lack expression by cancer cells. In this situation, binding to the common antigen "A" but blocking when binding to the antigen "B" on healthy cells would prevent CAR T cell activation against the healthy cells (Fig. 10.4). The inhibitory CAR with an ITIM (immunoreceptor tyrosine-based inhibitory motif) prevents signaling through the activating CAR as long as engaging its cognate antigen "B". Binding to "A" cancer cells that lack the antigen "B" allows CAR T cell activation due to a lack of signaling through the inhibitory CAR ("A-not-B" gating) [114, 115].

10.5 Universal CARs with Engrafted Multiple Antigen Specificities

Currently, the prototype CAR has one defined antigen specificity; changing specificity requires engineering T cells with a new CAR. In this situation, one "universal" CAR was developed that recognizes an antibody which in turn provides cancer specificity (Fig. 10.5). As an example, the immunoglobulin Fc region of an anticancer antibody is bound by a CAR with CD16 as a binding domain. CD16 CAR T cells bind the anticancer antibody and thereby gain specificity for cancer cells [116]. Alternatively, the CAR binds to a protein epitope linked to the cancer-targeting antibody [117]. T cells with a fluorescein isothiocyanate (FITC) specific CAR recognize FITC-labeled folate which binds to folate receptor-positive cancer cells [106]. Other epitopes linked to the bridging antibody can also be envisaged like avidin [118]. Simultaneous adding antibodies of different specificities allows redirecting the same CAR T cell product ("UniCAR") toward a plethora of antigens without the need for de novo CAR T cell manufacturing [119]. The advantage of the strategy is the simultaneous use of different antibodies for targeting a broad panel of antigens while applying only one manufactured CAR T cell product. The actual specificity is defined by the applied antibody mix in therapeutically sufficient concentrations. With the short half-life of applied antibodies, moreover, unexpected adverse events are shortly limited upon antibody withdrawal. Mechanistically, efficient tumor control requires that both the CAR T cells and the linking antibody infiltrate the tumor lesion; it is so far unclear whether this will be the case for advanced solid tumors. It is also unresolved whether

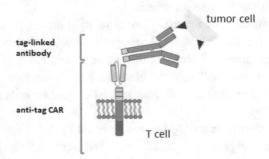

Fig. 10.5 The concept of "universal" CARs. T cells express a CAR with specificity for a tag that is linked to a bridging molecule, preferentially a natural antibody, scFv, or a BiTE, which recognizes the target antigen on the tumor cell. T cells can be equipped with different specificities by adding different bridging molecules without the need to engineer new CAR T cells. This is assumed to be also feasible after CAR T cell application to the patient; infusion of antibodies to different tumor antigens provides different specificities to the CAR T cell

lasting immunological memory against the cancer antigen can be established by "universal" CAR T cells which is thought to be required to control cancer in the long term.

10.6   Switch CARs Convert a Suppressor into an Activation Signal

The redirected T cell attack is often suppressed in tumor lesions by various means; inhibitory ligands like programmed cell death ligand-1 (PD-L1) and PD-L2 bind to programmed cell death-1 (PD-1) expressed by activated T cells. A CAR that binds to such an inhibitory ligand but transmits an activating signal to the engineered T cell would not only compete with the inhibitory receptor in binding but also overcome inhibition by providing the activating signal [120, 121] (Fig. 10.6a). Ideally, the PD-L1:CD28 "switch" CAR overruns the PD-1 mediated suppressive signal in the engineered T cell. In particular, a CAR with the PD-1 extracellular domain binding to PD-L1 and the CD28 intracellular signaling for activation converts the inhibitory to an activating signal resulting in increased ERK phosphorylation, the release of pro-inflammatory cytokines, T cell amplification, and cytolysis of the target cells [122]. A chimeric receptor composed of the exodomain of TIGIT (T cell immunoreceptor with Ig and ITIM domains) fused to the CD28 costimulatory domain enhances T cell function when binding to the TIGIT ligand CD155 [123]. Given the plethora of inhibitory signals provided by the tumor tissue, it remains challenging to identify a "key" suppressor that needs to be overcome in order to sustain a productive T cell response.

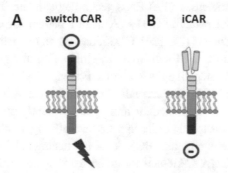

Fig. 10.6 Switch CAR and inhibitory CAR (iCAR). (**a**) The switch CAR binds to a ligand of a physiologically T cell inhibitory receptor, however, provides an activating signal to the engineered T cell. (**b**) The inhibitory CAR provides a negative signal by ITIM signaling to the engineered T cell upon binding to the antigen. The inhibitory signal overruns the activation through the TCR or through a co-expressed activating CAR

10.7   Inhibitory CARs: iCARs

While prototype CARs deliver an activating signal to the engineered T cell, CARs with an inhibitory signal are designed to repress CAR T cell activation when engaging healthy cells. Ideally, the inhibitory CAR mitigates off-tumor toxicities by overrunning the activating CAR signal when facing healthy tissues. A CAR with a dominant inhibitory signal recognizing healthy cells is co-expressed along with an activating, tumor-specific CAR; thereby the CAR T cells are only activated when engaging cancer cells, and the inhibitory CAR is not engaging healthy cells [114] (Fig. 10.6b). Such inhibitory CAR provides the PD-1 or CTLA-4 signal and overruns the activation signal by the cancer-specific CAR [114].

10.8   Allogeneic CAR T Cells and CAR NK Cells

So far, CAR T cell therapy is an individualized therapy by ex vivo manufacturing patient's own T cells. This is not only laborious, cost-intensive, and time-consuming but also often not applicable when patients with advanced disease cannot provide enough or sufficiently active T cells for engineering with a CAR. In this situation, allogeneic "off-the-shelf" CAR T cells will be of benefit for a broad cohort of patients. In order to avoid graft-versus-host disease (GvHD), the endogenous $\alpha\beta$ TCR and HLA molecules need to be deleted which can be achieved by the zinc finger nuclease [51] or CRISPR/Cas technology. Subsequent depleting of remaining TCR$\alpha\beta$ T cells reduces the GvHD risk through contaminating allogeneic TCR$^+$ cells [124, 125]. T cells were genetically edited by transcription activator-like effector nucleases (TALENs) technology in the TCRα and CD52 locus [124] for the treatment of pediatric CD19$^+$ ALL in a patient for whom autologous CAR T cells could not be produced [124, 126]. CD52, present on the patient's malignant B cells, needed to be deleted from CAR T cells to allow eliminating malignant lymphocytes while sparing the infused CD52-negative CAR T cells. For engineering allogeneic CAR T cells, CRISPR guide RNA and Cas9 are currently encoded by the viral vector for constitutive expression [127–129]; current research is aiming at transiently providing the gene-editing tools without persisting vectors in order to minimize off-target editing.

Allogeneic CAR T cells targeting BCMA for the treatment of multiple myeloma were engineered with the BCMA CAR and modified using gene-editing technology to limit TCR-mediated immune responses [130]. The safety profile of the allogencic BCMA CAR T cells was provided by incorporating a CD20 mimotope integrated into the extracellular CAR domain to enable CAR T elimination through rituximab.

With respect to use allogeneic effector cells, NK cells can also be used to induce a productive antitumor response due to their cytolytic potential and secretion of pro-inflammatory cytokines [52, 131, 132]. Apart from the use of allogeneic cells, there is a strong biological rationale to use NK cells for adoptive cell therapy of cancer; NK cells display a large repertoire of inhibitory and activating receptors to recognize foreign, infected or malignant cells. The balance of both signals finally induce or block the release of cytolytic granules for target cell killing by cytotoxic $CD56^{dim}$ NK cells. While a broad diversity of NK cell subsets likely exists based on the repertoire of activating and inhibitory receptors, engineering of NK cells with a CAR would provide predefined specificity for a particular cancer cell. The prototype CAR with CD28-CD3ζ signaling is also active in NK cells [52]; a CAR with the NK cell 2DS2 and DAP12 signaling proteins produced more potent antitumor activities [133]. CARs engineered for NK cells often harbor DAP10 or DAP12 for activating, alone or along with CD3ζ. A CAR with NKG2D as binder and DAP10-CD3ζ for activation efficiently activated primary NK cells against osteosarcoma in a mouse model [134]. CD3ζ outperformed DAP10 as the activating domain in NK cells [135] whereas DAP12 outperformed CD3ζ [136, 137].

The benefit of CAR redirected NK cells is their predefined targeting specificity and the potential use as allogeneic cell products making "off-the-shelf" manufacturing possible.

The established NK cell line NK92, which displays the activating NKp30, NKp46, and NKG2D receptors and a limited inhibitory repertoire including ILT-2, NKG2A, and KIR2DL4 at low levels, is engineered with a CAR for antigen-specific targeting. NK92 carry a number of genetic abnormalities, harbor Epstein–Barr virus, and are at risk to permanently engraft upon administration to the patient; engineered NK92 need to be irradiated before application to the patient resulting in short persistence, rapidly disappearance from circulation and lack of long-term memory which requires multiple rounds of administration. However, CAR NK92 cells showed efficacious upon local installation in a glioblastoma xenograft model [138] and in an orthotopic breast cancer model [139]. The therapeutic efficacy of CAR NK92 is currently evaluated in the treatment of $Her2^+$ glioblastoma (NCT03383978), ROBO1$^+$ pancreatic cancer (NCT03941457) and solid tumors (NCT03940820), and $BCMA^+$ multiple myeloma (NCT03940833). Anti-CD19 CAR engineered cord blood NK cells are explored in the treatment of B cell malignancies (NCT03056339) and peripheral blood NK cells engineered with an anti-NKG2D CAR for the treatment of metastatic carcinoma (NCT03415100).

10.9   Manufacturing CAR T Cells for Clinical Applications

For clinical application, CAR T cell products are manufactured from autologous T cells of the intended recipient patient according to the good manufacturing practice (GMP) rules (Fig. 10.7). In most cases blood is collected by leukapheresis, T cells are separated, genetically engineered by viral infection or electroporation and amplified to clinically relevant numbers [140]. Most clinical trials are using T cells modified by γ-retroviral or lentiviral gene transfer; transposon-based vectors like Sleeping Beauty and PiggyBac are increasingly applied for clinical applications with the benefit of avoiding cost- and time-intensive virus productions [141–144]. Viral and transposon-mediated gene transfer procedures are aiming at permanently modifying the patient's T cells for CAR expression; some trials are using T cells modified by RNA electroporation for transient CAR

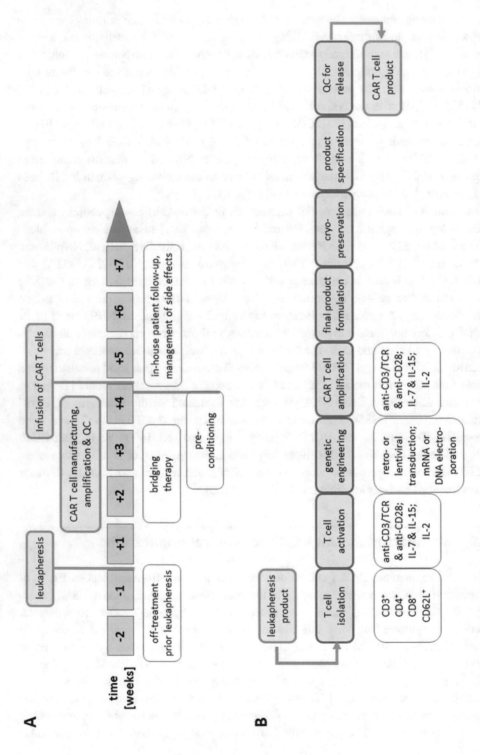

Fig. 10.7 CAR T cell manufacturing and patient treatment schedule. (**a**) T cells are collected from the patient by leukapheresis and processed ex vivo to obtain the final CAR T cell product which includes T cell engineering with the CAR, amplification to clinically relevant numbers, and final quality

expression. Upon permanent genetic modification, the risk of insertional mutagenesis and subsequent oncogenic transformation of mature T cells seems to be low and no oncogenic event was reported so far in this context. One caveat is the observation that upon lentiviral transfer the CAR gene inserted into the Tet2 locus leading to clonal expansion of the T cells that, however, contracted spontaneously [145]. However, malignant cells contaminating the apheresis product may also be accidentally modified by the CAR with the risk of CAR mediated tumor promotion after re-infusion to the patient [146].

Upon genetic modification, CAR T cells are extensively amplified ex vivo in the presence of cytokines to clinically relevant cell numbers; manually processed shaking reactors or bags as well as fully automated and supervised processing devices are currently in use. The aim of automation is to allow manufacturing with high reproducibility and quality and to produce cells from an increasing number of patients in parallel and with shortened vein-to-vein time.

The most suitable T cell population for CAR therapy is thought to be a naïve or young central memory cell with an acute inflammatory signature. CAR T cell amplification in the presence of IL-2 preferentially triggers effector T cells [147] while IL-7 or IL-2 plus IL-15 amplify central memory T cells with persistent capacities for cytokine release and antitumor activity [148, 149]. The addition of IL-21 keeps the CAR T cells in a less differentiated state [150, 151]. The mode of stimulation not only determines the yield of amplified CAR T cells but also the metabolic addiction; IL-15 increases the oxidative metabolism as well as carnitine palmitoyltransferase which is a rate-limiting step in fatty acid oxidation [67]. IL-7 increases Glut1 by STAT5 and Akt activation [152] and induces the glycerol transport and triglyceride synthesis [153] resulting in improved T cell persistence and survival. Akt inhibition during ex vivo amplification triggers a central memory phenotype with high levels of fatty acid oxidation and finally improved CAR T cell antitumor activities [154]. CD45RO$^+$ CD62L$^+$ memory CAR T cells provide a more durable antitumor response than effector T cells in more advanced stages of differentiation [155–157], making CD62L$^+$ enriched CAR T cells attractive for further exploration. Central memory CAR T cells, therefore, seem to be superior in clinical applications since responding patients did not accumulate T cells with an early memory and exhaustion signature while non-responding patients did [158]. The viability of engineered cells in the final product likely impacts the therapeutic efficacy. However, a first retrospective analysis

---

Fig. 10.7 (continued) control procedures (QC). Prior infusion of the CAR T cell product, the patient is subjected to non-myeloablative lymphodepletion ("pre-conditioning") in order to facilitate engraftment and in vivo amplification of CAR T cells. Subsequent in-house care is required to manage potential side effects like CRS or neurotoxicities. (b) The ex vivo CAR T cell manufacturing process usually starts from a leukapheresis product and involves a number of subsequent production steps including T cell isolation and activation, genetic engineering, CAR T cell amplification, formulation, and cryopreservation of the CAR T cell product, and final quality control for release. The entire process can be run manually or (semi)automated in a supervised manner

of CAR T cell therapy outcomes based on the viability of the manufactured cell products indicates no evidence for differences in CD19 CAR T cell expansion or patient survival when the patient received products of 70–80% viability compared products with at least 80% viability [159].

Manufacturing the T cell product is a personalized approach and the current procedure has some significant limitations:

1. A number of cancer patients are lymphopenic or have low numbers of naïve T cells resulting in delay or failure in production as experienced when producing for pediatric patients.
2. The manufacturing process is time-consuming which is detrimental for patients with acute or advanced stages of their disease.
3. Patient's T cells may be dysfunctional due to their disease as reported for CLL and some solid tumors [160–162] resulting in a product with poor response rates.

Engineering allogeneic T cells may be an alternative strategy in some situations. In patients who are subjected to hematopoietic stem cell transplantation, allogeneic T cells from the transplant were used to manufacture the CAR T cell product. Such allogeneic CD19 specific CAR T cells induced partial and complete remissions [163–165]; however, some patients were suffering from GvHD. Viral antigen-specific, partially HLA matched donor T cells lack alloreactivity and, engineered with a CAR, showed anti-lymphoma activity [166]. NK cells with their MHC-independent activity are also a source for third party donor cells for CAR engineering; however, they exhibit poor persistence capacities in patients [167, 168].

10.10 Adoptive Therapy with CAR T Cells Induces Lasting Remissions in Hematologic Malignancies

The therapeutic efficacy of adoptively transferred CAR T cells depends on multiple parameters including the CAR design, the CAR signaling, the binding affinity, the number of antigens on target cells, the spatial accessibility of the targeted antigen epitope, the maturation stage of T cells, persistence and amplification of CAR T cells, pre-conditioning of the patient's immune system and others. Current efforts are aiming at optimizing the conditions for certain diseases; some therapeutic regimens proved successful in the CAR T cell treatment of malignant diseases, in particular, B cell leukemia/lymphoma [169]; a minority of trials is aiming at treating solid cancer [170].

In the treatment of so far refractory B cell leukemia and lymphoma early clinical trials with "second generation" anti-CD19 CAR T cells achieved complete and lasting remissions [163, 166, 171–178]; treatments with "first generation" CAR T cells mostly failed due to rapid clearance of the CAR T cells from circulation. Currently, nearly 600 early phase trials using CAR T cells are in clinical exploration, mostly performed by academic centers (268 trials in China, 227 in the US, 67 in Europe as in early 2020). Anti-

CD19 CAR T cells for the treatment of pediatric B-ALL (Kymriah™, tisagenlecleucel, Novartis) and adult large B cell lymphoma (Yescarta™, axicabtagene ciloleucel, Kite-Gilead) have obtained FDA and EMA approval; others will follow soon.

The standard procedure in CAR T cell therapy starts with leukapheresis of the patient, isolation of the T cells, ex vivo genetic engineering with the respective CAR, CAR T cell amplification, and re-administration of the final cell product by i.v. infusion to the patient after non-myeloablative lymphodepletion and IL-2 support (Fig. 10.7). Since CAR T cell persistence is crucial for clinical efficacy [172], CARs with one or two costimulatory endodomains are currently used in trials, most of them providing CD28 or 4-1BB costimulation. With respect to persistence, CAR T cells with 4-1BB costimulation appear superior to CD28 CAR T cells [172] persisting for more than 4 years in circulation compared with 30 days of CD28 CAR T cells [179]. Since CD19 CAR T cells do not discriminate between healthy and malignant B cells, patients are experiencing lasting B cell depletion as long as CD19 CAR T cells persist. Although B cell depletion is clinically manageable, the situation asks for more selectivity in targeting the malignant B cells. Recently, the Fcμ (IgM) receptor was identified as a potentially more tumor-selective target in the treatment of chronic lymphocytic leukemia (CLL) sparing most healthy B cells from elimination due to lower levels of target expression [180]. Alternative targets are currently also explored.

During remission, most 4-1BB CAR T cell treated leukemia/lymphoma patients did not receive further cancer-specific treatment; patients with CD28 CAR therapy were frequently subjected to allogeneic stem cell transplantation which seems to be beneficial since CD28 CAR T cells persist less than 4-1BB CAR T cells [172, 179]. Further exploration needs to identify the more successful strategy in the long-term.

So far, patients with DLBCL, CLL, and B cell-acute lymphocytic leukemia (B-ALL) and follicular lymphoma were successfully treated by CD19 CAR T cells. Remarkably, patients with multiple myeloma also experienced remissions after CD19 CAR T cell therapy [181] although myeloma cells are CD19^{low} or CD19-negative as evaluated by flow cytometry. Alternative antigens are also targeted for the treatment of B cell malignancies, including CD20, CD22, the Igκ light chain, ROR-1, and CD30.

The success of CAR T cell therapy in various trials is difficult to compare due to the lack of any standardization with respect to the trial design, CAR composition, targeted antigen, pre-conditioning, and others. However, it becomes more and more evident that the CAR T cell dose and the degree of lymphodepletion or the lymphodepleting regimen are crucial to allow extensive T cell amplification, persistence, and finally therapeutic efficacy [61]. Repetitive re-stimulation of CAR T cells after application likely improves their persistence and the antitumor efficacy in the long term. This is the rationale to use virus-specific T cells for CAR engraftment which are repetitively re-stimulated through their TCR upon contact to viral antigens; for instance, Epstein–Barr virus (EBV) specific T cells were engineered with a CAR and triggered by EBV infected cells for amplification [182]. Indeed, EBV specific CAR T cells performed superior over CAR T cells without virus specificity [183].

10.11 The Challenges of CAR T Cells in Treating Solid Cancer

While the treatment of hematologic malignancies is increasingly successful, the treatment of solid cancer is still challenging (Table 10.2). CAR T cells as "living drugs" are basically ideal for the treatment of solid cancer and particular for the elimination of (micro-) metastases due to T cell-intrinsic migratory and tissue penetrating capabilities, penetrating the blood–brain barrier [184], and accumulating in virtually every tissue including privileged tissues like testes and eyes [185]. However, most solid tumors have a plethora of strategies to counteract CAR T cell trafficking, penetration, and activation in a very efficient way. Among those, an altered chemokine milieu impairs trafficking of CAR T cells to the tumor [186]; some adhesion factors are lost on tumor endothelia [187] hampering T cell penetration. The latter can be counteracted by locally depositing TNF-α that increases vascular adhesion molecules, such as vascular cell adhesion protein-1 and intracellular adhesion molecule-2 on endothelial cells [188]. Targeting vascular endothelial growth factor (VEGF) receptor-2 [189] or blocking migration inhibitory factors like the endothelin-B receptor [190] also improves endothelial cell adhesion and/or transmigration of T cells. Chemokine receptors, downregulated on T cells, were re-expressed, like CXCR2 (CXCL1 receptor), to facilitate targeting melanoma [191] or CCR2b for targeting neuroblastoma [192].

To circumvent trafficking barriers, CAR T cells were intratumorally installed [193] or injected pleurally or peritoneally for the treatment of mesothelioma and ovarian cancer, respectively [88]. Anti-CEA CAR T cells were applied by endoscopy into hepatic metastases [194]; anti-c-Met CAR T cells were injected into breast cancer metastases [195].

Once got through to the tumor lesion, CAR T cells need to break the barrier of stroma cells and extracellular matrix to get in the near vicinity to the cancer cells to execute their cytolytic attack. Degradation of heparan sulfate proteoglycans by transgenically released heparanase in the stroma improves T cell penetration [84]; high levels of IFN-γ are required to eliminate the stroma cells and to eradicate large established tumor lesions [196]. The overall antitumor activity was improved by targeting stroma cells directly by CAR T cells specific for fibroblast activation protein (FAP), a serine protease involved in extracellular matrix remodeling [197].

A number of processes make the tumor lesion hostile for the redirected CAR T cell turning them into an exhausted state or into apoptosis. Regulatory T (Treg) cells, myeloid-derived suppressor cells (MDSCs), and tumor-associated M2 macrophages release suppressive cytokines, like IL-4, IL-10, leukemia inhibitory factor (LIF) and TGF-β. MDSCs and macrophages decrease the intra-tumor tryptophan levels [198]; stroma cells are releasing indoleamine 2,3-dioxygenase (IDO) and kynurenine [198] inhibiting T cell effector functions and depriving the tissue of glucose and other nutrients. Profound acidosis moreover counteracts the antitumor activity of CAR T cells. CD3ζ expression and TCR signaling is repressed by low arginine levels [199] due to arginase released by MDSCs [200].

Table 10.2 CAR T cell therapy faces a plethora of challenges

Challenges	Consequences	Solutions for CAR T cell therapy
Lymphopenia of the tumor patient	Delay or failure in CAR T cell production	Allogeneic donor T cells
Dysfunctional patient's T cells	Poor response rate	Allogeneic donor T cells
Solid cancer lesions	Altered chemokine milieu	T cell engineering with chemokine receptors, e.g. CXCR2
	Migration inhibitory factors	Blocking endothelin-B receptor
	Lack of leukocyte adhesion factors	Locally depositing TNF-α
	Impaired homing to tumor lesions	Local CAR T cell administration
Altered extracellular matrix	Impaired T cell penetration into the tumor tissue	Release of transgenic heparanase
Stroma cells in solid cancer	Tumor sustaining milieu	Targeting FAP+ stroma cells
Immune suppressive tumor milieu	Impaired CAR T cell functions	T cells engineered with switch receptor
		Suppression of MDSCs and Tregs
		Checkpoint blockade
		Disruption of protein kinase A
Antigen loss cancer cells	Tumor relapse	Multi-antigen specific CAR T cells
		Local activation of the innate immune response
Systemic CAR T cell toxicities	Cytokine release syndrome	Anti-IL-6 receptor antibody tocilizumab
		Anti-IL-6 antibody siltuximab
		Corticosteroids
		Suicide switch to eliminate CAR T cells
		Off switches for CAR expression
		Transient CAR expression
		Depleting antibodies targeting tags integrated into the extracellular CAR domain
	Neurotoxicity (ICANS)	IL-6 blockade
		Steroids
	Anaphylaxis	Human or humanized scFvs
	Tumor lysis syndrome	Adjusting CAR T cell dose
On-target off-tumor toxicity	B cell depletion by anti-B cell CAR T cells	Immunoglobulin substitution
		Antibiotic and antifungal therapy
	Damage of healthy tissue	Local CAR T cell administration
		Increasing tumor selectivity by targeting neo-antigens or combinatorial antigen recognition

(continued)

Table 10.2 (continued)

Challenges	Consequences	Solutions for CAR T cell therapy
		CAR T cell inhibition by a co-expressed iCAR recognizing healthy tissues
Off-target off-tumor toxicity	Activation of innate immune cells	Deleting or mutating the FcγR binding site in the extracellular IgG CAR spacer

Adoptive cell therapy with CAR T cells for the treatment of malignant diseases is still challenging due to the biology of tumors that has consequences for the success of therapy. Some of these challenges can be addressed by altering the CAR design or treatment strategies

In this situation, several strategies are currently explored to make CAR T cells resistant and/or to eliminate the suppressive cells from the tumor tissue. MDSCs and Tregs can be suppressed by the multi-kinase inhibitor sunitinib; fludarabine and cyclophosphamide, used for patient pre-conditioning prior to adoptive cell therapy, deplete Treg cells, and decrease IDO levels. Disruption of membrane anchoring of protein kinase A (PKA), which is a key effector molecule in the downstream cascade of prostaglandin E2 and adenosine, increases CAR T cell infiltration, chemotaxis, persistence, and antitumor activity [201]. To convert resistance to inhibitory ligands, CAR T cells were engineered with suppressed PD-1 expression [71] or a PD-1:CD28 switch receptor which binds to PD-L1 and provides a costimulatory activating signal [121]. Checkpoint blockers like nivolumab, pembrolizumab, or ipilimumab are currently explored as adjuvant in CAR T cell trials [202, 203] (NCT02650999).

10.12 CAR T Cell Therapy Is Associated with Toxicities

Treatment with CAR T cells as "living drugs" achieved dramatic responses in patients in some trials; however, CAR T cells can cause severe side effects days and even weeks after initial application which requires intensive care interventions and can be fatal. Most toxicities are due to the specific biology of CAR T cells, other CAR T cell-associated toxicities are currently not fully understood. Mechanistically, the major toxicities are (1) general toxicities due to massive T cell activation, cytokine release and tumor lysis and (2) specific toxicities due to the CAR T cell interaction with healthy tissues ("on-target off-tumor" toxicities).

10.12.1 Systemic Toxicities

Cytokine release syndrome (CRS) Engagement of a high number of CAR T cells with their target cells results in massive cytokine release with super-physiologically high serum levels of pro-inflammatory cytokines including IFN-γ and TNF-α, IL-10 and IL-6 [175–

177, 204] which per se leads to cross-activation of other immune cells and to further cytokine release up to toxic levels. The toxicities are summarized as cytokine release syndrome (CRS) that comes clinically along with high fever, malaise, fatigue, myalgia, nausea, anorexia, tachycardia, hypotension, capillary leakage, cardiac dysfunction, renal impairment, hepatic failure, and disseminated intravascular coagulation [205]. CRS often occurs together with the vascular leakage syndrome (VLS) and is closely associated with macrophage activation syndrome (MAS) and macrophage activation, clinically resembling hemophagocytic lymphohistiocytosis (HLH). The risk of CRS somehow correlates with T cell activation and amplification which can, but must not, correlate with the overall tumor burden [175, 206]. To avoid CRS, lower CAR T cell doses are required in patients with high tumor load, a paradox that needs to be addressed by precisely defining the therapeutic window by adjusting T cell activation, cytokine release, and antitumor activity.

Diagnosis in the early stages of CRS and clinical management is facilitated by a newly established score and by treatment guidelines [176, 204, 206]. C-reactive protein, which is released by hepatocytes in response to increased IL-6 levels, currently serves as a laboratory marker of CRS onset and severity [176]; predictive markers in adults are also increase in soluble gp130 (sgp130), IFN-γ and IL-1Rα, in pediatric patients, IFN-γ, IL-13, and MIP1α. The current treatment of CRS is based on interrupting the IL-6/IL-6 receptor signaling axis by blocking antibodies like the anti-IL-6 receptor antibody tocilizumab or the anti-IL-6 antibody siltuximab which does not seem to interfere with CAR T cell efficacy [173, 206, 207]. Short-term systemic corticosteroid treatment also rapidly reversed CRS without compromising the initial antitumor response; however, it may impair the CAR T cell response in the long term [176, 177]. Neutralization of the pro-inflammatory cytokines during CRS contributes to reduce the severity of CRS. Activated monocytes and/or macrophages release high levels of IL-1 during CRS; neutralization by an IL-1 receptor antagonist ameliorates CRS and neurotoxicity in preclinical models [208, 209]. Neutralization of the macrophage-monocyte activating cytokine GM-CSF by administering lenzilumab antibody reduced the risk for CRS [210]. Reducing catecholamine by blocking tyrosine hydroxylase reduces cytokine levels in a mouse model implying a path for clinical intervention [211].

CAR T cell-associated HLH/MAS often shows CRS symptoms together with elevated ferritin and liver enzyme serum levels, haemophagocytosis, cytopenia, pulmonary edema, splenomegaly, renal failure, and loss of NK cells. Treatment includes systemic chemotherapy with etoposide or cytarabine which may be applied intrathecally if associated with neurotoxicity; IL-6/IL-6R blockade is mostly ineffective in this case.

Neurotoxicity (ICANS) CAR T cell therapy-related neurotoxicity is now referred to as immune effector cell-associated neurotoxicity syndrome (ICANS) [212]. The neurological effects are aphasia, hallucinations, confusion, delirium, expressive aphasia, myoclonus, delirium, seizure, and cerebral edema. Such neurotoxicity occurs in about 40% of patients, is reversible and frequent during CD19 CAR T cell therapy; other CAR T cells can also cause neurotoxicities pointing to a diffuse, but transient encephalopathy due to the infiltration by activated CAR T cells [175–177, 206]. The mechanism of neurotoxicity is so far unclear, however, is associated with disruption of the blood–brain barrier and increased cytokine levels in the cerebrospinal fluid [213]. Treatment with IL-6 blockade and steroids can be successful if in conjunction with CRS.

Anaphylaxis Since the CAR is a recombinant, foreign protein, in particular the antibody-derived binding domain, CAR T cells may trigger the development of antibodies as observed in one patient with elevated IgE levels after repeated doses of CAR T cells [214]. This can most likely be avoided by using human or humanized scFvs [215–217]. However, other neo-epitopes, like the linking regions between the CAR domains, will still be recognized as foreign. Along with reduced immunogenicity, the risk for inducing anti-CAR antibodies will likely be reduced.

Tumor Lysis Syndrome Rapid destructing bulk tumor mass causes the massive release of tumor cell components accompanied by disturbances in the electrolyte and metabolic balance with the risk of multi-organ failure. In the situation of high tumor mass, infusion of initially low doses of CAR T cells may paradoxically be beneficial.

10.12.2 "On-Target Off-Tumor" Toxicities

Since most targeted antigens are also expressed by healthy cells, CAR T cell treatment is associated with the risk of attacking healthy tissues causing tremendous "on-target off-tumor" toxicities. CD19 targeting CAR T cells to treat B cell leukemia also eliminate healthy B cells; a lasting B cell depletion requires immunoglobulin substitution and antibiotic and antifungal protection which is clinically manageable. The toxicity is more severe when the targeted antigen is expressed by vital tissues. In a trial for the treatment of renal carcinoma with carbonic anhydrase IX specific CAR T cells, patients developed liver toxicities due to low levels of carbonic anhydrase IX on bile duct epithelia; the severity of "off-tumor" toxicities could be reduced by pre-administering a blocking antibody [218].

Strategies to avoid "on-target off-tumor" toxicities are currently aiming at increasing tumor selectivity and limiting the CAR activity in a spatial and temporal fashion. Alternative routes of T cell administration may be an option to avoid systemic distribution of high numbers of CAR T cells through the bloodstream. Local CAR T cell administration by endoscopy or by intra-pleural or intra-peritoneal application provides high CAR T cell doses at the tumor site and avoids off-tumor toxicities to some extent [219, 220]. The application, however, is limited since in most cases the tumor lesions are not accessible or the tumor is broadly disseminated.

10.12.3 "Off-target Off-tumor" Toxicities

Most CARs engineered in the past harbor the IgG1 CH2CH3 domain as a spacer in the extracellular moiety which can potentially bind the Fc γ receptor (FcγR) (CD64), thereby activating innate cells like NK cells and macrophages. Although so far clinically not in the foreground, the risk can theoretically be reduced by deleting the IgG1 CH2 domain or mutating Asn297 within the FcγR binding site [37, 46]. Alternative spacers are the IgG4 Fc region or the extracellular CD8 domain.

Targeting ErB2 by third generation trastuzumab derived CAR T cells caused a fatal cardiopulmonary failure likely due to an acute respiratory distress syndrome and CRS resulting from immune activation independently of ErbB2 recognition in the lung [221]. Similarly, pulmonary edema was recorded upon treatment with EGFR variant-3 specific CAR T cells [222] and with CEA specific CAR T cells [223] without direct evidence for CAR T cells attacking the lung epithelia. The toxicity profile is even more complex since T cells with the second generation ErbB2 CAR derived from the FRP5 antibody in patients with sarcoma did not induce off-tumor toxicities [224]. Toxicity may be substantially different when using CARs targeting a different epitope of the same target or transmitting different signals, i.e., CAR T cells with another binding domain and of second generation costimulatory domain produced no dose-limiting toxicity when targeting ErbB2 [224, 225].

10.12.4 Strategies to Ameliorate CRS

Basically, CAR T cells need to be down-tuned or de-activated when CRS becomes clinically significant; transient CAR expression, "off-switches" and "suicide switches" are common tools.

Transient CAR expression may be beneficial when there is a substantial risk of severe, CAR T cell-related toxicities. RNA transfer produces transiently expressed CARs. The CAR is present on the T cell surface for several days and mediates efficient T cell activation upon target cell engagement [226], however, CAR expression declines with RNA degradation and RNA dilution upon T cell amplification. Such RNA modified T cells displayed some, although transient efficacy in trials [214, 227]. The drawback, however, is the short-term effect due to rapid CAR T cell elimination.

The use of permanently modified CAR T cells, however, requires strategies to actively eliminate those cells in case of uncontrolled toxicity. High dose pulse treatment with steroids stopped auto-immunity after the application of carbonic anhydrase IX-specific CAR T cells [228]. While this is a non-selective depletion of activated T cells, CAR T cells can be more selectively eliminated by specific targeting. The CAR binding domain can basically also be targeted by an anti-idiotypic antibody [229] to induce the lytic elimination of the CAR T cell. The depletion of CAR T cells can also be achieved by administering a depleting, FDA-approved antibody that recognizes an epitope integrated into the extracellular CAR domain. Examples are the use of rituximab to deplete CAR T cells with a CD20 epitope or mimotope [230–232] or cetuximab to deplete CAR T cells with the EGFR epitope [233]. Whether sufficient CAR T cell depletion can be achieved as rapidly as required to reverse ongoing CRS remains to be shown in a clinical situation.

Apoptosis-triggered CAR T cell elimination seems to be a more rapid solution in this situation. Transgenic caspase-9 (Cas9) linked to FKBP1A induces apoptosis in engineered CAR T cells upon systemic application of the otherwise inert small molecule AP1903 that promotes dimerization and activation of the inducible Cas9 (iCas9) [234]. The strategy is potent since donor T cell-mediated GvHD could be stopped within an hour by iCas9 mediated T cell apoptosis in an experimental system [235].

Instead of eliminating CAR T cells, the CAR on the cell surface can be controlled by integrated modules. The CAR is engineered as a fusion protein with an integrated self-cleaving degradation moiety reversely regulated by a protease that can be specifically blocked by a protease inhibitor. In presence of the protease the produced CAR fusion protein splits off the linked degradation domain and is expressed on the cell surface; inhibiting the protease results in CAR degradation and lack of surface expression [236].

The CAR can also be down-tuned in expression or modulated in its signaling strength. For instance, CAR downstream signaling can be inhibited by adding the tyrosine kinase inhibitor dasatinib that rapidly and reversibly blocks Lck activation; thereby CRS severity can be reduced in a mouse model [237]. The strategy has the benefit that CAR T cells are dampened in their signaling strength but not depleted as long as dasatinib is present; however, the strategy still needs clinical validation.

Further developments are needed to engineer CARs with a lower risk of cytokine-mediated toxicities. One key in modulating the signaling strength is CAR costimulation. CD28-ζ CARs more rapidly induce T cell activation and induce higher cytokine release, expansion, and subsequent exhaustion while 4-1BB-ζ CARs drive less acute expansion but are enduring T cell persistence with lower cytokine levels, thereby less associated with the risk of severe CRS [238]. As a consequence, 4-1BB-ζ CARs may be favorable in patients with high tumor load and high antigen density, while CD28-ζ CARs may be required for low tumor load.

Other strategies to reduce the CAR mediated T cell activation are, for instance, modifying the CD8 transmembrane and hinge region resulting in reduced risk of ICANS or CRS of higher than grade-1 [239]. As an alternative strategy, "universal" CARs allow dynamic control of T cell activation by titrating the adaptor molecule that links the CAR T cells to the target cells [107, 119, 240].

10.13 Challenges in CAR T Cell Therapy

During the last years, clinical trials are going to establish CAR T cell therapy in treatment of several types of B cell leukemia and lymphoma; however, in the treatment of other malignant diseases, major hurdles remain.

10.13.1 Which Is the Best Tumor-Selective Antigen Suitable for Targeting?

Selectivity in tumor targeting is crucial for the long-term compatibility of CAR T cell treatment. Most targetable tumor-associated antigens are not tumor-selective, which raises the risk of inducing on-target off-tumor toxicities in CAR T cell therapies. Tumor-specific mutations of surface proteins or tumor-associated glycosylation variants like Muc1 or Muc16 are rare [241]. However, there are some antigens that seem to be safe targets although expressed by healthy tissues. One example is carcinoembryonic antigen (CEA, CD66e) that is strictly expressed on the luminal side of healthy epithelia of the gastrointestinal tract and the lung while expressed in a depolarized fashion on cancer cells. Systemic application of CEA-specific CAR T cells (NCT01212887, NCT02349724) [223, 242] and local administration of anti-CEA CAR T cells by hepatic artery infusion (NCT01373047) did not induce treatment-related colitis while declining tumor progression [194]. Other antigens like ErbB2 may be suitable targets when tumor cells with high ErbB2 levels are selectively targeted while sparing medium and low expressing healthy cells.

In order to increase tumor selectivity, targeting a pattern of antigens instead of a single specific antigen is preferred. Technically, two CARs with different specificities are co-expressed in the same T cell to induce full T cell activation only when both CARs are simultaneously engaging their cognate antigen; one CAR is providing the primary signal, the other CAR the costimulatory signal, thereby complementing the signals for full T cell activation only in the presence of both antigens [103, 104, 109]. The strategy requires a weak primary signal in order to avoid some basic CAR T cell activation in the presence of one antigen and a strong costimulatory signal for driving full T cell activation.

In order to block CAR T cell activation when engaging healthy cells, an inhibitory CAR (iCAR) with intracellular PD-1 or CTLA-4 signaling is co-expressed together with an activating CAR [114]. The iCAR recognizes an antigen that is present on healthy cells but absent on cancer cells. Thereby, CAR T cell activation occurs only when engaging a target cell that lacks the antigen recognized by the iCAR.

Some antigens are expressed by cancer cells on higher levels than by healthy cells; "affinity tuning" of the CAR binding domain may help to discriminate between both tissues [11, 243, 244]. While feasibility is demonstrated in preclinical models, cancer cells may escape surveillance by down-regulating the respective antigen [11, 243, 244]. The therapeutic window in affinity tuning is narrow as increasing affinity may result in the recognition of healthy cells as observed with a GD2 specific CAR that induced fatal encephalitis in experimental models [245].

10.13.2 How to Optimize the CAR Architecture?

Most CARs currently in clinical evaluation are based on a prototype CAR architecture designed nearly two decades ago. Preclinical research has clearly shown that for each antigen the particular CAR needs to be optimized with respect to the targeted antigen epitope, the binding affinity, the spacer length, and the signaling and transmembrane domain among others. Due to the complex dependency between the moieties the optimal CAR composition for a specific antigen is currently not predictable. For instance, the binding affinity and the position of the targeted epitope provide a narrow affinity window for targeting cancer cells with high antigen while sparing healthy cells with low antigen load [10, 11, 14, 17].

10.13.3 How to Establish Long-Term Cancer Control After CAR T Cell Therapy?

Although complete remissions in pediatric B-ALL can be induced with high frequencies, about 40% of patients suffering from leukemia relapses despite persisting CAR T cells [173, 175, 177]. Several causes contribute to the high frequency of relapses including the loss of the entire CD19 molecule or the expression of a CD19 isoform lacking targeted exon-2 and thereby becoming invisible to CAR T cells [246]. Loss of CD19 by switching to a CD19-negative myeloid lineage occurred in two cases of B-ALL after CD19 CAR treatment [247]. In this situation, targeting an antigen mandatory for cancer cell survival would be an ideal strategy. Since such an antigen is frequently not available, co-targeting a second antigen, e.g., CD20 or CD22, by a bispecific CAR will likely increase the therapeutic pressure on leukemic cells. The situation is more relevant in solid tumors that often show a profound heterogeneity in antigen expression; early tumor relapse after initial tumor regression is likely due to selecting antigen-loss cancer cells. Those cancer cells may be recognized and eliminated in an antigen-independent fashion by an innate immune cell response. To attract and activate the innate cells and initiate a more broad immune response, CAR T cells were equipped with the inducible release of transgenic IL-12 in experimental models [77, 80], making an antitumor attack more effective. There is some indication that combining an antigen-dependent and -independent immune response may control the tumor in the long term.

10.13.4 Which T Cell Subset Provides Long-Term Tumor Control?

Preclinical models indicate that naïve or early central memory T cells with an enhanced capacity for amplification and long-lived persistence are the most suitable cells for adoptive therapy. Along this line CD62L$^+$ memory T cells seem to exhibit superior performance [248]. The discovery of central memory stem T cells (TSCMs), which are part of the CD62L$^+$ central memory T cell pool, with an extraordinary amplification potential [156] triggers the exploration of those cells for adoptive cell therapy [248]. There is growing interest in the identification of factors and regulatory networks orchestrating the formation and maintenance of stem cell-like T cells since these cells seem to be crucial for the long-term therapeutic efficacy upon adoptive transfer to patients.

An alternative strategy is based on inducing T cell maturation in a predicted fashion. For instance, 4-1BB costimulation initiates a central memory T cell response in young T cells, which is in contrast to CD28 CAR T cells that exhibit a more short-lived effector cell response [65]. On the other hand, cells in more matured stages show preferred accumulation in the periphery where most tumors are located. Those cells, however, require different costimuli, i.e., CCR7$^-$ T cells with preferred migration into the periphery benefit from combined CD28-OX40 costimulation [73]. The functional capacities of a given T cell subset can also be modulated by co-treatment with kinase inhibitors; ibrutinib, a Bruton tyrosine kinase (BTK) inhibitor, reduces PD-1 and exhaustion of CAR T cells and thereby increases persistence and antitumor activity [249].

10.13.5 How Can CAR T Cell Manufacturing Be Adapted to an Increasing Number of Patients?

Currently, most CAR T cells are produced by specialized GMP facilities, frozen and shipped to the patient's hospital. Manufacturing a growing number of cell products will bring the current procedures to their limits, in particular, when thousands of autologous cell products are demanded in due time. A de-centralized, in-hospital manufacturing by a (semi)automated, fully controlled, and entirely closed system is one of the favored concepts to address this situation. This will also require a high degree of standardization in the manufacturing process and would avoid sophisticated logistics in the transportation of blood and cell products. As an alternative concept, a highly specialized, centralized production facility with multiple automated manufacturing processes in parallel may act as a regional supplier of patient's cell products.

The other limitation is set by the need of the patient's autologous T cells for therapy; a "universal" donor for each patient would make the therapy much easier in multiple aspects. A cell bank of T cell products of a "universal" donor and with different engineered CARs would provide much flexibility in the clinical application according to the specific clinical situation of the patient. Moreover, upscaling the manufacturing would make the CAR T cell products available for a higher number of patients. Efforts are undertaken to generate such "universal" T cells that can be applied to a number of patients. This requires to make the CAR T cells "invisible" to the host immune cells and deficient in alloreactivity against the patient's healthy tissues while retaining CAR mediated antitumor activity [148]. In this line, MHC-deficient, CAR-modified virus-specific T cells and TCRα knock-out T cells, obtained by genome-editing techniques, are currently developed to act as an "off-the-shelf" or a "third-party" cell product with low risk of GvHD [250–252]. CAR gene insertion into the TCRα locus achieves both genetic engineering with a CAR and abolishing TCR expression in one step [253]. Editing MHC class I to prevent recognition and rejection of donor T cells by the host may induce elimination through NK cells and does not prevent MHC class II-mediated rejection. All currently explored genetic editing procedures, however, are prone to off-target modifications to some degrees and are thereby associated with some risk of unintended mutagenesis. The field is fast-moving and first trials using allogeneic CAR T cells are initiated for the treatment of B cell malignancies and solid tumors (reviewed by [254]).

10.13.6 How to Balance CAR T Cell Efficacy and Side Effects?

A number of CAR T cell treatments are currently associated with a high risk of severe side effects which may need intensive care hospitalization. Although the various CAR T cell protocols and treatment procedures differ and are not evaluated in a comparative clinical setting, some clinical management rules become visible. For instance, the early diagnosis, grading, and clinical management of CRS is going to be standardized and consensus treatment regimens established [176, 204, 206]. In the case of uncontrolled toxicity, CAR T cells can selectively and rapidly be eliminated by various means. Dimerization of the transgenic inducible caspase-9 (iCasp9) by adding the dimerizing agent AP1903 results in the elimination of >90% of T cells within 30 min [235, 255–257]. Alternatively, antibody-dependent cellular cytotoxicity (ADCC) induced by administering an antibody that targets a CAR domain or a co-expressed epitope clears CAR T cells from circulation [230, 232, 233]. The caveat of immune-mediated CAR T cell elimination still is that the potentially dysfunctional immune system of cancer patients may have limited capacities to eliminate CAR T cells by ADCC.

10.13.7 Do We Need Specific Pre-conditioning for Each Cancer Type?

In most trials, patients are subjected to a non-myeloablative lymphodepletion with fludarabine and cyclophosphamide prior to adoptive T cell therapy; the rationale for such pre-conditioning is to transiently provide "space" and cytokines for sufficient in vivo expansion of the transferred CAR T cells. The lymphodepletion also provides antitumor effects by immune modulation as it depletes suppressor cells and releases tumor-associated antigens to the immune system upon tumor destruction. Although the current regimens for pre-conditioning are basically effective, the basic protocol may require adaptation for each cancer entity. For instance, the conditioning regimen used in the treatment of Her2$^+$ tumors [221] was modified for cancers of the biliary tract and pancreas by applying nab-paclitaxel to deplete from desmoplastic stroma and to promote Her2 presentation and by applying cyclophosphamide that depletes Tregs and MDSCs [258]. On the other hand, pre-conditioning can be highly toxic in the context of CAR T cell therapy by inducing cerebral edema [175, 176, 259] and neurologic toxicities, in particular after fludarabine treatment as observed in a recent trial (NCT02535364).

10.13.8   How to Activate the Immune Network to Induce a Broad Inflammatory Antitumor Response?

CAR T cells as "living drugs" are becoming more and more "living factories" to produce and release bioactive compounds at the targeted organ. The rationale to use CAR T cells as factories is based on evidences from experimental tumor models that the host immune system is substantially involved and required for tumor rejection [260] making immune-stimulatory conditions within the targeted tumor tissue necessary. IL-12 releasing CAR T cells (IL-12 TRUCKs) can provoke an immune response in the tumor tissue by attracting and activating M1 macrophages that eliminate those cancer cells which are not recognized by CAR T cells [77]. Other cytokines with specific impact on the local immune environment can also be used; IL-18 CAR T cells increase the number of tumor-associated $CD206^-$ M1 macrophages and $NKG2D^+$ NK cells and reduce suppressor cells like Treg cells, $CD103^+$ dendritic cells, and M2 macrophages [80]. Checkpoint blockade like targeting PD-1 is currently explored in a trial in order to activate both CAR T cells and infiltrating lymphocytes (NCT02650999); other blockades of checkpoints or combinations thereof in order to re-activate the tumor intrinsic immune cells are likewise worthwhile for clinical exploration.

10.14   CAR T Cells for the Treatment of Other Diseases

CAR redirecting T cell activation is not limited to targeting cancer cells, but can also be used to target any cell defined by the targeted antigen. CAR T cells targeting memory B cells by an anti-Dsg3 CAR were designed to eliminate cells responsible for the pathology of pemphigus vulgaris [261]. Auto-reactive T cells were targeted by CAR T cells recognizing an MHC presented auto-antigen [262, 263]. CARs were engineered to target virus-infected cells like cells infected by hepatitis B virus [264], hepatitis C virus [265], cytomegalovirus [266], and HIV [267, 268]. Dectin-1 derived CAR T cells were used to disrupt the germination of the fungus *Aspergillus* [269]. Experimental models imply that CAR redirected Treg cells can be used to promote immune tolerance in the therapy of auto-immune diseases like colitis [270], allergic asthma [271], and graft-versus-host disease by targeting HLA [272–274].

Cardiac fibrosis was recently targeted by CAR T cells recognizing fibroblast activation protein (FAP) that is highly expressed by cardiac fibroblasts during remodeling the myocardium upon injury. Adoptive transfer of T cells with a FAP-specific CAR reduced cardiac fibrosis and restored the function after injury in mice [275]. The report implies CAR T cells as "living drugs" to ablate activated cardiac fibroblasts in the treatment of cardiac disease for which therapies are so far very limited. Other applications of CAR T cells for the treatment of non-malignant diseases are in preclinical development.

Take Home Messages
- Adoptive cell therapy with Chimeric Antigen Receptor (CAR) redirected T cells achieved remarkable responses in some hematologic malignancies; however, the therapeutic efficacy in other diseases is still limited.
- CARs are modularly composed, recombinant transmembrane molecules with a binding domain in the extracellular moiety, a transmembrane domain, and an intracellular signaling domain in order to activate the engineered cell. The optimal molecular design needs to be explored in each case through variations of the constituent domains.
- The prototype CAR design can be modified by using a variety of targeting and signaling moieties making CARs more universal. Also, various immune cells can be redirected by CARs to provide convenient therapeutic cells.
- CAR T cell therapy is frequently associated with toxicities which can be mitigated by concomitant therapies, altered CAR design, and targeting alternative antigens to reduce on-target off-tumor toxicities.
- Antitumor efficacy of CAR engineered immune cells can be improved by co-activating a broad immune response, co-targeting the tumor stroma, cancer stem cells, and other tumor-related structures.
- Manufacturing CAR T cell products is still challenging to provide individualized therapeutics for all patients; efforts are made to engineer an allogeneic cell product "off-the-shelf" for a broad number of patients.

References

1. Rosenberg SA, Packard BS, Aebersold PM, Solomon D, Topalian SL, Toy ST, et al. Use of tumor-infiltrating lymphocytes and interleukin-2 in the immunotherapy of patients with metastatic melanoma. N Engl J Med. 1988;319(25):1676–80.
2. Dudley ME, Wunderlich JR, Robbins PF, Yang JC, Hwu P, Schwartzentruber DJ, et al. Cancer regression and autoimmunity in patients after clonal repopulation with antitumor lymphocytes. Science. 2002;298(5594):850–4.
3. Eshhar Z, Waks T, Gross G, Schindler D. Specific activation and targeting of cytotoxic lymphocytes through chimeric single chains consisting of antibody-binding domains and the gamma or zeta subunits of the immunoglobulin and T-cell receptors. Proc Natl Acad Sci USA. 1993;90(2):720–4.
4. Bridgeman JS, Hawkins RE, Hombach AA, Abken H, Gilham DE. Building better chimeric antigen receptors for adoptive T cell therapy. Curr Gene Ther. 2010;10(2):77–90.
5. Gross G, Waks T, Eshhar Z. Expression of immunoglobulin-T-cell receptor chimeric molecules as functional receptors with antibody-type specificity. Proc Natl Acad Sci USA. 1989;86 (24):10024–8.
6. FDA approves tisagenlecleucel for B-cell ALL and tocilizumab for cytokine release syndrome. FDA. 2019. Available from: http://www.fda.gov/drugs/resources-information-approved-drugs/fda-approves-tisagenlecleucel-b-cell-all-and-tocilizumab-cytokine-release-syndrome

7. FDA approves axicabtagene ciloleucel for large B-cell lymphoma. FDA. 2019. Available from: http://www.fda.gov/drugs/resources-information-approved-drugs/fda-approves-axicabtagene-ciloleucel-large-b-cell-lymphoma

8. Hudecek M, Lupo-Stanghellini M-T, Kosasih PL, Sommermeyer D, Jensen MC, Rader C, et al. Receptor affinity and extracellular domain modifications affect tumor recognition by ROR1-specific chimeric antigen receptor T cells. Clin Cancer Res Off J Am Assoc Cancer Res. 2013;19 (12):3153–64.

9. Lynn RC, Feng Y, Schutsky K, Poussin M, Kalota A, Dimitrov DS, et al. High-affinity FRβ-specific CAR T cells eradicate AML and normal myeloid lineage without HSC toxicity. Leukemia. 2016;30(6):1355–64.

10. Liu X, Jiang S, Fang C, Yang S, Olalere D, Pequignot EC, et al. Affinity-tuned ErbB2 or EGFR chimeric antigen receptor T cells exhibit an increased therapeutic index against tumors in mice. Cancer Res. 2015;75(17):3596–607.

11. Caruso HG, Hurton LV, Najjar A, Rushworth D, Ang S, Olivares S, et al. Tuning sensitivity of CAR to EGFR density limits recognition of normal tissue while maintaining potent antitumor activity. Cancer Res. 2015;75(17):3505–18.

12. Watanabe K, Terakura S, Uchiyama S, Martens A, Meerten T, Kiyoi H, et al. Excessively high-affinity single-chain fragment variable region in a chimeric antigen receptor can counteract T-cell proliferation. Blood. 2014;124:4799.

13. Ghorashian S, Kramer AM, Onuoha S, Wright G, Bartram J, Richardson R, et al. Enhanced CAR T cell expansion and prolonged persistence in pediatric patients with ALL treated with a low-affinity CD19 CAR. Nat Med. 2019;25(9):1408–14.

14. Chmielewski M, Hombach A, Heuser C, Adams GP, Abken H. T cell activation by antibody-like immunoreceptors: increase in affinity of the single-chain fragment domain above threshold does not increase T cell activation against antigen-positive target cells but decreases selectivity. J Immunol. 2004;173(12):7647–53.

15. Smith EL, Harrington K, Staehr M, Masakayan R, Jones J, Long TJ, et al. GPRC5D is a target for the immunotherapy of multiple myeloma with rationally designed CAR T cells. Sci Transl Med. 2019;11:485.

16. Ajina A, Maher J. Strategies to address chimeric antigen receptor tonic signaling. Mol Cancer Ther. 2018;17(9):1795–815.

17. Hombach AA, Schildgen V, Heuser C, Finnern R, Gilham D, Abken H. T cell activation by antibody-like immunoreceptors: the position of the binding epitope within the target molecule determines the efficiency of activation of redirected T cells. Abstract – Europe PMC. J Immunol. 2007;178(7):4650–7.

18. Zhang G, Wang L, Cui H, Wang X, Zhang G, Ma J, et al. Anti-melanoma activity of T cells redirected with a TCR-like chimeric antigen receptor. Sci Rep. 2014;4:3571.

19. Oren R, Hod-Marco M, Haus-Cohen M, Thomas S, Blat D, Duvshani N, et al. Functional comparison of engineered T cells carrying a native TCR versus TCR-like antibody-based chimeric antigen receptors indicates affinity/avidity thresholds. J Immunol. 2014;193 (11):5733–43.

20. Inaguma Y, Akahori Y, Murayama Y, Shiraishi K, Tsuzuki-Iba S, Endoh A, et al. Construction and molecular characterization of a T-cell receptor-like antibody and CAR-T cells specific for minor histocompatibility antigen HA-1H. Gene Ther. 2014;21(6):575–84.

21. Rafiq S, Purdon TJ, Daniyan AF, Koneru M, Dao T, Liu C, et al. Optimized T-cell receptor-mimic chimeric antigen receptor T cells directed toward the intracellular Wilms tumor 1 antigen. Leukemia. 2017;31(8):1788–97.

22. Ma Q, Garber HR, Lu S, He H, Tallis E, Ding X, et al. A novel TCR-like CAR with specificity for PR1/HLA-A2 effectively targets myeloid leukemia in vitro when expressed in human adult peripheral blood and cord blood T cells. Cytotherapy. 2016;18(8):985–94.

23. Stewart-Jones G, Wadle A, Hombach A, Shenderov E, Held G, Fischer E, et al. Rational development of high-affinity T-cell receptor-like antibodies. Proc Natl Acad Sci USA. 2009;106(14):5784–8.

24. Xie YJ, Dougan M, Jailkhani N, Ingram J, Fang T, Kummer L, et al. Nanobody-based CAR T cells that target the tumor microenvironment inhibit the growth of solid tumors in immunocompetent mice. Proc Natl Acad Sci USA. 2019;116(16):7624–31.

25. Faitschuk E, Nagy V, Hombach AA, Abken H. A dual chain chimeric antigen receptor (CAR) in the native antibody format for targeting immune cells towards cancer cells without the need of an scFv. Gene Ther. 2016;23(10):718–26.

26. Hammill JA, VanSeggelen H, Helsen CW, Denisova GF, Evelegh C, Tantalo DGM, et al. Designed ankyrin repeat proteins are effective targeting elements for chimeric antigen receptors. J Immunother Cancer. 2015;3:55.

27. Siegler E, Li S, Kim YJ, Wang P. Designed ankyrin repeat proteins as Her2 targeting domains in chimeric antigen receptor-engineered T cells. Hum Gene Ther. 2017;28(9):726–36.

28. Han X, Cinay GE, Zhao Y, Guo Y, Zhang X, Wang P. Adnectin-based design of chimeric antigen receptor for T cell engineering. Mol Ther. 2017;25(11):2466–76.

29. Kahlon KS, Brown C, Cooper LJN, Raubitschek A, Forman SJ, Jensen MC. Specific recognition and killing of glioblastoma multiforme by interleukin 13-zetakine redirected cytolytic T cells. Cancer Res. 2004;64(24):9160–6.

30. Kong S, Sengupta S, Tyler B, Bais AJ, Ma Q, Doucette S, et al. Suppression of human glioma xenografts with second-generation IL13R-specific chimeric antigen receptor-modified T cells. Clin Cancer Res Off J Am Assoc Cancer Res. 2012;18(21):5949–60.

31. Krebs S, Chow KK, Yi Z, Rodriguez-Cruz T, Hegde M, Gerken C, et al. T cells redirected to IL13Rα2 with IL13 mutein-CARs have antiglioma activity but also recognize IL13Rα1. Cytotherapy. 2014;16(8):1121–31.

32. Brown CE, Alizadeh D, Starr R, Weng L, Wagner JR, Naranjo A, et al. Regression of glioblastoma after chimeric antigen receptor T-cell therapy. N Engl J Med. 2016;375 (26):2561–9.

33. Lee L, Draper B, Chaplin N, Philip B, Chin M, Galas-Filipowicz D, et al. An APRIL-based chimeric antigen receptor for dual targeting of BCMA and TACI in multiple myeloma. Blood. 2018;131(7):746–58.

34. Wang Y, Xu Y, Li S, Liu J, Xing Y, Xing H, et al. Targeting FLT3 in acute myeloid leukemia using ligand-based chimeric antigen receptor-engineered T cells. J Hematol Oncol. 2018;11 (1):60.

35. Nakazawa Y, Matsuda K, Kurata T, Sueki A, Tanaka M, Sakashita K, et al. Anti-proliferative effects of T cells expressing a ligand-based chimeric antigen receptor against CD116 on CD34+ cells of juvenile myelomonocytic leukemia. J Hematol Oncol. 2016;9:27.

36. Baumeister SH, Murad J, Werner L, Daley H, Trebeden-Negre H, Gicobi JK, et al. Phase I trial of autologous CAR T cells targeting NKG2D ligands in patients with AML/MDS and multiple myeloma. Cancer Immunol Res. 2019;7(1):100–12.

37. Hudecek M, Sommermeyer D, Kosasih PL, Silva-Benedict A, Liu L, Rader C, et al. The nonsignaling extracellular spacer domain of chimeric antigen receptors is decisive for in vivo antitumor activity. Cancer Immunol Res. 2015;3(2):125–35.

38. Hombach A, Heuser C, Gerken M, Fischer B, Lewalter K, Diehl V, et al. T cell activation by recombinant FcepsilonRI gamma-chain immune receptors: an extracellular spacer domain

impairs antigen-dependent T cell activation but not antigen recognition. Gene Ther. 2000;7 (12):1067–75.

39. Jensen MC, Riddell SR. Designing chimeric antigen receptors to effectively and safely target tumors. Curr Opin Immunol. 2015;33:9–15.

40. Grakoui A, Bromley SK, Sumen C, Davis MM, Shaw AS, Allen PM, et al. The immunological synapse: a molecular machine controlling T cell activation. Science. 1999;285(5425):221–7.

41. James SE, Greenberg PD, Jensen MC, Lin Y, Wang J, Till BG, et al. Antigen sensitivity of CD22-specific chimeric TCR is modulated by target epitope distance from the cell membrane. J Immunol. 2008;180(10):7028–38.

42. Haso W, Lee DW, Shah NN, Stetler-Stevenson M, Yuan CM, Pastan IH, et al. Anti-CD22-chimeric antigen receptors targeting B-cell precursor acute lymphoblastic leukemia. Blood. 2013;121(7):1165–74.

43. Qin L, Zhao R, Li P. Incorporation of functional elements enhances the antitumor capacity of CAR T cells. Exp Hematol Oncol. 2017;6:28.

44. Watanabe N, Bajgain P, Sukumaran S, Ansari S, Heslop HE, Rooney CM, et al. Fine-tuning the CAR spacer improves T-cell potency. Onco Targets Ther. 2016;5(12):e1253656.

45. Srivastava S, Riddell SR. Engineering CAR-T cells: design concepts. Trends Immunol. 2015;36 (8):494–502.

46. Hombach A, Hombach AA, Abken H. Adoptive immunotherapy with genetically engineered T cells: modification of the IgG1 Fc "spacer" domain in the extracellular moiety of chimeric antigen receptors avoids "off-target" activation and unintended initiation of an innate immune response. Gene Ther. 2010;17(10):1206–13.

47. Jonnalagadda M, Mardiros A, Urak R, Wang X, Hoffman LJ, Bernanke A, et al. Chimeric antigen receptors with mutated IgG4 fc spacer avoid fc receptor binding and improve T cell persistence and antitumor efficacy. Mol Ther. 2015;23(4):757–68.

48. Bridgeman JS, Hawkins RE, Bagley S, Blaylock M, Holland M, Gilham DE. The optimal antigen response of chimeric antigen receptors harboring the CD3zeta transmembrane domain is dependent upon incorporation of the receptor into the endogenous TCR/CD3 complex. J Immunol. 2010;184(12):6938–49.

49. Dotti G, Gottschalk S, Savoldo B, Brenner MK. Design and development of therapies using chimeric antigen receptor-expressing T cells. Immunol Rev. 2014;257(1):107–26.

50. Alabanza L, Pegues M, Geldres C, Shi V, Wiltzius JJW, Sievers SA, et al. Function of novel anti-CD19 chimeric antigen receptors with human variable regions is affected by hinge and transmembrane domains. Mol Ther. 2017;25(11):2452–65.

51. Torikai H, Reik A, Liu P-Q, Zhou Y, Zhang L, Maiti S, et al. A foundation for universal T-cell based immunotherapy: T cells engineered to express a CD19-specific chimeric-antigen-receptor and eliminate expression of endogenous TCR. Blood. 2012;119(24):5697–705.

52. Kruschinski A, Moosmann A, Poschke I, Norell H, Chmielewski M, Seliger B, et al. Engineering antigen-specific primary human NK cells against HER-2 positive carcinomas. Proc Natl Acad Sci USA. 2008;105(45):17481–6.

53. Hombach AA, Abken H. Costimulation by chimeric antigen receptors revisited the T cell antitumor response benefits from combined CD28-OX40 signaling. Int J Cancer. 2011;129 (12):2935–44.

54. Shen C-J, Yang Y-X, Han EQ, Cao N, Wang Y-F, Wang Y, et al. Chimeric antigen receptor containing ICOS signaling domain mediates specific and efficient antitumor effect of T cells against EGFRvIII expressing glioma. J Hematol Oncol. 2013;6:33.

55. Guedan S, Chen X, Madar A, Carpenito C, McGettigan SE, Frigault MJ, et al. ICOS-based chimeric antigen receptors program bipolar TH17/TH1 cells. Blood. 2014;124(7):1070–80.

56. Song D-G, Ye Q, Poussin M, Harms GM, Figini M, Powell DJ. CD27 costimulation augments the survival and antitumor activity of redirected human T cells in vivo. Blood. 2012;119 (3):696–706.
57. Foster AE, Mahendravada A, Shinners NP, Chang W-C, Crisostomo J, Lu A, et al. Regulated expansion and survival of chimeric antigen receptor-modified T cells using small molecule-dependent inducible MyD88/CD40. Mol Ther. 2017;25(9):2176–88.
58. Collinson-Pautz MR, Chang W-C, Lu A, Khalil M, Crisostomo JW, Lin P-Y, et al. Constitutively active MyD88/CD40 costimulation enhances expansion and efficacy of chimeric antigen receptor T cells targeting hematological malignancies. Leukemia. 2019;33(9):2195–207.
59. Cheadle EJ, Rothwell DG, Bridgeman JS, Sheard VE, Hawkins RE, Gilham DE. Ligation of the CD2 co-stimulatory receptor enhances IL-2 production from first-generation chimeric antigen receptor T cells. Gene Ther. 2012;19(11):1114–20.
60. Altvater B, Landmeier S, Pscherer S, Temme J, Juergens H, Pule M, et al. 2B4 (CD244) signaling via chimeric receptors costimulates tumor-antigen specific proliferation and in vitro expansion of human T cells. Cancer Immunol Immunother. 2009;58(12):1991–2001.
61. Zhang T, Cao L, Xie J, Shi N, Zhang Z, Luo Z, et al. Efficiency of CD19 chimeric antigen receptor-modified T cells for treatment of B cell malignancies in phase I clinical trials: a meta-analysis. Oncotarget. 2015;6(32):33961–71.
62. Hombach A, Abken H. Costimulation tunes tumor-specific activation of redirected T cells in adoptive immunotherapy. Cancer Immunol Immunother. 2007;56(5):731–7.
63. Wang X, Popplewell LL, Wagner JR, Naranjo A, Blanchard MS, Mott MR, et al. Phase 1 studies of central memory-derived CD19 CAR T-cell therapy following autologous HSCT in patients with B-cell NHL. Blood. 2016;127(24):2980–90.
64. Frauwirth KA, Riley JL, Harris MH, Parry RV, Rathmell JC, Plas DR, et al. The CD28 signaling pathway regulates glucose metabolism. Immunity. 2002;16(6):769–77.
65. Kawalekar OU, O'Connor RS, Fraietta JA, Guo L, McGettigan SE, Posey AD, et al. Distinct signaling of coreceptors regulates specific metabolism pathways and impacts memory development in CAR T cells. Immunity. 2016;44(2):380–90.
66. van der Windt GJW, Pearce EL. Metabolic switching and fuel choice during T-cell differentiation and memory development. Immunol Rev. 2012;249(1):27–42.
67. van der Windt GJW, Everts B, Chang C-H, Curtis JD, Freitas TC, Amiel E, et al. Mitochondrial respiratory capacity is a critical regulator of CD8+ T cell memory development. Immunity. 2012;36(1):68–78.
68. Pearce EL, Walsh MC, Cejas PJ, Harms GM, Shen H, Wang L-S, et al. Enhancing CD8 T-cell memory by modulating fatty acid metabolism. Nature. 2009;460(7251):103–7.
69. Sukumar M, Liu J, Ji Y, Subramanian M, Crompton JG, Yu Z, et al. Inhibiting glycolytic metabolism enhances CD8+ T cell memory and antitumor function. J Clin Invest. 2013;123 (10):4479–88.
70. Gattinoni L, Zhong X-S, Palmer DC, Ji Y, Hinrichs CS, Yu Z, et al. Wnt signaling arrests effector T cell differentiation and generates CD8+ memory stem cells. Nat Med. 2009;15 (7):808–13.
71. Cherkassky L, Morello A, Villena-Vargas J, Feng Y, Dimitrov DS, Jones DR, et al. Human CAR T cells with cell-intrinsic PD-1 checkpoint blockade resist tumor-mediated inhibition. J Clin Invest. 2016;126(8):3130–44.
72. Golumba-Nagy V, Kuehle J, Hombach AA, Abken H. CD28-ζ CAR T cells resist TGF-β repression through IL-2 signaling, which can be mimicked by an engineered IL-7 autocrine loop. Mol Ther. 2018;26(9):2218–30.

73. Hombach AA, Chmielewski M, Rappl G, Abken H. Adoptive immunotherapy with redirected T cells produces CCR7- cells that are trapped in the periphery and benefit from combined CD28-OX40 costimulation. Hum Gene Ther. 2013;24(3):259–69.

74. Chmielewski M, Hombach AA, Abken H. Of CARs and TRUCKs: chimeric antigen receptor (CAR) T cells engineered with an inducible cytokine to modulate the tumor stroma. Immunol Rev. 2014;257(1):83–90.

75. Chmielewski M, Abken H. TRUCKs: the fourth generation of CARs. Expert Opin Biol Ther. 2015;15(8):1145–54.

76. Pegram HJ, Park JH, Brentjens RJ. CD28z CARs and armored CARs. Cancer J Sudbury Mass. 2014;20(2):127–33.

77. Chmielewski M, Kopecky C, Hombach AA, Abken H. IL-12 release by engineered T cells expressing chimeric antigen receptors can effectively muster an antigen-independent macrophage response on tumor cells that have shut down tumor antigen expression. Cancer Res. 2011;71(17):5697–706.

78. Pegram HJ, Lee JC, Hayman EG, Imperato GH, Tedder TF, Sadelain M, et al. Tumor-targeted T cells modified to secrete IL-12 eradicate systemic tumors without need for prior conditioning. Blood. 2012;119(18):4133–41.

79. Hu B, Ren J, Luo Y, Keith B, Young RM, Scholler J, et al. Augmentation of antitumor immunity by human and mouse CAR T cells secreting IL-18. Cell Rep. 2017;20(13):3025–33.

80. Chmielewski M, Abken H. CAR T cells releasing IL-18 convert to T-Bethigh FoxO1low effectors that exhibit augmented activity against advanced solid tumors. Cell Rep. 2017;21(11):3205–19.

81. Kunert A, Chmielewski M, Wijers R, Berrevocts C, Abken H, Debets R. Intra-tumoral production of IL18, but not IL12, by TCR-engineered T cells is non-toxic and counteracts immune evasion of solid tumors. Onco Targets Ther. 2017;7(1):e1378842.

82. Koneru M, Purdon TJ, Spriggs D, Koneru S, Brentjens RJ. IL-12 secreting tumor-targeted chimeric antigen receptor T cells eradicate ovarian tumors in vivo. Onco Targets Ther. 2015;4(3):e994446.

83. Ligtenberg MA, Mougiakakos D, Mukhopadhyay M, Witt K, Lladser A, Chmielewski M, et al. Coexpressed catalase protects chimeric antigen receptor-redirected T cells as well as bystander cells from oxidative stress-induced loss of antitumor activity. J Immunol. 2016;196(2):759–66.

84. Caruana I, Savoldo B, Hoyos V, Weber G, Liu H, Kim ES, et al. Heparanase promotes tumor infiltration and antitumor activity of CAR-redirected T lymphocytes. Nat Med. 2015;21(5):524–9.

85. Yeku OO, Brentjens RJ. Armored CAR T-cells: utilizing cytokines and pro-inflammatory ligands to enhance CAR T-cell anti-tumour efficacy. Biochem Soc Trans. 2016;44(2):412–8.

86. Kaczanowska S, Joseph AM, Davila E. TLR agonists: our best frenemy in cancer immunotherapy. J Leukoc Biol. 2013;93(6):847–63.

87. Geng D, Kaczanowska S, Tsai A, Younger K, Ochoa A, Rapoport AP, et al. TLR5 ligand-secreting T cells reshape the tumor microenvironment and enhance antitumor activity. Cancer Res. 2015;75(10):1959–71.

88. Koneru M, O'Cearbhaill R, Pendharkar S, Spriggs DR, Brentjens RJ. A phase I clinical trial of adoptive T cell therapy using IL-12 secreting MUC-16(ecto) directed chimeric antigen receptors for recurrent ovarian cancer. J Transl Med. 2015;13:102.

89. Xu A, Bhanumathy KK, Wu J, Ye Z, Freywald A, Leary SC, et al. IL-15 signaling promotes adoptive effector T-cell survival and memory formation in irradiation-induced lymphopenia. Cell Biosci. 2016;6:30.

90. Hoyos V, Savoldo B, Quintarelli C, Mahendravada A, Zhang M, Vera J, et al. Engineering CD19-specific T lymphocytes with interleukin-15 and a suicide gene to enhance their anti-lymphoma/leukemia effects and safety. Leukemia. 2010;24(6):1160–70.
91. Hsu C, Jones SA, Cohen CJ, Zheng Z, Kerstann K, Zhou J, et al. Cytokine-independent growth and clonal expansion of a primary human CD8+ T-cell clone following retroviral transduction with the IL-15 gene. Blood. 2007;109(12):5168–77.
92. Vera JF, Hoyos V, Savoldo B, Quintarelli C, Giordano Attianese GM, Leen AM, et al. Genetic manipulation of tumor-specific cytotoxic T lymphocytes to restore responsiveness to IL-7. Mol Ther. 2009;17(5):880–8.
93. Perna SK, Pagliara D, Mahendravada A, Liu H, Brenner M, Savoldo B, et al. Interleukin-7 mediates selective expansion of tumor-redirected cytotoxic T lymphocytes without enhancement of regulatory T-cell inhibition. Clin Cancer Res. 2014;20(1):131–9.
94. Mohammed S, Sukumaran S, Bajgain P, Watanabe N, Heslop HE, Rooney CM, et al. Improving chimeric antigen receptor-modified T cell function by reversing the immunosuppressive tumor microenvironment of pancreatic cancer. Mol Ther. 2017;25(1):249–58.
95. Zhang L, Yu Z, Muranski P, Palmer DC, Restifo NP, Rosenberg SA, et al. Inhibition of TGF-β signaling in genetically engineered tumor antigen-reactive T cells significantly enhances tumor treatment efficacy. Gene Ther. 2013;20(5):575–80.
96. Boice M, Salloum D, Mourcin F, Sanghvi V, Amin R, Oricchio E, et al. Loss of the HVEM tumor suppressor in lymphoma and restoration by modified CAR-T cells. Cell. 2016;167(2):405–418.e13.
97. Grada Z, Hegde M, Byrd T, Shaffer DR, Ghazi A, Brawley VS, et al. TanCAR: a novel bispecific chimeric antigen receptor for cancer immunotherapy. Mol Ther Nucleic Acids. 2013;2:e105.
98. Zah E, Lin M-Y, Silva-Benedict A, Jensen MC, Chen YY. T cells expressing CD19/CD20 bispecific chimeric antigen receptors prevent antigen escape by malignant B cells. Cancer Immunol Res. 2016;4(6):498–508.
99. Ruella M, Barrett DM, Kenderian SS, Shestova O, Hofmann TJ, Perazzelli J, et al. Dual CD19 and CD123 targeting prevents antigen-loss relapses after CD19-directed immunotherapies. J Clin Invest. 2016;126(10):3814–26.
100. Hudecek M, Schmitt TM, Baskar S, Lupo-Stanghellini MT, Nishida T, Yamamoto TN, et al. The B-cell tumor-associated antigen ROR1 can be targeted with T cells modified to express a ROR1-specific chimeric antigen receptor. Blood. 2010;116(22):4532–41.
101. Vera J, Savoldo B, Vigouroux S, Biagi E, Pule M, Rossig C, et al. T lymphocytes redirected against the kappa light chain of human immunoglobulin efficiently kill mature B lymphocyte-derived malignant cells. Blood. 2006;108(12):3890–7.
102. Zhao W-H, Liu J, Wang B-Y, Chen Y-X, Cao X-M, Yang Y, et al. A phase 1, open-label study of LCAR-B38M, a chimeric antigen receptor T cell therapy directed against B cell maturation antigen, in patients with relapsed or refractory multiple myeloma. J Hematol Oncol. 2018;11(1):141.
103. Wilkie S, van Schalkwyk MCI, Hobbs S, Davies DM, van der Stegen SJC, Pereira ACP, et al. Dual targeting of ErbB2 and MUC1 in breast cancer using chimeric antigen receptors engineered to provide complementary signaling. J Clin Immunol. 2012;32(5):1059–70.
104. Lanitis E, Poussin M, Klattenhoff AW, Song D, Sandaltzopoulos R, June CH, et al. Chimeric antigen receptor T cells with dissociated signaling domains exhibit focused antitumor activity with reduced potential for toxicity in vivo. Cancer Immunol Res. 2013;1(1):43–53.
105. He X, Feng Z, Ma J, Ling S, Cao Y, Gurung B, et al. Bi-specific and split CAR T cells targeting CD13 and TIM3 eradicate acute myeloid leukemia. Blood. 2020;135(10):713–23.

106. Kim MS, Ma JSY, Yun H, Cao Y, Kim JY, Chi V, et al. Redirection of genetically engineered CAR-T cells using bifunctional small molecules. J Am Chem Soc. 2015;137(8):2832–5.
107. Rodgers DT, Mazagova M, Hampton EN, Cao Y, Ramadoss NS, Hardy IR, et al. Switch-mediated activation and retargeting of CAR-T cells for B-cell malignancies. Proc Natl Acad Sci USA. 2016;113(4):E459–68.
108. Wu C-Y, Roybal KT, Puchner EM, Onuffer J, Lim WA. Remote control of therapeutic T cells through a small molecule-gated chimeric receptor. Science. 2015;350(6258):aab4077.
109. Kloss CC, Condomines M, Cartellieri M, Bachmann M, Sadelain M. Combinatorial antigen recognition with balanced signaling promotes selective tumor eradication by engineered T cells. Nat Biotechnol. 2013;31(1):71–5.
110. Morsut L, Roybal KT, Xiong X, Gordley RM, Coyle SM, Thomson M, et al. Engineering customized cell sensing and response behaviors using synthetic notch receptors. Cell. 2016;164 (4):780–91.
111. Roybal KT, Rupp LJ, Morsut L, Walker WJ, McNally KA, Park JS, et al. Precision tumor recognition by T cells with combinatorial antigen-sensing circuits. Cell. 2016;164(4):770–9.
112. Srivastava S, Salter AI, Liggitt D, Yechan-Gunja S, Sarvothama M, Cooper K, et al. Logic-gated ROR1 chimeric antigen receptor expression rescues T cell-mediated toxicity to normal tissues and enables selective tumor targeting. Cancer Cell. 2019;35(3):489–503.e8.
113. Fisher J, Abramowski P, Wisidagamage Don ND, Flutter B, Capsomidis A, Cheung GW-K, et al. Avoidance of on-target off-tumor activation using a co-stimulation-only chimeric antigen receptor. Mol Ther. 2017;25(5):1234–47.
114. Fedorov VD, Themeli M, Sadelain M. PD-1- and CTLA-4-based inhibitory chimeric antigen receptors (iCARs) divert off-target immunotherapy responses. Sci Transl Med. 2013;5 (215):215ra172.
115. Yu S, Yi M, Qin S, Wu K. Next generation chimeric antigen receptor T cells: safety strategies to overcome toxicity. Mol Cancer. 2019;18(1):125.
116. Kudo K, Imai C, Lorenzini P, Kamiya T, Kono K, Davidoff AM, et al. T lymphocytes expressing a CD16 signaling receptor exert antibody-dependent cancer cell killing. Cancer Res. 2014;74(1):93–103.
117. Cartellieri M, Feldmann A, Koristka S, Arndt C, Loff S, Ehninger A, et al. Switching CAR T cells on and off: a novel modular platform for retargeting of T cells to AML blasts. Blood Cancer J. 2016;6(8):e458.
118. Urbanska K, Lanitis E, Poussin M, Lynn RC, Gavin BP, Kelderman S, et al. A universal strategy for adoptive immunotherapy of cancer through use of a novel T-cell antigen receptor. Cancer Res. 2012;72(7):1844–52.
119. Bachmann M. The UniCAR system: a modular CAR T cell approach to improve the safety of CAR T cells. Immunol Lett. 2019;211:13–22.
120. Kobold S, Grassmann S, Chaloupka M, Lampert C, Wenk S, Kraus F, et al. Impact of a new fusion receptor on PD-1-mediated immunosuppression in adoptive T cell therapy. J Natl Cancer Inst. 2015;107:8.
121. Liu X, Ranganathan R, Jiang S, Fang C, Sun J, Kim S, et al. A chimeric switch-receptor targeting PD1 augments the efficacy of second-generation CAR T cells in advanced solid tumors. Cancer Res. 2016;76(6):1578–90.
122. Prosser ME, Brown CE, Shami AF, Forman SJ, Jensen MC. Tumor PD-L1 co-stimulates primary human CD8(+) cytotoxic T cells modified to express a PD1:CD28 chimeric receptor. Mol Immunol. 2012;51(3–4):263–72.
123. Hoogi S, Eisenberg V, Mayer S, Shamul A, Barliya T, Cohen CJ. A TIGIT-based chimeric co-stimulatory switch receptor improves T-cell anti-tumor function. J Immunother Cancer. 2019;7(1):243.

124. Poirot L, Philip B, Schiffer-Mannioui C, Le Clerre D, Chion-Sotinel I, Derniame S, et al. Multiplex genome-edited T-cell manufacturing platform for "off-the-shelf" adoptive T-cell immunotherapies. Cancer Res. 2015;75(18):3853–64.
125. Bertaina A, Merli P, Rutella S, Pagliara D, Bernardo ME, Masetti R, et al. HLA-haploidentical stem cell transplantation after removal of αβ+ T and B cells in children with nonmalignant disorders. Blood. 2014;124(5):822–6.
126. Qasim W, Zhan H, Samarasinghe S, Adams S, Amrolia P, Stafford S, et al. Molecular remission of infant B-ALL after infusion of universal TALEN gene-edited CAR T cells. Sci Transl Med. 2017;9:374.
127. Shalem O, Sanjana NE, Hartenian E, Shi X, Scott DA, Mikkelson T, et al. Genome-scale CRISPR-Cas9 knockout screening in human cells. Science. 2014;343(6166):84–7.
128. Ren J, Zhang X, Liu X, Fang C, Jiang S, June CH, et al. A versatile system for rapid multiplex genome-edited CAR T cell generation. Oncotarget. 2017;8(10):17002–11.
129. Ren J, Liu X, Fang C, Jiang S, June CH, Zhao Y. Multiplex genome editing to generate universal CAR T cells resistant to PD1 inhibition. Clin Cancer Res. 2017;23(9):2255–66.
130. Sommer C, Boldajipour B, Kuo TC, Bentley T, Sutton J, Chen A, et al. Preclinical evaluation of allogeneic CAR T cells targeting BCMA for the treatment of multiple myeloma. Mol Ther. 2019;27(6):1126–38.
131. Klingemann H. Are natural killer cells superior CAR drivers? Onco Targets Ther. 2014;3: e28147.
132. Huenecke S, Zimmermann SY, Kloess S, Esser R, Brinkmann A, Tramsen L, et al. IL-2-driven regulation of NK cell receptors with regard to the distribution of CD16+ and CD16- subpopulations and in vivo influence after haploidentical NK cell infusion. J Immunother. 2010;33(2):200–10.
133. Wang E, Wang L-C, Tsai C-Y, Bhoj V, Gershenson Z, Moon E, et al. Generation of potent T-cell immunotherapy for cancer using DAP12-based, multichain, chimeric immunoreceptors. Cancer Immunol Res. 2015;3(7):815–26.
134. Chang Y-H, Connolly J, Shimasaki N, Mimura K, Kono K, Campana D. A chimeric receptor with NKG2D specificity enhances natural killer cell activation and killing of tumor cells. Cancer Res. 2013;73(6):1777–86.
135. Imai C, Iwamoto S, Campana D. Genetic modification of primary natural killer cells overcomes inhibitory signals and induces specific killing of leukemic cells. Blood. 2005;106(1):376–83.
136. Xiao L, Cen D, Gan H, Sun Y, Huang N, Xiong H, et al. Adoptive transfer of NKG2D CAR mRNA-engineered natural killer cells in colorectal cancer patients. Mol Ther. 2019;27 (6):1114–25.
137. Töpfer K, Cartellieri M, Michen S, Wiedemuth R, Müller N, Lindemann D, et al. DAP12-based activating chimeric antigen receptor for NK cell tumor immunotherapy. J Immunol. 2015;194 (7):3201–12.
138. Zhang C, Burger MC, Jennewein L, Genßler S, Schönfeld K, Zeiner P, et al. ErbB2/HER2-specific NK cells for targeted therapy of glioblastoma. J Natl Cancer Inst. 2016;108:5.
139. Schönfeld K, Sahm C, Zhang C, Naundorf S, Brendel C, Odendahl M, et al. Selective inhibition of tumor growth by clonal NK cells expressing an ErbB2/HER2-specific chimeric antigen receptor. Mol Ther. 2015;23(2):330–8.
140. Köhl U, Arsenieva S, Holzinger A, Abken H. CAR T cells in trials: recent achievements and challenges that remain in the production of modified T cells for clinical applications. Hum Gene Ther. 2018;29(5):559–68.
141. Singh H, Figliola MJ, Dawson MJ, Olivares S, Zhang L, Yang G, et al. Manufacture of clinical-grade CD19-specific T cells stably expressing chimeric antigen receptor using sleeping beauty system and artificial antigen presenting cells. PLoS One. 2013;8(5):e64138.

142. Singh H, Moyes JSE, Huls MH, Cooper LJN. Manufacture of T cells using the sleeping beauty system to enforce expression of a CD19-specific chimeric antigen receptor. Cancer Gene Ther. 2015;22(2):95–100.
143. Manuri PVR, Wilson MH, Maiti SN, Mi T, Singh H, Olivares S, et al. piggyBac transposon/transposase system to generate CD19-specific T cells for the treatment of B-lineage malignancies. Hum Gene Ther. 2010;21(4):427–37.
144. Monjezi R, Miskey C, Gogishvili T, Schleef M, Schmeer M, Einsele H, et al. Enhanced CAR T-cell engineering using non-viral sleeping beauty transposition from minicircle vectors. Leukemia. 2017;31(1):186–94.
145. Fraietta JA, Nobles CL, Sammons MA, Lundh S, Carty SA, Reich TJ, et al. Disruption of TET2 promotes the therapeutic efficacy of CD19-targeted T cells. Nature. 2018;558(7709):307–12.
146. Ruella M, Xu J, Barrett DM, Fraietta JA, Reich TJ, Ambrose DE, et al. Induction of resistance to chimeric antigen receptor T cell therapy by transduction of a single leukemic B cell. Nat Med. 2018;24(10):1499–503.
147. Pipkin ME, Sacks JA, Cruz-Guilloty F, Lichtenheld MG, Bevan MJ, Rao A. Interleukin-2 and inflammation induce distinct transcriptional programs that promote the differentiation of effector cytolytic T cells. Immunity. 2010;32(1):79–90.
148. Kaneko S, Mastaglio S, Bondanza A, Ponzoni M, Sanvito F, Aldrighetti L, et al. IL-7 and IL-15 allow the generation of suicide gene-modified alloreactive self-renewing central memory human T lymphocytes. Blood. 2009;113(5):1006–15.
149. Butler MO, Lee J-S, Ansén S, Neuberg D, Hodi FS, Murray AP, et al. Long-lived antitumor CD8+ lymphocytes for adoptive therapy generated using an artificial antigen-presenting cell. Clin Cancer Res. 2007;13(6):1857–67.
150. Li Y, Bleakley M, Yee C. IL-21 influences the frequency, phenotype, and affinity of the antigen-specific CD8 T cell response. J Immunol. 2005;175(4):2261–9.
151. Hinrichs CS, Spolski R, Paulos CM, Gattinoni L, Kerstann KW, Palmer DC, et al. IL-2 and IL-21 confer opposing differentiation programs to CD8+ T cells for adoptive immunotherapy. Blood. 2008;111(11):5326–33.
152. Wofford JA, Wieman HL, Jacobs SR, Zhao Y, Rathmell JC. IL-7 promotes Glut1 trafficking and glucose uptake via STAT5-mediated activation of Akt to support T-cell survival. Blood. 2008;111(4):2101–11.
153. Cui G, Staron MM, Gray SM, Ho P-C, Amezquita RA, Wu J, et al. IL-7-induced glycerol transport and TAG synthesis promotes memory CD8+ T cell longevity. Cell. 2015;161(4):750–61.
154. van der Waart AB, van de Weem NMP, Maas F, Kramer CSM, Kester MGD, Falkenburg JHF, et al. Inhibition of Akt signaling promotes the generation of superior tumor-reactive T cells for adoptive immunotherapy. Blood. 2014;124(23):3490–500.
155. Klebanoff CA, Gattinoni L, Restifo NP. Sorting through subsets: which T-cell populations mediate highly effective adoptive immunotherapy? J Immunother. 2012;35(9):651–60.
156. Gattinoni L, Lugli E, Ji Y, Pos Z, Paulos CM, Quigley MF, et al. A human memory T cell subset with stem cell-like properties. Nat Med. 2011;17(10):1290–7.
157. Singh N, Perazzelli J, Grupp SA, Barrett DM. Early memory phenotypes drive T cell proliferation in patients with pediatric malignancies. Sci Transl Med. 2016;8(320):320ra3.
158. O'Rourke DM, Nasrallah MP, Desai A, Melenhorst JJ, Mansfield K, Morrissette JJD, et al. A single dose of peripherally infused EGFRvIII-directed CAR T cells mediates antigen loss and induces adaptive resistance in patients with recurrent glioblastoma. Sci Transl Med. 2017;9:399.
159. Chong EA, Levine BL, Grupp SA, Davis MM, Siegel DL, Maude SL, et al. CAR T cell viability release testing and clinical outcomes: is there a lower limit? Blood. 2019;134(21):1873–5.

160. Tötterman T, Carlsson M, Simonsson B, Bengtsson M, Nilsson K. T-cell activation and subset patterns are altered in B-CLL and correlate with the stage of the disease. Blood. 1989;74 (2):786–92.
161. Fraietta JA, Lacey SF, Orlando EJ, Pruteanu-Malinici I, Gohil M, Lundh S, et al. Determinants of response and resistance to CD19 chimeric antigen receptor (CAR) T cell therapy of chronic lymphocytic leukemia. Nat Med. 2018;24(5):563–71.
162. Philip M, Fairchild L, Sun L, Horste EL, Camara S, Shakiba M, et al. Chromatin states define tumour-specific T cell dysfunction and reprogramming. Nature. 2017;545(7655):452–6.
163. Kochenderfer JN, Dudley ME, Carpenter RO, Kassim SH, Rose JJ, Telford WG, et al. Donor-derived CD19-targeted T cells cause regression of malignancy persisting after allogeneic hematopoietic stem cell transplantation. Blood. 2013;122(25):4129–39.
164. Kebriaei P, Singh H, Huls MH, Figliola MJ, Bassett R, Olivares S, et al. Phase I trials using sleeping beauty to generate CD19-specific CAR T cells. J Clin Invest. 2016;126(9):3363–76.
165. Brudno JN, Somerville RPT, Shi V, Rose JJ, Halverson DC, Fowler DH, et al. Allogeneic T cells that express an anti-CD19 chimeric antigen receptor induce remissions of B-cell malignancies that progress after allogeneic hematopoietic stem-cell transplantation without causing graft-versus-host disease. J Clin Oncol. 2016;34(10):1112–21.
166. Cruz CRY, Micklethwaite KP, Savoldo B, Ramos CA, Lam S, Ku S, et al. Infusion of donor-derived CD19-redirected virus-specific T cells for B-cell malignancies relapsed after allogeneic stem cell transplant: a phase 1 study. Blood. 2013;122(17):2965–73.
167. Rotolo R, Leuci V, Donini C, Cykowska A, Gammaitoni L, Medico G, et al. CAR-based strategies beyond T lymphocytes: integrative opportunities for cancer adoptive immunotherapy. Int J Mol Sci. 2019;20(11):2839.
168. Hu W, Wang G, Huang D, Sui M, Xu Y. Cancer immunotherapy based on natural killer cells: current progress and new opportunities. Front Immunol. 2019;10:1205.
169. Holzinger A, Barden M, Abken H. The growing world of CAR T cell trials: a systematic review. Cancer Immunol Immunother. 2016;65(12):1433–50.
170. Abken H. Driving CARs on the highway to solid cancer: some considerations on the adoptive therapy with CAR T cells. Hum Gene Ther. 2017;28(11):1047–60.
171. Porter DL, Levine BL, Kalos M, Bagg A, June CH. Chimeric antigen receptor-modified T cells in chronic lymphoid leukemia. N Engl J Med. 2011;365(8):725–33.
172. Porter DL, Hwang W-T, Frey NV, Lacey SF, Shaw PA, Loren AW, et al. Chimeric antigen receptor T cells persist and induce sustained remissions in relapsed refractory chronic lympho-cytic leukemia. Sci Transl Med. 2015;7(303):303ra139.
173. Grupp SA, Kalos M, Barrett D, Aplenc R, Porter DL, Rheingold SR, et al. Chimeric antigen receptor-modified T cells for acute lymphoid leukemia. N Engl J Med. 2013;368(16):1509–18.
174. Brentjens RJ, Davila ML, Riviere I, Park J, Wang X, Cowell LG, et al. CD19-targeted T cells rapidly induce molecular remissions in adults with chemotherapy-refractory acute lymphoblas-tic leukemia. Sci Transl Med. 2013;5(177):177ra38.
175. Maude SL, Frey N, Shaw PA, Aplenc R, Barrett DM, Bunin NJ, et al. Chimeric antigen receptor T cells for sustained remissions in leukemia. N Engl J Med. 2014;371(16):1507–17.
176. Davila ML, Riviere I, Wang X, Bartido S, Park J, Curran K, et al. Efficacy and toxicity management of 19-28z CAR T cell therapy in B cell acute lymphoblastic leukemia. Sci Transl Med. 2014;6(224):224ra25.
177. Lee DW, Kochenderfer JN, Stetler-Stevenson M, Cui YK, Delbrook C, Feldman SA, et al. T cells expressing CD19 chimeric antigen receptors for acute lymphoblastic leukaemia in children and young adults: a phase 1 dose-escalation trial. Lancet. 2015;385(9967):517–28.
178. Kochenderfer JN, Dudley ME, Kassim SH, Somerville RPT, Carpenter RO, Stetler-Stevenson-M, et al. Chemotherapy-refractory diffuse large B-cell lymphoma and indolent B-cell

malignancies can be effectively treated with autologous T cells expressing an anti-CD19 chimeric antigen receptor. J Clin Oncol. 2015;33(6):540–9.

179. Brentjens RJ, Rivière I, Park JH, Davila ML, Wang X, Stefanski J, et al. Safety and persistence of adoptively transferred autologous CD19-targeted T cells in patients with relapsed or chemotherapy refractory B-cell leukemias. Blood. 2011;118(18):4817–28.

180. Faitschuk E, Hombach AA, Frenzel LP, Wendtner C-M, Abken H. Chimeric antigen receptor T cells targeting Fc μ receptor selectively eliminate CLL cells while sparing healthy B cells. Blood. 2016;128(13):1711–22.

181. Garfall AL, Maus MV, Hwang W-T, Lacey SF, Mahnke YD, Melenhorst JJ, et al. Chimeric antigen receptor T cells against CD19 for multiple myeloma. N Engl J Med. 2015;373 (11):1040–7.

182. Savoldo B, Rooney CM, Di Stasi A, Abken H, Hombach A, Foster AE, et al. Epstein Barr virus specific cytotoxic T lymphocytes expressing the anti-CD30zeta artificial chimeric T-cell receptor for immunotherapy of Hodgkin disease. Blood. 2007;110(7):2620 30.

183. Louis CU, Savoldo B, Dotti G, Pule M, Yvon E, Myers GD, et al. Antitumor activity and long-term fate of chimeric antigen receptor-positive T cells in patients with neuroblastoma. Blood. 2011;118(23):6050–6.

184. Pule MA, Savoldo B, Myers GD, Rossig C, Russell HV, Dotti G, et al. Virus-specific T cells engineered to coexpress tumor-specific receptors: persistence and antitumor activity in individuals with neuroblastoma. Nat Med. 2008;14(11):1264–70.

185. Brudno JN, Kochenderfer JN. Toxicities of chimeric antigen receptor T cells: recognition and management. Blood. 2016;127(26):3321–30.

186. Franciszkiewicz K, Boissonnas A, Boutet M, Combadière C, Mami-Chouaib F. Role of chemokines and chemokine receptors in shaping the effector phase of the antitumor immune response. Cancer Res. 2012;72(24):6325–32.

187. Bouzin C, Brouet A, De Vriese J, Dewever J, Feron O. Effects of vascular endothelial growth factor on the lymphocyte-endothelium interactions: identification of caveolin-1 and nitric oxide as control points of endothelial cell anergy. J Immunol. 2007;178(3):1505–11.

188. Calcinotto A, Grioni M, Jachetti E, Curnis F, Mondino A, Parmiani G, et al. Targeting TNF-α to neoangiogenic vessels enhances lymphocyte infiltration in tumors and increases the therapeutic potential of immunotherapy. J Immunol. 2012;188(6):2687–94.

189. Chinnasamy D, Yu Z, Theoret MR, Zhao Y, Shrimali RK, Morgan RA, et al. Gene therapy using genetically modified lymphocytes targeting VEGFR-2 inhibits the growth of vascularized syngenic tumors in mice. J Clin Invest. 2010;120(11):3953–68.

190. Kandalaft LE, Facciabene A, Buckanovich RJ, Coukos G. Endothelin B receptor, a new target in cancer immune therapy. Clin Cancer Res. 2009;15(14):4521–8.

191. Kershaw MH, Wang G, Westwood JA, Pachynski RK, Tiffany HL, Marincola FM, et al. Redirecting migration of T cells to chemokine secreted from tumors by genetic modification with CXCR2. Hum Gene Ther. 2002;13(16):1971–80.

192. Craddock JA, Lu A, Bear A, Pule M, Brenner MK, Rooney CM, et al. Enhanced tumor trafficking of GD2 chimeric antigen receptor T cells by expression of the chemokine receptor CCR2b. J Immunother. 2010;33(8):780–8.

193. Adusumilli PS, Cherkassky L, Villena-Vargas J, Colovos C, Servais E, Plotkin J, et al. Regional delivery of mesothelin-targeted CAR T cell therapy generates potent and long-lasting CD4-dependent tumor immunity. Sci Transl Med. 2014;6(261):261ra151.

194. Katz SC, Burga RA, McCormack E, Wang LJ, Mooring W, Point GR, et al. Phase I hepatic immunotherapy for metastases study of intra-arterial chimeric antigen receptor-modified T-cell therapy for CEA+ liver metastases. Clin Cancer Res. 2015;21(14):3149–59.

195. Tchou J, Zhao Y, Levine BL, Zhang PJ, Davis MM, Melenhorst JJ, et al. Safety and efficacy of intratumoral injections of chimeric antigen receptor (CAR) T cells in metastatic breast cancer. Cancer Immunol Res. 2017;5(12):1152–61.
196. Textor A, Listopad JJ, Wührmann LL, Perez C, Kruschinski A, Chmielewski M, et al. Efficacy of CAR T-cell therapy in large tumors relies upon stromal targeting by IFNγ. Cancer Res. 2014;74(23):6796–805.
197. Kakarla S, Chow KKH, Mata M, Shaffer DR, Song X-T, Wu M-F, et al. Antitumor effects of chimeric receptor engineered human T cells directed to tumor stroma. Mol Ther. 2013;21 (8):1611–20.
198. Ninomiya S, Narala N, Huye L, Yagyu S, Savoldo B, Dotti G, et al. Tumor indoleamine 2,3-dioxygenase (IDO) inhibits CD19-CAR T cells and is downregulated by lymphodepleting drugs. Blood. 2015;125(25):3905–16.
199. Rodriguez PC, Quiceno DG, Ochoa AC. L-arginine availability regulates T-lymphocyte cell-cycle progression. Blood. 2007;109(4):1568–73.
200. Rodriguez PC, Zea AH, Culotta KS, Zabaleta J, Ochoa JB, Ochoa AC. Regulation of T cell receptor CD3zeta chain expression by L-arginine. J Biol Chem. 2002;277(24):21123–9.
201. Newick K, O'Brien S, Sun J, Kapoor V, Maceyko S, Lo A, et al. Augmentation of CAR T-cell trafficking and antitumor efficacy by blocking protein kinase a localization. Cancer Immunol Res. 2016;4(6):541–51.
202. John LB, Devaud C, Duong CPM, Yong CS, Beavis PA, Haynes NM, et al. Anti-PD-1 antibody therapy potently enhances the eradication of established tumors by gene-modified T cells. Clin Cancer Res. 2013;19(20):5636–46.
203. Chong EA, Melenhorst JJ, Lacey SF, Ambrose DE, Gonzalez V, Levine BL, et al. PD-1 blockade modulates chimeric antigen receptor (CAR)-modified T cells: refueling the CAR. Blood. 2017;129(8):1039–41.
204. Maude SL, Barrett D, Teachey DT, Grupp SA. Managing cytokine release syndrome associated with novel T cell-engaging therapies. Cancer J. 2014;20(2):119–22.
205. Lee DW, Gardner R, Porter DL, Louis CU, Ahmed N, Jensen M, et al. Current concepts in the diagnosis and management of cytokine release syndrome. Blood. 2014;124(2):188–95.
206. Teachey DT, Lacey SF, Shaw PA, Melenhorst JJ, Maude SL, Frey N, et al. Identification of predictive biomarkers for cytokine release syndrome after chimeric antigen receptor T-cell therapy for acute lymphoblastic leukemia. Cancer Discov. 2016;6(6):664–79.
207. Chen F, Teachey DT, Pequignot E, Frey N, Porter D, Maude SL, et al. Measuring IL-6 and sIL-6R in serum from patients treated with tocilizumab and/or siltuximab following CAR T cell therapy. J Immunol Methods. 2016;434:1–8.
208. Giavridis T, van der Stegen SJC, Eyquem J, Hamieh M, Piersigilli A, Sadelain M. CAR T cell-induced cytokine release syndrome is mediated by macrophages and abated by IL-1 blockade. Nat Med. 2018;24(6):731–8.
209. Norelli M, Camisa B, Barbiera G, Falcone L, Purevdorj A, Genua M, et al. Monocyte-derived IL-1 and IL-6 are differentially required for cytokine-release syndrome and neurotoxicity due to CAR T cells. Nat Med. 2018;24(6):739–48.
210. Sterner RM, Sakemura R, Cox MJ, Yang N, Khadka RH, Forsman CL, et al. GM-CSF inhibition reduces cytokine release syndrome and neuroinflammation but enhances CAR-T cell function in xenografts. Blood. 2019;133(7):697–709.
211. Staedtke V, Bai R-Y, Kim K, Darvas M, Davila ML, Riggins GJ, et al. Disruption of a self-amplifying catecholamine loop reduces cytokine release syndrome. Nature. 2018;564 (7735):273–7.

212. Lee DW, Santomasso BD, Locke FL, Ghobadi A, Turtle CJ, Brudno JN, et al. ASTCT consensus grading for cytokine release syndrome and neurologic toxicity associated with immune effector cells. Biol Blood Marrow Transplant. 2019;25(4):625–38.
213. Santomasso BD, Park JH, Salloum D, Riviere I, Flynn J, Mead E, et al. Clinical and biological correlates of neurotoxicity associated with CAR T-cell therapy in patients with B-cell acute lymphoblastic leukemia. Cancer Discov. 2018;8(8):958–71.
214. Maus MV, Haas AR, Beatty GL, Albelda SM, Levine BL, Liu X, et al. T cells expressing chimeric antigen receptors can cause anaphylaxis in humans. Cancer Immunol Res. 2013;1 (1):26–31.
215. Sommermeyer D, Hill T, Shamah SM, Salter AI, Chen Y, Mohler KM, et al. Fully human CD19-specific chimeric antigen receptors for T-cell therapy. Leukemia. 2017;31(10):2191–9.
216. Hombach A, Koch D, Sircar R, Heuser C, Diehl V, Kruis W, et al. A chimeric receptor that selectively targets membrane-bound carcinoembryonic antigen (mCEA) in the presence of soluble CEA. Gene Ther. 1999;6(2):300–4.
217. Hombach A, Heuser C, Sircar R, Tillmann T, Diehl V, Pohl C, et al. An anti-CD30 chimeric receptor that mediates CD3-zeta-independent T-cell activation against Hodgkin's lymphoma cells in the presence of soluble CD30. Cancer Res. 1998;58(6):1116–9.
218. Lamers CH, Sleijfer S, van Steenbergen S, van Elzakker P, van Krimpen B, Groot C, et al. Treatment of metastatic renal cell carcinoma with CAIX CAR-engineered T cells: clinical evaluation and management of on-target toxicity. Mol Ther. 2013;21(4):904–12.
219. Parente-Pereira AC, Burnet J, Ellison D, Foster J, Davies DM, van der Stegen S, et al. Trafficking of CAR-engineered human T cells following regional or systemic adoptive transfer in SCID beige mice. J Clin Immunol. 2011;31(4):710–8.
220. Katz SC, Point GR, Cunetta M, Thorn M, Guha P, Espat NJ, et al. Regional CAR-T cell infusions for peritoneal carcinomatosis are superior to systemic delivery. Cancer Gene Ther. 2016;23(5):142–8.
221. Morgan RA, Yang JC, Kitano M, Dudley ME, Laurencot CM, Rosenberg SA. Case report of a serious adverse event following the administration of T cells transduced with a chimeric antigen receptor recognizing ERBB2. Mol Ther. 2010;18(4):843–51.
222. Goff SL, Morgan RA, Yang JC, Sherry RM, Robbins PF, Restifo NP, et al. Pilot trial of adoptive transfer of chimeric antigen receptor-transduced T cells targeting EGFRvIII in patients with glioblastoma. J Immunother. 2019;42(4):126–35.
223. Thistlethwaite FC, Gilham DE, Guest RD, Rothwell DG, Pillai M, Burt DJ, et al. The clinical efficacy of first-generation carcinoembryonic antigen (CEACAM5)-specific CAR T cells is limited by poor persistence and transient pre-conditioning-dependent respiratory toxicity. Cancer Immunol Immunother CII. 2017;66(11):1425–36.
224. Ahmed N, Brawley VS, Hegde M, Robertson C, Ghazi A, Gerken C, et al. Human epidermal growth factor receptor 2 (HER2) -specific chimeric antigen receptor-modified T cells for the immunotherapy of HER2-positive sarcoma. J Clin Oncol. 2015;33(15):1688–96.
225. Feng K, Liu Y, Guo Y, Qiu J, Wu Z, Dai H, et al. Phase I study of chimeric antigen receptor modified T cells in treating HER2-positive advanced biliary tract cancers and pancreatic cancers. Protein Cell. 2018;9(10):838–47.
226. Birkholz K, Hombach A, Krug C, Reuter S, Kershaw M, Kämpgen E, et al. Transfer of mRNA encoding recombinant immunoreceptors reprograms CD4+ and CD8+ T cells for use in the adoptive immunotherapy of cancer. Gene Ther. 2009;16(5):596–604.
227. Beatty GL, Haas AR, Maus MV, Torigian DA, Soulen MC, Plesa G, et al. Mesothelin-specific chimeric antigen receptor mRNA-engineered T cells induce anti-tumor activity in solid malignancies. Cancer Immunol Res. 2014;2(2):112–20.

228. Lamers CHJ, Sleijfer S, Vulto AG, Kruit WHJ, Kliffen M, Debets R, et al. Treatment of metastatic renal cell carcinoma with autologous T-lymphocytes genetically retargeted against carbonic anhydrase IX: first clinical experience. J Clin Oncol. 2006;24(13):e20–2.

229. Jena B, Maiti S, Huls H, Singh H, Lee DA, Champlin RE, et al. Chimeric antigen receptor (CAR)-specific monoclonal antibody to detect CD19-specific T cells in clinical trials. PLoS One. 2013;8(3):e57838.

230. Serafini M, Manganini M, Borleri G, Bonamino M, Imberti L, Biondi A, et al. Characterization of CD20-transduced T lymphocytes as an alternative suicide gene therapy approach for the treatment of graft-versus-host disease. Hum Gene Ther. 2004;15(1):63–76.

231. Griffioen M, van Egmond EHM, Kester MGD, Willemze R, Falkenburg JHF, Heemskerk MHM. Retroviral transfer of human CD20 as a suicide gene for adoptive T-cell therapy. Haematologica. 2009;94(9):1316–20.

232. Philip B, Kokalaki E, Mekkaoui L, Thomas S, Straathof K, Flutter B, et al. A highly compact epitope-based marker/suicide gene for easier and safer T-cell therapy. Blood. 2014;124 (8):1277–87.

233. Wang X, Chang W-C, Wong CW, Colcher D, Sherman M, Ostberg JR, et al. A transgene-encoded cell surface polypeptide for selection, in vivo tracking, and ablation of engineered cells. Blood. 2011;118(5):1255–63.

234. Straathof KC, Pulè MA, Yotnda P, Dotti G, Vanin EF, Brenner MK, et al. An inducible caspase 9 safety switch for T-cell therapy. Blood. 2005;105(11):4247–54.

235. Di Stasi A, Tey S-K, Dotti G, Fujita Y, Kennedy-Nasser A, Martinez C, et al. Inducible apoptosis as a safety switch for adoptive cell therapy. N Engl J Med. 2011;365(18):1673–83.

236. Juillerat A, Tkach D, Busser BW, Temburni S, Valton J, Duclert A, et al. Modulation of chimeric antigen receptor surface expression by a small molecule switch. BMC Biotechnol. 2019;19(1):44.

237. Mestermann K, Giavridis T, Weber J, Rydzek J, Frenz S, Nerreter T, et al. The tyrosine kinase inhibitor dasatinib acts as a pharmacologic on/off switch for CAR T cells. Sci Transl Med. 2019;11:499.

238. Salter AI, Ivey RG, Kennedy JJ, Voillet V, Rajan A, Alderman EJ, et al. Phosphoproteomic analysis of chimeric antigen receptor signaling reveals kinetic and quantitative differences that affect cell function. Sci Signal. 2018;11:544.

239. Ying Z, Huang XF, Xiang X, Liu Y, Kang X, Song Y, et al. A safe and potent anti-CD19 CAR T cell therapy. Nat Med. 2019;25(6):947–53.

240. Lee YG, Chu H, Lu Y, Leamon CP, Srinivasarao M, Putt KS, et al. Regulation of CAR T cell-mediated cytokine release syndrome-like toxicity using low molecular weight adapters. Nat Commun. 2019;10(1):2681.

241. Posey AD, Schwab RD, Boesteanu AC, Steentoft C, Mandel U, Engels B, et al. Engineered CAR T cells targeting the cancer-associated Tn-glycoform of the membrane mucin MUC1 control adenocarcinoma. Immunity. 2016;44(6):1444–54.

242. Zhang C, Wang Z, Yang Z, Wang M, Li S, Li Y, et al. Phase I escalating-dose trial of CAR-T therapy targeting CEA+ metastatic colorectal cancers. Mol Ther. 2017;25(5):1248–58.

243. Liu K, Liu X, Peng Z, Sun H, Zhang M, Zhang J, et al. Retargeted human avidin-CAR T cells for adoptive immunotherapy of EGFRvIII expressing gliomas and their evaluation via optical imaging. Oncotarget. 2015;6(27):23735–47.

244. Song D-G, Ye Q, Poussin M, Liu L, Figini M, Powell DJ. A fully human chimeric antigen receptor with potent activity against cancer cells but reduced risk for off-tumor toxicity. Oncotarget. 2015;6(25):21533–46.

245. Richman SA, Nunez-Cruz S, Moghimi B, Li LZ, Gershenson ZT, Mourelatos Z, et al. High-affinity GD2-specific CAR T cells induce fatal encephalitis in a preclinical neuroblastoma model. Cancer Immunol Res. 2018;6(1):36–46.
246. Sotillo E, Barrett DM, Black KL, Bagashev A, Oldridge D, Wu G, et al. Convergence of acquired mutations and alternative splicing of CD19 enables resistance to CART-19 immuno-therapy. Cancer Discov. 2015;5(12):1282–95.
247. Gardner R, Wu D, Cherian S, Fang M, Hanafi L-A, Finney O, et al. Acquisition of a CD19-negative myeloid phenotype allows immune escape of MLL-rearranged B-ALL from CD19 CAR-T-cell therapy. Blood. 2016;127(20):2406–10.
248. Sabatino M, Hu J, Sommariva M, Gautam S, Fellowes V, Hocker JD, et al. Generation of clinical-grade CD19-specific CAR-modified CD8+ memory stem cells for the treatment of human B-cell malignancies. Blood. 2016;128(4):519–28.
249. Ruella M, Kenderian SS, Shestova O, Fraietta JA, Qayyum S, Zhang Q, et al. The addition of the BTK inhibitor ibrutinib to anti-CD19 chimeric antigen receptor T cells (CART19) improves responses against mantle cell lymphoma. Clin Cancer Res. 2016;22(11):2684–96.
250. Osborn MJ, Webber BR, Knipping F, Lonetree C, Tennis N, DeFeo AP, et al. Evaluation of TCR gene editing achieved by TALENs, CRISPR/Cas9, and megaTAL nucleases. Mol Ther. 2016;24(3):570–81.
251. Georgiadis C, Preece R, Nickolay L, Etuk A, Petrova A, Ladon D, et al. Long terminal repeat CRISPR-CAR-coupled "universal" T cells mediate potent anti-leukemic effects. Mol Ther. 2018;26(5):1215–27.
252. Liu X, Zhang Y, Cheng C, Cheng AW, Zhang X, Li N, et al. CRISPR-Cas9-mediated multiplex gene editing in CAR-T cells. Cell Res. 2017;27(1):154–7.
253. Eyquem J, Mansilla-Soto J, Giavridis T, van der Stegen SJC, Hamieh M, Cunanan KM, et al. Targeting a CAR to the TRAC locus with CRISPR/Cas9 enhances tumour rejection. Nature. 2017;543(7643):113–7.
254. Bailey SR, Maus MV. Gene editing for immune cell therapies. Nat Biotechnol. 2019;37 (12):1425–34.
255. Thomis DC, Marktel S, Bonini C, Traversari C, Gilman M, Bordignon C, et al. A Fas-based suicide switch in human T cells for the treatment of graft-versus-host disease. Blood. 2001;97 (5):1249–57.
256. Tey S-K, Dotti G, Rooney CM, Heslop HE, Brenner MK. Inducible caspase 9 suicide gene to improve the safety of allodepleted T cells after haploidentical stem cell transplantation. Biol Blood Marrow Transplant. 2007;13(8):913–24.
257. Zhou X, Di Stasi A, Tey S-K, Krance RA, Martinez C, Leung KS, et al. Long-term outcome after haploidentical stem cell transplant and infusion of T cells expressing the inducible caspase 9 safety transgene. Blood. 2014;123(25):3895–905.
258. Von Hoff DD, Ramanathan RK, Borad MJ, Laheru DA, Smith LS, Wood TE, et al. Gemcitabine plus nab-paclitaxel is an active regimen in patients with advanced pancreatic cancer: a phase I/II trial. J Clin Oncol. 2011;29(34):4548–54.
259. Hu Y, Sun J, Wu Z, Yu J, Cui Q, Pu C, et al. Predominant cerebral cytokine release syndrome in CD19-directed chimeric antigen receptor-modified T cell therapy. J Hematol Oncol. 2016;9 (1):70.
260. Sampson JH, Choi BD, Sanchez-Perez L, Suryadevara CM, Snyder DJ, Flores CT, et al. EGFRvIII mCAR-modified T-cell therapy cures mice with established intracerebral glioma and generates host immunity against tumor-antigen loss. Clin Cancer Res. 2014;20(4):972–84.
261. Ellebrecht CT, Bhoj VG, Nace A, Choi EJ, Mao X, Cho MJ, et al. Reengineering chimeric antigen receptor T cells for targeted therapy of autoimmune disease. Science. 2016;353 (6295):179–84.

262. Jyothi MD, Flavell RA, Geiger TL. Targeting autoantigen-specific T cells and suppression of autoimmune encephalomyelitis with receptor-modified T lymphocytes. Nat Biotechnol. 2002;20 (12):1215–20.
263. Margalit A, Fishman S, Berko D, Engberg J, Gross G. Chimeric beta2 microglobulin/CD3zeta polypeptides expressed in T cells convert MHC class I peptide ligands into T cell activation receptors: a potential tool for specific targeting of pathogenic CD8(+) T cells. Int Immunol. 2003;15(11):1379–87.
264. Krebs K, Böttinger N, Huang L-R, Chmielewski M, Arzberger S, Gasteiger G, et al. T cells expressing a chimeric antigen receptor that binds hepatitis B virus envelope proteins control virus replication in mice. Gastroenterology. 2013;145(2):456–65.
265. Sautto GA, Wisskirchen K, Clementi N, Castelli M, Diotti RA, Graf J, et al. Chimeric antigen receptor (CAR)-engineered T cells redirected against hepatitis C virus (HCV) E2 glycoprotein. Gut. 2016;65(3):512–23.
266. Full F, Lehner M, Thonn V, Goetz G, Scholz B, Kaufmann KB, et al. T cells engineered with a cytomegalovirus-specific chimeric immunoreceptor. J Virol. 2010;84(8):4083–8.
267. Romeo C, Seed B. Cellular immunity to HIV activated by CD4 fused to T cell or Fc receptor polypeptides. Cell. 1991;64(5):1037–46.
268. Deeks SG, Wagner B, Anton PA, Mitsuyasu RT, Scadden DT, Huang C, et al. A phase II randomized study of HIV-specific T-cell gene therapy in subjects with undetectable plasma viremia on combination antiretroviral therapy. Mol Ther. 2002;5(6):788–97.
269. Kumaresan PR, Manuri PR, Albert ND, Maiti S, Singh H, Mi T, et al. Bioengineering T cells to target carbohydrate to treat opportunistic fungal infection. Proc Natl Acad Sci USA. 2014;111 (29):10660–5.
270. Elinav E, Waks T, Eshhar Z. Redirection of regulatory T cells with predetermined specificity for the treatment of experimental colitis in mice. Gastroenterology. 2008;134(7):2014–24.
271. Skuljec J, Chmielewski M, Happle C, Habener A, Busse M, Abken H, et al. Chimeric antigen receptor-redirected regulatory T cells suppress experimental allergic airway inflammation, a model of asthma. Front Immunol. 2017;8:1125.
272. MacDonald KG, Hoeppli RE, Huang Q, Gillies J, Luciani DS, Orban PC, et al. Alloantigen-specific regulatory T cells generated with a chimeric antigen receptor. J Clin Invest. 2016;126 (4):1413–24.
273. Boardman DA, Philippeos C, Fruhwirth GO, Ibrahim MA, Hannen RF, Cooper D, et al. Expression of a chimeric antigen receptor specific for donor HLA class I enhances the potency of human regulatory T cells in preventing human skin transplant rejection. Am J Transplant. 2017;17(4):931–43.
274. Noyan F, Zimmermann K, Hardtke-Wolenski M, Knoefel A, Schulde E, Geffers R, et al. Prevention of allograft rejection by use of regulatory T cells with an MHC-specific chimeric antigen receptor. Am J Transplant. 2017;17(4):917–30.
275. Aghajanian H, Kimura T, Rurik JG, Hancock AS, Leibowitz MS, Li L, et al. Targeting cardiac fibrosis with engineered T cells. Nature. 2019;573(7774):430–3.

Improvement of Key Characteristics of Antibodies

11

Neil Brewis

Contents

Keywords

Affinity maturation · Antibody fragments · Biophysical properties · Fc receptors ·
Immunogenicity · Serum half-life

N. Brewis (✉)
F-star Therapeutics, Cambridge, UK
e-mail: neil.brewis@f-star.com

© Springer Nature Switzerland AG 2021
F. Rüker, G. Wozniak-Knopp (eds.), *Introduction to Antibody Engineering*,
Learning Materials in Biosciences,
https://doi.org/10.1007/978-3-030-54630-4_11

What You Will Learn in This Chapter

Recombinant antibodies have been engineered for several decades to make highly effective medicines that have transformed the lives of patients. Researchers have optimised key characteristics of the natural activities of antibodies, including affinity and Fc-mediated activities, to address the disease-causing agent. Moreover, antibodies have been engineered to have desirable manufacturability properties, low immunogenicity and good tolerability. These improvements will be covered in this chapter.

11.1 Improving the Natural Functions of Antibodies to Create Therapeutics

The natural function of an antibody is to bind to foreign microbes and toxins, to neutralise their activity and facilitate clearance from the body. Binding to the foreign target is performed by the Fab region of the antibody. This can neutralise infectivity and, through the activity of the Fc region, facilitate phagocytosis and antibody-mediated killing via the processes of cellular cytotoxicity (ADCC) and complement-mediated lysis (CDC).

11.2 Antigen Binding

A fundamental property that contributes to the success of therapeutic antibodies is their high specificity and affinity for targeted antigens. This binding is mediated through the Fab proportion of the antibody, which in turn contains domains with a high degree of sequence variation, termed variable domains. Within these variable domains, antigen binding is mediated by the complementarity-determining regions (CDRs). The most common form of a therapeutic antibody, IgG_1, is a Y-shaped molecule with two identical Fab regions, each with six CDRs and a Fc region (Fig. 11.1). It is the high degree of sequence variation, the large surface area and the bivalent (two binding sites) which enables high specificity and affinity.

The traditional approach to discover a therapeutic antibody is by immunisation of animals, mainly mice, with a target antigen to generate a mixed population of B lymphocytes producing antibodies against the target. In the mouse, during the course of an immune response, multiple rounds of somatic hypermutation and clonal selection of the B cells increases the activity of the polyclonal antibody pool against a particular antigen. The ground-breaking technology developed by Köhler and Milstein was to immortalise the B lymphocytes by fusion with myeloma cells to produce monoclonal hybridoma cell lines that secrete mixtures of antibody candidates [1]. These monoclonal antibodies while highly specific for their target, were of murine origin and had a high potential for being

Fig. 11.1 Schematic representation of an IgG$_1$ monoclonal antibody. The variable domains (V_H and V_k) are shown in blue and the constant regions (CH1, CH2 and CH3) in grey

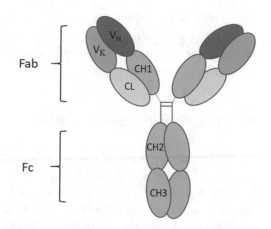

immunogenic in humans, with the patient making anti-drug antibodies. This immunogenicity has both a safety liability and likely lead to a loss of effectiveness through increased clearance. To address this, technologies were developed to make chimeric and humanised antibodies by grafting the murine variable regions or the CDR residues to a human antibody framework. Both approaches have led to a dramatic decrease in immunogenicity and subsequently increased clinical usefulness [2, 3].

An alternative approach for the discovery of therapeutic antibodies is to create transgenic mice with the endogenous murine antibody sequences for the heavy and light chain of the antibody inactivated and replaced with the equivalent human germline sequences [4, 5]. The XenoMouse and HuMab-Mouse were developed in parallel. Injection of antigens into these mice led to the clonal selection of B cells providing 'fully human' high-affinity antibodies that had undergone somatic hypermutation and selection. A number of these antibodies have entered clinical development and have gone on to create important medicines targeting epidermal growth factor receptor [6], RANK ligand [7] and tumour necrosis factor [8].

An in vitro method for generating human antibodies has also been successfully accomplished by creating large libraries of human antibody genes and display technologies. The concept is to have individual antibody clones expressed and displayed on vectors such as bacteria phage viruses where each individual phage encodes and expresses a different antibody variant. By creating large libraries, that are either cloned from B cell antibody repertoires or made synthetically, high-affinity clones can be isolated using recombinant target antigen to 'pull out' the desired binders. This technology was first demonstrated using a single-chain variable fragment of an antibody (scFv) to isolate a high-affinity antagonist of tumour necrosis factor to create the medicine adalimumab [9]. Numerous phage display-derived candidates have entered clinical development and demonstrated the established and reliable nature of this technology.

Next-generation sequencing has been successfully used to augment the approaches above. It can be used to analyse both in vitro- and in vivo-derived libraries of binding clones and thereby increase the depth of analysis of the binders. This can enable readily identifying related clones with a range of properties such as affinities [10]. These clones can be expressed and screened for the desired properties and improve the success of identifying the therapeutic antibody candidate.

11.2.1 Binding Epitope and Affinity

Identifying clones that bind to the optimal region or epitope on the target antigen is essential. Whether the desired biology is to block and antagonise a receptor–ligand interaction or crosslink and agonise a membrane receptor, identifying the appropriate binder is performed using functional assays. These assays may include purified recombinant ligand and receptors, cell-based physiological assays or in vivo models. It is preferable to screen the repertoire of binders from a library of clones for function as soon as possible in the drug discovery process.

Antibody clones can have a very wide range of affinities to their target (from femtomolar to micromolar). It is typical to optimise the binding affinities of antibodies to best address the biology of the target and typically therapeutic antibodies have a strong affinity that is single digit nanomolar or less. Increasing the affinity may increase the biological effectiveness of a therapy and can also decrease the injection dose, enable less frequent dosing and decrease the cost of therapy [11].

Engineering antibodies to optimise binding affinity has relied on experimental affinity maturation but this has now been complemented by in silico modelling. Experimental in vitro maturation can be performed on an antibody random- or targeted-mutagenesis. In the random method, saturation or error-prone polymerase chain reaction, PCR, are used to introduce mutations in the CDR sequences. In the targeted approach, alanine-scanning or small libraries with diversified specific regions are created. In both cases, the resulting variants are selected for affinity, often using phage or yeast display, and then individual clones are isolated, expressed and screened. Improvements in affinity can be detected at medium screening throughput using binding methods such as surface plasmon residence, or SPR. Several rounds of maturation may be required to achieve the desired affinity; however, it should be noted that during this process, while the affinity may improve, there is a potential for a reduction in specificity and stability of the antibody. Highly interactive residues such as aromatic amino acids could increase nonspecific binding of the purified antibody to itself and to non-target antigens. These undesirable properties need to be identified using biophysical methods and such clones deselected.

The use of in silico or computer-aided maturation gives the ability to search for the desired antibody in large virtual libraries and at the same time better understand the antibody–antigen interaction. When the three-dimensional crystallographic structure is available, it is possible to identify residues that are involved in the antigen interaction. Different approaches can be employed to stabilise the interaction using calculations of the hydrophobic, electrostatic, non-polar and binding energies. Internal stabilising energetics and dynamics for an anti-VEGF antibody modelled using the GROMACS software correlated with experimental affinity [12]. Electrostatic optimisation has been used successfully to increase the affinity of an anti-synuclein antibody by more than 100-fold [13].

11.2.2 Creating a Therapeutic Antibody with pH Sensitivity or 'pH Switch'

Clinical benefit can be created with an antibody that is sensitive to changes in pH. One application is to enable dissociation from the antigen in the acidified late endosomal compartment. This results in the ligand, dissociated from the antibody, entering the degradation pathways while the antibody is recycled by binding to the neonatal Fc receptor (FcRn). The potential is for increased longevity of the antibody in circulation in the patient's body and active degradation of the target. Sensitivity of binding to changes in pH can be created by introducing ionisable histidine residues for example into the CDRs and then screening in neutral and acidic conditions for differential antigen binding. Antibodies to IL-6, IL-6R and proprotein convertase subtilisin kexin type 9 (PCSK9) were all successfully engineered to retain high affinity at pH 7.4 and show decreased binding at acidic pH [14]. This resulted in more favourable dosing regimens.

11.3 Improving and Optimising the Interactions with Fc Receptors

Antibody-dependent cellular cytotoxicity (ADCC) is an important function of an antibody and part of the humoral immune response to limit and contain the infection. It is also an important mechanism for immunotherapy particularly for the treatment of cancers using therapeutic antibodies where the ADCC activity can augment other mechanisms such as blocking growth factor receptor signalling and acting as a targeting agent for toxic payloads.

ADCC is the process by which an antibody coats a target cell and recruits effector cells to induce target cell death via non-phagocytic mechanisms. Several classes of antibodies can mediate ADCC, including IgG, IgA and IgE. IgG_1 is the most used isotype for therapeutic antibodies. To carry out ADCC an antibody must bind simultaneously to its target and Fc receptors. There are several classes of FcR and the most important for tumour cell clearance by myeloid cells are the FcγRs. These are comprised of activating FcγR I (CD64), FcγR IIA (CD32A), FcγR IIIA (CD16A), and inhibitory FcγR IIB (CD32B) receptors [15]. Antibody engagement of these receptors triggers receptor clustering and downstream signalling via their tyrosine-based activation and inhibitory motifs. Myeloid cells capable of being ADCC effectors are NK cells, monocytes, macrophages, neutrophils, eosinophils and dendritic cells. Once activated, effector cells exocytose granules with perforin and granzyme B inducing the death of the target cell. In addition, Fas ligand expression is upregulated in activated effector cells and this can also lead to apoptosis in the target cell through a process mediated by Fas signalling [16].

It has been proposed that a number of therapeutic antibodies use ADCC as a prominent mechanism of action in their clinical activity. The role of ADCC has been demonstrated in the case of the anti-CD20 therapeutic antibody rituximab, where polymorphisms in the activatory FcγR gene correlated with clinical survival in patients. These polymorphisms were shown to be associated with higher affinity for Fc-mediated ADCC [17]. Similar correlations have been found in neuroblastoma patients treated with anti-GD2 antibodies [18]. Given the evidence that ADCC can be an important mechanism to kill tumours cells in cancer patients, new technologies have emerged to enhance ADCC activity. This includes co-administration of drugs including IL-2 to enhance ADCC in the patient, although this is beyond the scope of this chapter. Approaches have been developed that change the antibody structure and the most prominent is to increase, by site mutagenesis, the binding affinity to activating FcγR IIIA and by changing the type and level of Fc glycosylation. Modification of residues in the Fc region to increase the affinity to the Fc receptors has been successful and in addition, asymmetrical engineering of the Fc portion to create heterodimers with different heavy chains resulted in more stable antibodies with enhanced ADCC functionality [19]. Of the sugars that modify the Fc region, fucose appears to play the largest role in changing the binding to the Fc receptors. Removal of fucose, either through mutagenesis or production in non-fucosylating cell lines, improved Fc receptor affinity dramatically and resulted in higher levels of ADCC. This is being explored in the clinical development of a number of antibodies including a defucosylated anti-CD20 antibody therapeutic antibody [20].

Finally, for some antibody targets, it is beneficial to reduce binding to the Fc receptors for both efficacy and safety considerations. It has been shown that binding to the Fc receptors in organs, such as the liver, can lead to aberrant activation by therapeutic antibodies targeting costimulatory receptors, for example those targeting CD137, also termed 4-1BB. This can induce dose-limiting hepato-toxicities [21]. Reducing this interaction, by disabling the interaction with the FcγR decreases this toxicity. Furthermore, when using antibodies to activate immune cells, such as exhausted tumour T cells in the setting of immuno-oncology, by targeting checkpoint inhibitory receptors, it is useful to prevent the subsequent destruction of the T cell by ADCC [22]. There are several strategies to reduce the ADCC activity of an antibody and these include using IgG subclasses with lower ADCC activity or specific point mutation in the Fc that reduce affinity for the FcγRs. Some IgG subclasses, such as IgG_2 and IgG_4, have naturally reduced FcγR interactions compared with IgG_1. Additionally, a variety of Fc mutations have been created to achieve this reduced affinity and including removal of native Fc N-linked glycosylation including leucine to glutamic acid substitutions at position 235 of IgG_1. The modification of the Fc with two point mutations (L234A/L235A), the so-called LALA Fc mutations, were first described several decades ago [23] and more recently further Fc mutations have been described that are devoid of FcγR-mediated effector function [24]. It should be noted that these Fc changes increase the potential for the introduction of new T cell epitopes with subsequent immunogenicity and poor manufacturability as the IgG becomes less stable [25].

11.4 Complement Activation and Inactivation

Antibodies can engage the humoral immune response through interactions with complement component 1q (C1q) and the initiation of complement cascade by a series of proteolytic cleavage events [26]. Potent complement activation has been demonstrated in vitro for a number of approved antibodies including ofatumumab and rituximab. Furthermore, to enhance this potentially potent mechanism of action for cell killing, a variety of Fc engineering approaches have been reported. These generally use the approach of increasing the antibody affinity for the C1q either by point mutations or using sections of the IgG_3. IgG_3 has the most potent CDC activity of the four antibody subclasses but has both a relatively short serum half-life and manufacturing challenges. To overcome these issues, using IgG_3 CH2/CH3 regions were substituted into an IgG_1, and enhanced CDC more than 50-fold compared with the IgG_1 parental molecule.

As with liabilities with full ADCC activity, to increase the safety of administered antibodies, mutations have been introduced to diminish C1q binding and complement activation. These mutations include L234A/L235A [26] and V234A/G237A/P238S/H268A [24] and can lead to undetectable CDC and ADCC activities.

11.5 Improving Serum Half-Life to Decrease Dosing Frequency

Antibodies typically have a half-life of 7–21 days in humans (depending on the IgG subclass) and the effects of target-mediated clearance. The binding to the neonatal Fc receptor (FcRn)) plays a central role in both cellular trafficking and serum half-life [27]. This long half-life is largely due to the ability of the Fc to bind to the FcRn in endosomes following pinocytosis and avoid lysosomal degradation. IgG has a low affinity for FcRn at physiological pH however they become protonated in the acidic endosomal pH (6–6.5) leading to pH-dependent binding to the FcRn. The IgG avoids degradation by binding to the FcRn and is recycled into the bloodstream. Key contact residues in the Fc with the FcRn have been identified through elucidating the co-crystal structure. Histidine at position 310 is a significant contributor to pH-dependent binding [28]. Mutations of H310 to other amino acids lead to undetectable binding to FcRn at pH 6.0.

The understanding of the mechanism of recycling at the molecular interaction level, facilitated approaches to attempt to improve the natural IgG half-life by changing the interactions with the FcRn. This offered the potential for lower therapeutic doses, less frequent dosing and ultimately improving the cost of therapy. Using phage display to screen Fc variant libraries, mutations were identified that enhanced binding to FcRn at pH 6.0 [29]. These mutations, termed YTE (M252Y/S254T/T256E) increased the binding to the human and cynomolgus FcRn by tenfold and increased the half-life in a cynomolgus pharmacokinetic study by fourfold [30]. In a phase I study, testing motavizumab-YTE, an antibody targeting respiratory syncytial virus with the engineered increase in affinity to FcRn, resulted in a two- to fourfold increase in half-life relative to the parental motavizumab IgG$_1$ [31]. This demonstrated the clinical utility of this approach. Another set of mutations, M428L/N434S, was identified to result in an 11-fold increase in the affinity for FcRn at pH 6.0 and using both anti-EGFR and anti-VEGF antibodies in preclinical studies, increased the efficacies in mouse tumour models [32].

11.6 Immunogenicity of Therapeutic Antibodies

The first generation of monoclonal antibodies to enter clinical development came from the hybridoma technology developed by Köhler and Milstein [33]. The first approved therapeutic antibody, OKT-3, for the prevention of allograft rejection after transplantation, was of murine origin and like other early therapeutics, was highly immunogenic in patients [34]. The resulting anti-drug antibodies increased the clearance of the therapeutic antibodies, neutralised their binding and ultimately limited efficacy [35]. Additionally, therapeutic antibody re-administration was associated with severe infusion reactions, which may in part be mediated by the activation of the complement system, a major humoral defence system of innate immunity as well as an adaptive antibody response [36].

Several innovations have been introduced to isolate therapeutic antibodies that have a higher degree of human sequences (covered in a previous section). The creation of murine/human chimeric therapeutic antibodies by exchanging the murine constant regions with their human counterparts resulted in a dramatic reduction of immunogenicity with the loss of murine 'non-self' epitopes. The next generation of therapeutic antibodies was humanised antibodies created by grafting murine CDRs into the human monoclonal backbone. The amount of further benefit that was derived from this humanisation is more debatable and this may be due to the homology between variable regions from human and mouse being higher than between the constant regions. Indeed humanised antibodies can demonstrate more differences from natural human diversity than the differences with murine variable regions [35, 37].

The advent of fully human therapeutic antibodies using both phage and transgenic systems (where the sequences were derived from the human repertoire) gave potential for a lack of immunogenicity. However even fully human sequence-derived antibodies can induce notable immune responses. Adalimumab, which was first approved by the US Food and Drug Administration (FDA) for treatment of rheumatoid arthritis, induces high titre, neutralising antibody responses in a subset of patients that varies depending on disease setting (5–89%) [38–40].

A major driver of the immunogenicity for therapeutic antibodies is the presence of human T cell epitopes within the amino acid sequence. This activates helper T cells resulting in the production of antibodies against the therapeutic antibodies and potent neutralisation of the therapeutic effect. Antibodies can be 'de-immunised' by creating mutations to remove T cell epitopes without significantly reducing the binding affinity of the antibody [41].

While primary amino acid sequences can drive strong immunogenicity, other aspects can contribute. Examples include different patterns of glycosylation of the Fc region where antibody responses can be reduced by removal of N-linked glycosylation [42]. Furthermore, non-human glycosylation, for example galactose-α-1,3-galactose, from the production in Chinese hamster ovary (CHO) cell lines, can create hypersensitivity reactions due to pre-existing IgE antibodies in the patient that were created by prior exposure to the glycosylation in a tick infection [43]. The regional nature of the exposure to the tick infection, explains the regional nature of the hypersensitivity.

The preparation of the antibody and route of administration can also impact immunogenicity. The formation of aggregates during production or storage of the therapeutic antibody product increases the risk for immunogenicity [44]. Moreover, it has been demonstrated that it is the type, rather than the quantity or size, of the antibody aggregates which has more impact on the development of an immune response [45]. Considerable antibody discovery and manufacturing development time is spent identifying therapeutic antibodies with good biophysical properties and removing aggregates during production and preventing their formation during long term storage, for example by using different formulation buffers.

11.7 Antibody Fragments

The modular structure of an antibody, with multiple domains, enables the creation of smaller antigen-binding formats. Such smaller units include the fragment antigen-binding (Fab), the single-chain fragment variable (scFv) and single-domain antibodies (dAb). They can be used in their own right or combined to create a large variety of different therapeutics including coupled to antibody–drug conjugates (ADC) and multi-specific antibodies which can bind to more than one antigen. While ADC and bispecific antibodies both represent exciting opportunities, they are outside the scope of this chapter.

Antibody fragments offer several advantages over conventional antibodies including lack of Fc functionality, increased tissue penetration, greater stability, and large-scale production using cost-effective microbial systems. Moreover, their smaller structure facilitates binding to difficult epitopes for example those in a deep protein groove which would be more challenging with classical monoclonal antibody approaches. While promoting tissue penetration, their smaller size and lack of FcRn binding results in faster kidney-mediated excretion. This can be overcome by high dosing, local dosing or extending serum residency by half-life extension technologies including conjugation to polyethylene glycol (PEG) or albumin-binding domains.

One of the first antibody fragments to gain market approval was a PEGylated Fab, Cimzia, targeting tumour necrosis factor-alpha for the treatment of the Crohns' disease (reviewed in [46]). It was subsequently approved for other major autoimmune diseases including rheumatoid arthritis. Without the Fc activity, Cimzia had a clear mechanism of action difference compared with anti-TNF therapeutic antibodies.

As an alternative to systemic administration, the smaller size and greater stability of antibody fragments have enabled clinical studies using local delivery to the diseased tissue. This enables high concentrations at the site of delivery and low systemic exposure due to subsequent rapid renal clearance. Ranibizumab (Lucentis), an anti-VEGF Fab antibody fragment, was approved in 2006 for local administration into the eyes of patients with wet age-related macular degeneration (AMD). Controversy ensued however, as the parental therapeutic antibody, bevacizumab, was reported to be equally safe and efficacious but at approximately 1/20th of the price [47–49].

Creating antibody fragments for inhalation has also been tested in the clinic including an anti-tumour necrosis factor receptor I domain antibody for the treatment of acute respiratory distress syndrome [50] and an anti-respiratory syncytial virus (RSV) domain antibody for the treatment of RSV infection [51].

11.8 Improving Biophysical Properties of Therapeutic Antibodies

The biophysical properties of an antibody can dramatically affect clinical development and ultimately commercial success. Progression to a commercial manufacturing scale is lengthy, scientifically complex and costly. Therapeutic antibodies must be produced to a very high purity and be sterile. Levels of host cell proteins must be reduced to parts per million (ppm), DNA to parts per billion (ppb) and virus levels to less than one virus particle per million doses [52]. Even the first step of making a batch of Phase I clinical-grade material can take 18–24 months and cost multimillion euros.

The majority of the production of full-length antibodies is in Chinese hamster ovary cells (CHO) as antibodies require complex folding, assembly and post-translational modifications. One advantage of antibody fragments is they can be made in microbial systems, including bacteria and yeast, reducing costs and timelines. However, the complexity of production of both full-length and fragments of antibodies is ultimately dependent on the primary sequence of the antibody, and therefore, understanding factors which can affect antibody production and stability will be reviewed here.

During manufacture and storage, antibodies are at risk of degradation via a number of pathways. Each therapeutic antibody candidate has unique characteristics and risks. The identification of degradation prone or unstable regions in the antibody discovery process would permit re-engineering of problematic areas without the loss of biological activity.

11.8.1 Aggregation

Antibody aggregation is a common and significant type of physical degradation and can lead to reduced activity and the formation of immunogenic products [44]. For higher concentration formulations, such as those for subcutaneous injection at >100 mg/ml, protein–protein interactions become more frequent and the opportunity for aggregation formation increases as the concentration goes up. Changes in extrinsic conditions such as temperature, pH, salt, light, shaking and viscosity can transiently expose hydrophobic residues which promote protein–protein interactions that can lead to aggregation.

11.8.2 Thermal Denaturation

Protein denaturation is the partial or complete unfolding of the native-folded protein structure. Unfolding can lead to loss of antigen-binding and expose aggregation-prone regions. Thermal unfolding can be observed by differential scanning calorimetry (DSC) and a typical DSC profile for an IgG antibody molecule contains three peaks identified as Fab, CH2, and CH3 domains [53]. The CH2 domain is typically the least stable. The pH of the solution and the presence of salt can increase or decrease the melting temperature and therefore affect stability.

11.8.3 Fragmentation, Deamidation and Oxidation

Fragmentation of antibodies can occur following enzymatic or non-enzymatic hydrolysis of the peptide backbone and at a number of regions including the hinge, CH2/CH3 interface and regions containing aspartic acid and tryptophan residues. Deamidation is the most common chemical degradation pathway and results from the hydrolysis of glutamine and asparagine residues. There is a strong association with surrounding amino acid residues, and it can also be affected by pH, temperature and buffer composition. Deamidation can affect both antibody activity and stability. Oxidation of antibodies is also common and affects particularly methionine, cysteine, histidine, tyrosine and tryptophan residues. Oxidation can affect stability and biological activity and can reduce serum half-life and affinity for purification matrixes such as protein A.

11.8.4 Computational Design Tools

Computational tools allow for the rapid identification of specific amino acid sequences and regions that are prone to aggregation, degradation and chemical modification. The current tools and emerging trends are reviewed by [54]. Such computational design can also be used in the creation of discovery libraries where improvements have been built in and result in primary clones with improved developability properties [55].

Take Home Messages
- Antibodies can make complex but highly effective therapeutics
- Many key characteristics that control their intrinsic biological activities can be optimised to increase both the efficacy and safety of the therapeutic antibody
- Manufacturability of the therapeutic antibody is important and can be improved by directed mutagenesis of sequences generating biophysical liabilities

References

1. Köhler G, Milstein C. Derivation of specific antibody-producing tissue culture and tumor lines by cell fusion. Eur J Immunol. 1976;6(7):511–9. https://doi.org/10.1002/eji.1830060713.
2. Jones PT, Dear PH, Foote J, Neuberger MS, Winter G. Replacing the complementarity-determining regions in a human antibody with those from a mouse. Nature. 1986;321 (6069):522–5. https://doi.org/10.1038/321522a0.
3. Riechmann L, Clark M, Waldmann H, Winter G. Reshaping human antibodies for therapy. Nature. 1988;332(6162):323–7. https://doi.org/10.1038/332323a0.
4. Lonberg N. Human antibodies from transgenic animals. Nat Biotechnol. 2005;23(9):1117–25. https://doi.org/10.1038/nbt1135.
5. Scott CT. Mice with a human touch. Nat Biotechnol. 2007;25(10):1075–7. https://doi.org/10.1038/nbt1007-1075.
6. Jakobovits A, Amado RG, Yang X, Roskos L, Schwab G. From XenoMouse technology to panitumumab, the first fully human antibody product from transgenic mice. Nat Biotechnol. 2007;25(10):1134–43. https://doi.org/10.1038/nbt1337.
7. Elgundi Z, Reslan M, Cruz E, Sifniotis V, Kayser V. The state-of-play and future of antibody therapeutics. Adv Drug Deliv Rev. 2017;122:2–19. https://doi.org/10.1016/j.addr.2016.11.004.
8. Zhou H, Jang H, Fleischmann RM, et al. Pharmacokinetics and safety of golimumab, a fully human anti-TNF-alpha monoclonal antibody, in subjects with rheumatoid arthritis. J Clin Pharmacol. 2007;47(3):383–96. https://doi.org/10.1177/0091270006298188.
9. Weinblatt ME, Keystone EC, Furst DE, et al. Adalimumab, a fully human anti-tumor necrosis factor α monoclonal antibody, for the treatment of rheumatoid arthritis in patients taking concomitant methotrexate: the ARMADA trial. Arthritis Rheum. 2003;48(1):35–45. https://doi.org/10.1002/art.10697.
10. Ravn U, Didelot G, Venet S, et al. Deep sequencing of phage display libraries to support antibody discovery. Methods. 2013;60(1):99–110. https://doi.org/10.1016/j.ymeth.2013.03.001.
11. Hoogenboom HR. Selecting and screening recombinant antibody libraries. Nat Biotechnol. 2005;23(9):1105–16. https://doi.org/10.1038/nbt1126.
12. Corrada D, Colombo G. Energetic and dynamic aspects of the affinity maturation process: characterizing improved variants from the bevacizumab antibody with molecular simulations. J Chem Inf Model. 2013;53(11):2937–50. https://doi.org/10.1021/ci400416e.
13. Mahajan SP, Meksiriporn B, Waraho-Zhmayev D, et al. Computational affinity maturation of camelid single-domain intrabodies against the nonamyloid component of alpha-synuclein. Sci Rep. 2018;8(1):17611. https://doi.org/10.1038/s41598-018-35464-7.

14. Igawa T, Mimoto F, Hattori K. pH-dependent antigen-binding antibodies as a novel therapeutic modality. Biochim Biophys Acta – Proteins Proteomics. 2014;1844(11):1943–50. https://doi.org/10.1016/j.bbapap.2014.08.003.

15. Chenoweth AM, Wines BD, Anania JC, Mark HP. Harnessing the immune system *via* FcγR function in immune therapy: a pathway to next-gen mAbs. Immunol Cell Biol. 2020;98 (4):287–304. https://doi.org/10.1111/imcb.12326.

16. Bryceson YT, Chiang SCC, Darmanin S, et al. Molecular mechanisms of natural killer cell activation. J Innate Immun. 2011;3(3):216–26. https://doi.org/10.1159/000325265.

17. Rezvani AR, Maloney DG. Rituximab resistance. Best Pract Res Clin Haematol. 2011;24 (2):203–16. https://doi.org/10.1016/j.beha.2011.02.009.

18. Delgado DC, Hank JA, Kolesar J, et al. Genotypes of NK cell KIR receptors, their ligands, and Fc receptors in the response of neuroblastoma patients to Hu14.18-IL2 immunotherapy. Cancer Res. 2010;70(23):9554–61. https://doi.org/10.1158/0008-5472.CAN-10-2211.

19. Liu Z, Gunasekaran K, Wang W, et al. Asymmetrical Fc engineering greatly enhances antibody-dependent cellular cytotoxicity (ADCC) effector function and stability of the modified antibodies. J Biol Chem. 2014;289(6):3571–90. https://doi.org/10.1074/jbc.M113.513366.

20. Robak T. GA-101, a third-generation, humanized and glyco-engineered anti-CD20 mAb for the treatment of B-cell lymphoid malignancies. Curr Opin Investig Drugs. 2009;10(6):588–96. http://www.ncbi.nlm.nih.gov/pubmed/19513948. Accessed 25 Apr 2020

21. Bartkowiak T, Jaiswal AR, Ager CR, et al. Activation of 4-1BB on liver myeloid cells triggers hepatitis via an Interleukin-27–dependent pathway. Clin Cancer Res. 2018;24(5):1138–51. https://doi.org/10.1158/1078-0432.CCR-17-1847.

22. Schlothauer T, Herter S, Koller CF, et al. Novel human IgG1 and IgG4 Fc-engineered antibodies with completely abolished immune effector functions. Protein Eng Des Sel. 2016;29(10):457–66. https://doi.org/10.1093/protein/gzw040.

23. Lund J, Winter G, Jones PT, et al. Human Fc gamma RI and Fc gamma RII interact with distinct but overlapping sites on human IgG. J Immunol. 1991;147(8):2657–62. http://www.ncbi.nlm.nih.gov/pubmed/1833457. Accessed 25 April 2020

24. Vafa O, Gilliland GL, Brezski RJ, et al. An engineered Fc variant of an IgG eliminates all immune effector functions via structural perturbations. Methods. 2014;65(1):114–26. https://doi.org/10.1016/j.ymeth.2013.06.035.

25. Sifniotis V, Cruz E, Eroglu B, Kayser V. Current advancements in addressing key challenges of therapeutic antibody design, manufacture, and formulation. Antibodies (Basel, Switzerland). 2019;8(2):36. https://doi.org/10.3390/antib8020036.

26. Wang X, Mathieu M, Brezski RJ. IgG Fc engineering to modulate antibody effector functions. Protein Cell. 2018;9(1):63–73. https://doi.org/10.1007/s13238-017-0473-8.

27. Rath T, Kuo TT, Baker K, et al. The immunologic functions of the neonatal Fc receptor for IgG. J Clin Immunol. 2013;33(S1):9–17. https://doi.org/10.1007/s10875-012-9768-y.

28. Oganesyan V, Damschroder MM, Cook KE, et al. Structural insights into neonatal Fc receptor-based recycling mechanisms. J Biol Chem. 2014;289(11):7812–24. https://doi.org/10.1074/jbc.M113.537563.

29. Dall'Acqua WF, Woods RM, Ward ES, et al. Increasing the affinity of a human IgG1 for the neonatal Fc receptor: biological consequences. J Immunol. 2002;169(9):5171–80. https://doi.org/10.4049/jimmunol.169.9.5171.

30. Dall'Acqua WF, Kiener PA, Wu H. Properties of human IgG1s engineered for enhanced binding to the neonatal Fc receptor (FcRn). J Biol Chem. 2006;281(33):23514–24. https://doi.org/10.1074/jbc.M604292200.

31. Robbie GJ, Criste R, Dall'acqua WF, et al. A novel investigational Fc-modified humanized monoclonal antibody, motavizumab-YTE, has an extended half-life in healthy adults. Antimicrob Agents Chemother. 2013;57(12):6147–53. https://doi.org/10.1128/AAC.01285-13.

32. Zalevsky J, Chamberlain AK, Horton HM, et al. Enhanced antibody half-life improves in vivo activity. Nat Biotechnol. 2010;28(2):157–9. https://doi.org/10.1038/nbt.1601.

33. Köhler G, Milstein C. Continuous cultures of fused cells secreting antibody of predefined specificity. Nature. 1975;256(5517):495–7. https://doi.org/10.1038/256495a0.

34. Zlabinger GJ, Ulrich W, Pohanka E, Kovarik J. OKT 3 treatment of kidney transplant recipients. Wien Klin Wochenschr. 1990;102(5):142–7. http://www.ncbi.nlm.nih.gov/pubmed/2108521. Accessed 25 Apr 2020

35. Hwang WYK, Foote J. Immunogenicity of engineered antibodies. Methods. 2005;36(1):3–10. https://doi.org/10.1016/j.ymeth.2005.01.001.

36. Fülöp T, Mészáros T, Kozma G, Szebeni J, Józsi M. Infusion reactions associated with the medical application of monoclonal antibodies: the role of complement activation and possibility of inhibition by factor H. Antibodies. 2018;7(1):14. https://doi.org/10.3390/antib7010014.

37. Clark M. Antibody humanization: a case of the "Emperor's new clothes"? Immunol Today. 2000;21(8):397–402. https://doi.org/10.1016/s0167-5699(00)01680-7.

38. Harding FA, Stickler MM, Razo J, DuBridge R. The immunogenicity of humanized and fully human antibodies. MAbs. 2010;2(3):256–65. https://doi.org/10.4161/mabs.2.3.11641.

39. Bender NK, Heilig CE, Dröll B, Wohlgemuth J, Armbruster F-P, Heilig B. Immunogenicity, efficacy and adverse events of adalimumab in RA patients. Rheumatol Int. 2006;27(3):269–74. https://doi.org/10.1007/s00296-006-0183-7.

40. Radstake TRDJ, Svenson M, Eijsbouts AM, et al. Formation of antibodies against infliximab and adalimumab strongly correlates with functional drug levels and clinical responses in rheumatoid arthritis. Ann Rheum Dis. 2009;68(11):1739–45. https://doi.org/10.1136/ard.2008.092833.

41. Jones TD, Crompton LJ, Carr FJ, Baker MP. Deimmunization of monoclonal antibodies. Methods Mol Biol. 2009;525:405–23, xiv. https://doi.org/10.1007/978-1-59745-554-1_21.

42. Zhou Q, Qiu H. The mechanistic impact of N-glycosylation on stability, pharmacokinetics, and immunogenicity of therapeutic proteins. J Pharm Sci. 2019;108(4):1366–77. https://doi.org/10.1016/j.xphs.2018.11.029.

43. Chung CH, Mirakhur B, Chan E, et al. Cetuximab-induced anaphylaxis and IgE specific for galactose-α-1,3-galactose. N Engl J Med. 2008;358(11):1109–17. https://doi.org/10.1056/NEJMoa074943.

44. Bessa J, Boeckle S, Beck H, et al. The immunogenicity of antibody aggregates in a novel transgenic mouse model. Pharm Res. 2015;32(7):2344–59. https://doi.org/10.1007/s11095-015-1627-0.

45. Filipe V, Jiskoot W, Basmeleh AH, Halim A, Schellekens H, Brinks V. Immunogenicity of different stressed IgG monoclonal antibody formulations in immune tolerant transgenic mice. MAbs. 2012;4(6):740–52. https://doi.org/10.4161/mabs.22066.

46. Goel N, Stephens S. Certolizumab pegol. MAbs. 2010;2(2):137–47. https://doi.org/10.4161/mabs.2.2.11271.

47. Hutton D, Newman-Casey PA, Tavag M, Zacks D, Stein J. Switching to less expensive blindness drug could save medicare part B $18 billion over a ten-year period. Health Aff. 2014;33 (6):931–9. https://doi.org/10.1377/hlthaff.2013.0832.

48. Schmucker C, Ehlken C, Agostini HT, et al. A safety review and meta-analyses of bevacizumab and ranibizumab: off-label versus goldstandard. PLoS One. 2012;7(8):e42701. https://doi.org/10.1371/journal.pone.0042701.

49. Moja L, Lucenteforte E, Kwag KH, et al. Systemic safety of bevacizumab versus ranibizumab for neovascular age-related macular degeneration. Cochrane Database Syst Rev. 2014;9:CD011230. https://doi.org/10.1002/14651858.CD011230.pub2.
50. Proudfoot A, Bayliffe A, O'Kane CM, et al. Novel anti-tumour necrosis factor receptor-1 (TNFR1) domain antibody prevents pulmonary inflammation in experimental acute lung injury. Thorax. 2018;73(8):723–30. https://doi.org/10.1136/thoraxjnl-2017-210305.
51. Wilken L, McPherson A. Application of camelid heavy-chain variable domains (VHHs) in prevention and treatment of bacterial and viral infections. Int Rev Immunol. 2018;37(1):69–76. https://doi.org/10.1080/08830185.2017.1397657.
52. Birch JR, Racher AJ. Antibody production. Adv Drug Deliv Rev. 2006;58(5–6):671–85. https://doi.org/10.1016/j.addr.2005.12.006.
53. Vermeer AWP, Norde W. The thermal stability of immunoglobulin: unfolding and aggregation of a multi-domain protein. Biophys J. 2000;78(1):394–404. https://doi.org/10.1016/S0006-3495(00)76602-1.
54. Norman RA, Ambrosetti F, Bonvin AMJJ, et al. Computational approaches to therapeutic antibody design: established methods and emerging trends. Brief Bioinform. 2019. https://doi.org/10.1093/bib/bbz095
55. Amimeur T, Shaver JM, Ketchem RR, et al. Designing feature-controlled humanoid antibody discovery libraries using generative adversarial networks. bioRxiv. 2020, 04.12.024844. https://doi.org/10.1101/2020.04.12.024844

Engineering Therapeutic Antibodies for Development

12

Henri Kornmann and Björn Hock

Contents

H. Kornmann · B. Hock (✉)
Biologics Technologies and Development, Ferring International Center S.A., Saint-Prex, Switzerland
e-mail: henri.kornmann@ferring.com; bjorn.hock@ferring.com

© Springer Nature Switzerland AG 2021
F. Rüker, G. Wozniak-Knopp (eds.), *Introduction to Antibody Engineering*,
Learning Materials in Biosciences,
https://doi.org/10.1007/978-3-030-54630-4_12

Keywords

Antibody manufacturability · Immunogenicity of biotherapeutics · Minimization of
unwanted side effects · Antibody product homogeneity · mAb formulation into product

What You Will Learn from This Chapter
Most chapters in this book describe methods to select antibodies and to engineer their
properties for desired profiles in in-vitro assays or animal studies, focusing on
binding and engaging the target. To become drugs, such early antibodies require
further sequence optimization to ensure that the molecules can consistently be
produced in high quantities fulfilling the necessary quality requirements and that
they can be applied in humans with minimal unwanted side effects. In this chapter,
we will discuss the nature of some side effects caused by monoclonal antibodies
(mAbs) and how protein engineering methodologies are applied to reduce unwanted
side effects. In addition, we will cover the challenges of manufacturing and process
development with a focus on product homogeneity and explain how sequence
optimization and process controls are applied to minimize product heterogeneity
and ensure batch-to-batch consistency.

12.1   Introduction

Protein therapeutics including mAbs comprise a very successful class of drugs generating
significant revenue and provide valuable treatment options for patients with devastating
conditions such as metastatic cancer, autoimmune diseases, cardiovascular diseases, and
many others. mAbs are considered relatively safe drugs due to their high target specificity.
Antibody therapy started more than 100 years ago using passive immunization with sera
from immunized animals to treat infectious diseases such as diphtheria. In 1975, Köhler
and Milstein developed the hybridoma technology and this milestone paved the way for
modern mAb therapeutics [1]. The success of mAbs in the market place is a result of their
biology: Antibodies are components of the human immune system where they provide key
functions for the humoral host defense. They bind to surface structures of pathogens—so-
called epitopes—very specifically and connect them to other components of the immune
system such as macrophages, natural killer cells, and components of the complement
system. Antibodies are highly abundant in the bloodstream and share high structural
similarity. At the same time; however, they have high sequence diversity in their
so-called variable regions. Hence, antibodies—although structurally very similar—can
bind to all kinds of different epitopes including proteins, lipids, glycostructures, etc.,
opening a wide target space. Whereas antibodies in the bloodstream are polyclonal,

meaning a mixture from different sequences targeting several epitopes, mAbs are antibodies derived from a single cell clone with a defined sequence and binding to one specific epitope, hence the term "monoclonal antibody." Since a large surface area of a mAb is involved in binding of the epitope, very high selectivity is ensured, and "unwanted" targets are not affected. Thus, safety profiles of mAbs are favorable compared to, e.g., small molecules, which very often show activity on unwanted targets.

Nevertheless, the complex structure and biology of mAbs come with its own challenges in terms of development: mAbs are proteins and can cause immunogenicity in subsets of patients leading to the development of anti-drug antibodies (ADA), impairing drug efficacy and pharmacokinetics. In addition, antibodies naturally engage components of the immune system mediated by their Fc-part, which can result in unwanted side effects. Also, aggregation of antibodies and the formation of immune complexes or impurities during manufacturing or non-human posttranslational modifications can result in further unwanted side effects.

The structural similarity of mAbs allows the implementation of platform processes for development and production: Today, most marketed antibodies are produced by fed-batch processes using Chinese Hamster Ovary (CHO) cells as the expression host and affinity purification using Protein A resins as the key enrichment step. Production processes share high similarities for different mAb products, thus reducing the technology risk, enabling faster process-development timelines, and lowering investments. Consequently, mAbs have substantially higher success rates and shorter cycle times compared to small molecules in discovery and development. However, as mAbs are complex glycoproteins with a molecular weight of roughly 150 kDa consisting of four polypeptide chains. Hence, many unwanted structural modifications can occur during manufacturing, impairing the homogeneity of the final product. Some of those modifications may change their efficacy on target or their serum half-life, while others do not. As the final product is always a mixture of millions of variants, it is critical to ensure that modifications impairing the activity of the antibody in humans is minimal and that batch-to-batch consistency is guaranteed.

Often, very high doses of antibodies are required, thus triggering the requirement for exceedingly large-scale manufacturing (>700 kg per annum). In addition, the final drug product must be highly concentrated if the drug is to be administered through the subcutaneous route. Many of the aspects described can be modulated using protein engineering and process technologies and some examples are described below in this chapter.

12.2 Safety Considerations

Although mAbs are generally better tolerated than conventional chemotherapeutic compounds, there is a broad variety of side effects that have been reported over the years. These range from rather mild effects including: nausea, diarrhea, headache, mild gastrointestinal symptoms or itching, to more serious conditions, such as hypersensitivity reactions including: anaphylaxis, injection site reactions, cytopenia, cardiac toxicity, infections or cytokine release syndrome [2]. In addition, immunogenicity and immune complex formation is always a safety concern for all biotherapeutics including mAbs. Some of the side effects of mAbs are clearly related to their specific target, while others appear to relate to attributes of the antibody itself and their ability to interact with various components of the host immune system.

12.3   Examples of Target-Dependent Side Effects of mAbs

An example of target-dependent side effects is cardiotoxicity mediated by Trastuzumab, a chimeric IgG1 targeting the Her-2 receptor in breast cancer. As Her-2 is also expressed on cardiomyocytes, blocking Her-2 signal transduction with trastuzumab led to cardiotoxicity in 4% of treated patients [3].

Another target-mediated side effect is skin rash induced by exposure of patients to anti-EGFR antibodies such as Cetuximab. EGFR is a prominent cancer target and cetuximab is used for the treatment of metastatic colon cancer, as well as head and neck cancer. EGFR is widely expressed in epithelial cells including keratinocytes. Thus, blocking EGFR signal transduction with cetuximab or other EGFR blocking agents including small molecules often causes skin rash in the face and on the upper torso. There is a correlation between the occurrence of a skin rash and a positive response to the drug [4].

Progressive multifocal leukoencephalopathy (PML) is a rare demyelinating disease of the central nervous system caused by the *John Cunningham Polyomavirus* (JCV). Although the pathophysiology is not fully understood, it appears that immunosuppression plays a role. Several antibodies applied for the treatment of lymphoproliferative- or autoimmune-disorders appear to increase the incidence of PML infections [5]. Natalizumab, a mAb used for the treatment of relapsing-remitting multiple sclerosis, was suspended in 2005 because of the occurrence of some PML cases and reintroduced in 2006 with specific warnings on the label [6]. Besides natalizumab, cases of PML were described for several other antibodies including rituximab, ofatumumab, obinutuzumab, brentuximab vedotin, alemtuzumab, and others [2] (and references therein).

Another common side effect of antibodies modulating the immune response is the so-called "cytokine storm" (cytokine release syndrome, CRS) that has been described, for example, for Rituximab, Alemtuzumab or Muromonab [2] (and references therein). In 2006, a clinical phase 1 study with an experimental drug, TGN1412, a CD28 superagonist, had a dramatic effect as treated patients developed CRS within minutes after dosing, resulting in systemic inflammation symptoms and multiorgan failure requiring intensive care [7].

12.4   Immunogenicity of Biotherapeutics

The first mAb to be approved by the US Food and Drug Administration (FDA) was Muromonab (OKT3) for the treatment of kidney transplant rejection in 1985. Muromonab represents a mouse IgG2a isotype [8] and caused immune reactions which led to alterations in efficacy, safety, and pharmacokinetics [9]: Mouse antibodies can be detected as "non-self" by the human immune system resulting in an "anti-drug-antibody" (ADA) response. Those early findings triggered the discovery of antibody engineering technologies leading to less immunogenic formats [10]: In the years following, protein engineering technologies were applied to develop new antibodies displaying a more favorable profile (Fig. 12.1): In the mid-80s the first chimeric antibody was described [11] followed a few years later by the first humanized antibody [12]. Chimeric antibodies comprise an antigen-binding region (Fab) of rodent origin which is fused to a human constant region (Fc). Humanization means grafting the CDR regions of rodent origin into a human antibody scaffold, further reducing the presence of rodent-derived sequences in a therapeutic molecule. Nowadays, fully human antibodies derived from transgenic animals or from phage display technologies are state-of-the-art. Those molecules can be considered "fully human" and do not contain any sequence patches of rodent origin.

As shown by [13], dosing of murine antibodies causes marked ADA responses, resulting in impaired clinical outcomes and ultimately in discontinuation of clinical studies. Substitution of the rodent Fc-regions to those of human origin by chimerization causes a reduction of ADA responses, which can be further reduced by using humanized or fully human antibodies. Although the use of humanized and fully human antibodies reduces the overall immunogenicity risk compared to chimeric and mouse antibodies, there is still a considerable patient population that develops ADA responses: In the case of adalimumab (Humira)), a fully human antibody targeting the cytokine TNFα, 12% of patients in a clinical trial developed an ADA response. Whereas infliximab (Remicade), a chimeric

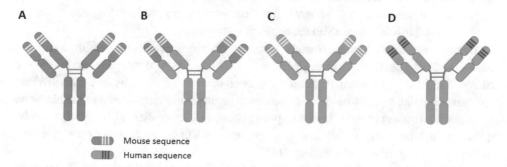

Fig. 12.1 Antibody engineering for chimeric and humanized antibodies to reduce ADA response. Variable regions of a mouse antibody (**a**) are fused to the constant regions of a human antibody to build a chimeric antibody (**b**). Grafting of mouse CDRs into a human backbone results in a humanized antibody (**c**). Fully human antibodies are now state of the art and can be generated by display technologies or by using transgenic mice or rats harboring the human antibody repertoire (**d**)

antibody targeting the same cytokine, led to an ADA response in 75% of patients. In both cases, the ADA response was found to be further reduced by coadministration of immuno-suppressive drugs.

12.5   What Causes Immunogenicity?

It is easy to understand that patients receiving repeatedly high doses of antibodies containing sequence patches from other species would be detected as "non-self" by the host and consequently trigger an immune response. However, unlike one would expect, even fully human antibodies have still proven to be immunogenic, causing an ADA response in a subset of individuals.

One example can be anti-allotype reactions for human IgG-Fc and kappa light chains. Allotypes comprise minor deviations in the amino acid sequence of IgGs of the same subclass that occur throughout a species. Those allotypes are inherited and different allelic forms are expressed by individuals. IgG1, for example, exists in four different allotypes located in the heavy chain (G1m (1–3, 17)) and three different allotypes in the kappa light chain (Km (1–3)). There are no allotypes for the lambda light chain reported [14]. Hence, IgG1-based therapeutic antibodies may induce an anti-allotype response in a subset of patients. It must also be taken into consideration that paratopes of therapeutic antibodies are usually not part of the natural antibody repertoire. Consequently, an anti-idiotypic response directed against the paratope may be triggered when a therapeutic mAb is selected using hybridoma or transgenic animal technology, or when changes in the amino acid sequence of the variable regions occur through rearrangement and affinity maturation of the variable regions [10] (and references therein). Imprecision of the rearrangement of genomic sequences (VL-JL and VH-DH-JH recombination) contributes to the sequence diversity of antibodies. In addition, somatic hypermutations during affinity maturation may intro-duce new amino acid variations deviating from germline sequences. Those novel sequences can lead to an anti-idiotypic immune response, as the host immune system is unable to establish tolerance to all those possible variations.

Besides sequence attributes, immune responses to posttranslational modifications have been described, e.g., for cetuximab (see above). Cetuximab is N-glycosylated on N88 in the heavy chain [15] and when produced in mouse-derived SP2/0 cells, the carbohydrate structure contains a galactose-α-1,3-galactose glycostructure, causing severe anaphylactic reactions in a subset of patients due to pre-existing IgE antibodies. This problem can be circumvented by using a cell line (for example, CHO cells) that does not perform galactose-α-1,3-galactose connections [15].

Besides those so-called intrinsic factors that can be attributed to the amino acid sequence and the structure of biopharmaceuticals, some external factors such as purification- and formulation-dependent impurities are important as well. Furthermore, factors including route and frequency of administration, immune status of the patient, or comedications play a role [10]. Amongst those extrinsic factors contributing to the immunogenicity of biopharmaceutical drugs, aggregation propensity appears to be very important [16], as many of the side effects mediated by ADAs require the formation of drug–ADA immune complexes (IC) leading to Fc-mediated downstream effects causing hypersensitivity reactions (discussed below).

12.6   Mechanisms Leading to an ADA Response

12.6.1 T Cell-Dependent Immune Response

In many cases, a single injection of a non-self-antigen is sufficient to induce a high level of ADAs with high affinity that can persist for an extended period of time in the body. In addition, memory cells are induced which leads to a booster reaction if the host organism is re-challenged with the same non-self-antigen. This "vaccination-like" adaptive immune response involves antigen-presenting cells (APCs) and CD4+ T helper cells. Following administration, foreign proteins are phagocytosed by APCs such as dendritic cells (DC) or macrophages. After degradation by the immunoproteasome, peptides derived from the antigen are presented by the major histocompatibility complex-2 (MHC-2) [17]. CD4+ T cells recognize linear peptides presented by the MHC-2 complex via their T cell receptors and secrete cytokines promoting differentiation of B cells, isotype switching, and affinity maturation, resulting in the production of high-affinity IgG-type ADAs in plasma cells. Until this differentiation occurs, the immune response is driven by IgM-expressing B cells with low affinity and specificity. In conclusion, the presence of high-affinity ADAs of the IgG subclass is usually associated with CD4+ T helper cell activity. Such responses exist for biotherapeutics of microbial origin [18] but have also been described for the V-regions of antibodies [19]. This process, leading to the production of high-affinity IgG1, takes a few days and requires co-stimulation by additional factors such as CD28 and CD80 [17], as without co-stimulation, T cells will become anergic. Inhibitory DCs (iDC) present peptides without co-stimulation, playing a key role in maintaining immune tolerance [20]. During development, CD4+ T cells are tolerized in the thymus and later anergized or deleted upon contact with a self-antigen.

12.7   T Cell-Independent Immune Response

Modern antibody therapeutics are humanized or fully human and hence it is counterintuitive that they would cause immune responses in humans. Antibodies occur in mg/ml concentrations in human serum and there is a high level of immunotolerance. However, prolonged exposure to humanized or human antibodies can result in tolerance breaking, and aggregates of therapeutic proteins appear to play an important role. Aggregates can directly interact with B cells and activate them, presumably by oligomerizing and activating B cell receptors, a process that does not discriminate between self and non-self. The ADA responses resulting from this process are often milder compared to the CD4+ T cell-driven response described above and disappear after treatment as there is no memory effect [16].

The distinction between both types of immunogenicity is not absolute, but rather patient-dependent, and sometimes both types of responses can be observed in a patient population exposed to the same biopharmaceutical.

The correlation between aggregation of biotherapeutics and immunogenicity is well established, and it appears that breaking tolerance is the main mechanism for ADA formation for modern therapeutic antibodies. Besides the B cell induction described above, mAbs—depending on their subtype—modulate immune reactions such as complement—or macrophage activation with their Fc-part that may contribute to their immunogenicity. It has been shown that deglycosylation of IgG1 reduces those reactions [21] but human antibodies devoid of Fc-mediated effector functions can still be immunogenic.

12.8   Clinical Consequences of Immunogenicity

It is difficult to compare the consequences of immunogenicity of different studies due to the lack of standardization with respect to study design, analytical methods, or data interpretation. In addition, side effects, including those which are induced by immunogenicity such as ADA responses, vary significantly between individual patients within a given study and when comparing studies of different drugs for different indications [16]. Hence, some of the mechanisms are poorly understood and it is difficult to discriminate target-dependent mechanisms from those induced by the antibody itself.

Persisting high levels of neutralizing antibodies may lead to a complete loss of efficacy. ADAs to biotherapeutics that have an endogenous counterpart can have devastating effects if the ADA is cross-reactive to the endogenous factor. For example, healthy volunteers exposed to recombinant thrombopoietin or erythropoietin developed thrombocytopenia or pure red cell aplasia, respectively [22, 23].

The effect of ADAs to mAb therapeutics may result in loss of efficacy, either by directly interfering with target binding or by impairing the serum half-life. Other clinical effects that are observed with therapeutic antibodies and correlate with the occurrence of ADAs are injection site reactions and serum sickness [24].

Type 1 hypersensitivity reactions caused by pre-existing IgE antibodies directed to glycostructures have been described above for cetuximab. Anaphylactic hypersensitivities have been described for other mAbs as well (mostly mAbs containing rodent sequences), although the incidence of such an event is relatively small.

Infusion of mAbs may result in infusion site reactions, typically a few hours after the first administration, characterized by rather moderate "flu-like"-symptoms such as nausea, fever, and headache [25]. However, life-threatening events such as bronchospasm or cardiac arrest occur occasionally. The biological mechanism leading to infusion site reactions is not fully understood but pre-existing antibodies to impurities or glycostructures are thought to play a role.

mAb therapies often require high doses that may result in the formation of circulating immune-complexes (IC) between the ADAs and the therapeutic mAb, causing side effects. Various parameters including size, distribution, polyclonal diversity, and Fc-mediated effects contribute to the clinical outcome. Whereas larger and insoluble ICs are cleared by the mononuclear phagocytic system in the liver, smaller ICs stay in circulation and trigger Immune responses via Fc-receptors. Fcγ-receptors (Fcγ-R), which are expressed widely in the hematopoietic system, appear to play a key role in the pathological effects of ICs. Low-affinity Fcγ-R, such as Fcγ-RIII (CD16) or Fcγ-RII (CD32), that interact with ICs exceeding a certain size threshold can trigger proinflammatory responses.

Deposition of ICs in the capillary network can lead to vascular thrombosis and a local inflammatory response. In addition, those ICs can activate the classical complement cascade by binding to C1q via the Fc-region of the antibody. Proximity of IgG1-Fc effectively leads to the generation of C3-convertase, which is a danger factor inducing various inflammatory effects [26]. Therapeutic mAbs can also attract ICs to cell surface targets causing symptoms such as vasculitis and serum sickness [27].

12.9 Engineering mAbs for Safety and Minimal Immunogenicity

As described above, the immunogenicity of mAbs can be influenced by a variety of extrinsic and intrinsic factors and is difficult to predict. Several protein engineering strategies are applied to reduce the risk of immunogenicity. Immune reactions with murine mAbs led to the development of chimeric antibodies and humanized mAbs, and later to fully human antibodies derived from transgenic rodents or combinatorial library approaches (described elsewhere in this book). In addition, posttranslational modifications can result in immunogenicity, as described in this chapter for the example of cetuximab and should be avoided. Moreover, the choice of the isotype influences immunogenic potential as the engagement of effector cells and the complement system varies for different isotopes. Drug aggregation is another factor that induces an immune response and should be generally avoided.

12.10 Humanization

In the past decades, several strategies for the humanization of mAbs have been reported. However, CDR grafting onto a homologous human antibody framework of high sequence homology remains the most common technique [28]. CDR grafting often results in a reduction or loss of antigen recognition. This problem can be resolved by back mutating critical framework sites to the corresponding amino acid residues in the initial rodent sequence. Such critical framework residues are selected from structural data on antibodies or from homology-modeled structures revealing framework positions important for structural integrity. In some cases, this approach does not fully restore affinity to the target, and affinity maturation technologies must be applied (described elsewhere in this book).

Guided selection is a method to convert murine antibodies into fully human antibodies with similar binding characteristics [29]; Mouse VH and VL domains are sequentially or in parallel replaced by human VH and VL domains, respectively, followed by phage selection to derive human antibodies with higher affinities. A potential disadvantage of the guided selection approach is that shuffling one or both antibody chains can result in an epitope drift [30]. To maintain the epitope recognized by the source non-human antibody, CDRs can be conserved. In this alternative method, one or both non-human CDR3s are commonly retained as they tend to play a critical role in the recognition of the antigen.

12.11 Selection of Isotypes

Therapeutic antibodies are usually of the IgG subclass as IgGs are highly abundant in serum with g/L concentrations. IgGs are well understood as a drug class and the pharmaceutical industry has implemented platform processes for manufacturing. Four isotypes of IgGs exist in humans with slightly different properties, and only IgG3 is not exploited due to its complex structure. IgG1, IgG2, and IgG4 differ in their affinity to Fcγ receptors and the complement receptor C1q. Consequently, the immunologic effector functions of IgG Isotypes vary in strength [14].

IgG1 is a commonly used isotype and considered to have the strongest effect on "antibody-dependent cellular cytotoxicity" (ADCC) and "complement-dependent cytotoxicity" (CDC), and both effects can contribute to efficacy. If the engagement of the immune system is not necessary for therapeutic efficacy, IgG2 and IgG4 are used as isotypes, because these isotypes do not bind to Fcγ-RIII, the only Fc-receptor expressed on natural killer cells. Nevertheless, cytotoxic activity has been described for some IgG2- and IgG4-based antibodies, which likely resulted from the engagement of neutrophil granulocytes [31].

The use of the IgG2 isotype brings in additional challenges as structural heterogeneity occurs due to connectivity of disulfide bonds within the hinge region, resulting in at least three variants with differing disulfide bonds [32]. Nonetheless, IgG2 hinge region homogeneity with uniform disulfide bonds can be readily achieved using a modified IgG1 hinge [33].

The hinge region of the IgG4 heavy chain comprises two cysteine residues that can form either an intrachain disulfide bond or two interchain disulfide bonds, resulting in the abundance of "half antibodies" comprising only one heavy and one light chain. Therapeutic IgG4 antibodies can engage in Fab-arm exchange with endogenous human IgG4 in vivo, resulting in the functional monovalency of the therapeutic mAb, with the loss of binding avidity and cross-linking effects. Gemtuzumab is an IgG4 mAb and contains a hinge modification, whereas natalizumab, another IgG4, does not [34]. Natalizumab showed significant levels of Fab-arm exchange in 15 of 16 patients, as early as 1 h after infusion [35]. Therefore, therapeutic IgG4 antibodies should be designed to prevent the Fab-arm exchange by introducing the S228P mutation to stabilize the hinge region, and, more importantly, the R409K mutation to stabilize the CH3 region [36].

Further approaches to modulate effector functions of mAbs are based on glycoengineering and mutagenesis. Approaches to enhance effector functions have been described elsewhere [37].

Several mutations to abrogate ADCC and/or CDC have been described in the literature. One common strategy is to incorporate natural motifs from IgG2 and IgG4 to decrease binding to Fcγ-Receptors (Fcγ-R) and complement, respectively, while keeping potential immunogenicity of the changes to a minimum. For example, Armour et al. [38] replaced residues 233–236 in the lower hinge region of IgG1 responsible for Fcγ-R binding with the corresponding residues from IgG2, and introduced IgG4 residues at position 327, 330, and 331 to abolish binding to C1q. Another variant that is widely used is the double mutation L234A, L235A ("LALA") that has been shown to abrogate binding to Fcγ-RI, Fcγ-RIIa and Fcγ-RIIIa [39]. Several other mutations have been described disrupting the interaction with FcγR and with C1q [40].

Glycoengineering is another method to modulate effector functions. Antibodies contain a biantennary N-glycosylation on N297 and it has been shown that a mutation to alanine of glutamine disrupts the interaction with FcγR and C1q [40].

12.12 Heterogeneity of Therapeutic mAbs

mAbs can be theoretically described by a defined primary structure of amino acids of around 150 kDa. In reality, a single dose of a therapeutic mAb product represents a mixture of numerous mAb variants. Those variants are inherent to the biotechnological processes used to manufacture drug products. Some variants are formed at a transcriptional level leading to misincorporation of amino acids and ultimately delivering a different amino acid sequence than originally planned [41]. Yet, most of the variants are caused by posttranslational modifications (PTM)) or are generated after secretion during the manufacturing process. A description of the major quality modifications leading to multiple mAb variants are listed in Table 12.1.

12.13 Quality Attributes of mAbs Can Impact Safety and Efficacy

Some of the quality attributes listed in Table 12.1 may impact the safety and efficacy of the therapeutic mAbs. As described earlier in this chapter, *aggregation* triggers potential immune responses, especially neutralizing antibodies that limit efficacy, as well as severe immediate hypersensitivity responses such as anaphylaxis [42]. *Fragmentation and clipping* impact the function of a mAb [43]. In the case of fragmentation in the CDRs or the

Table 12.1 Quality modifications leading to multiple mAb variants

Quality attribute	Definition
N-terminal pyroglutamate	N-terminal glutamine of heavy chains is partially or completely derivatized to pyroglutamate resulting in more acidic antibodies
Deamidation	Asparagine residues are, via succinimide intermediates, converted into aspartate and iso-aspartate residues. Glutamine residues can also undergo deamidation resulting in glutamate residues. The kinetics of this reaction is slower than with asparagine
C-terminal lysine truncation	Truncation of the C-terminal lysine of antibodies due to carboxypeptidase activity, resulting in charge heterogeneity since the number of charged terminal lysine residues per antibody molecule maybe 0, 1, or 2 (K0, K1, K2 respectively). Loss of lysine residues results in a decrease in positive charge and more acidic antibodies
Oxidation	Oxidation is a covalent modification of an amino acid that is induced by reactive oxygen. mAb oxidation occurs predominantly on methionine residues. Oxidation of methionine to methionine sulfoxide makes the side chain of methionine more polar
Glycation	Glycation results in non-enzymatic glycosylation at the protein amine group on lysine side chains. It occurs without the controlling action of an enzyme when the protein is incubated in the presence of sugars
Glycosylation	For most of the recombinant mAbs, glycosylation represents N-linked oligosaccharides (biantennary complexes) with a core fucose with 0, 1, or 2 terminal galactose residues, and various truncated structures with the loss of the core fucose, N-acetyl-glucosamine (GlcNAc), or both (Fig. 12.2). This allows for a possible total of 32 different oligosaccharides and potentially more than 400 glycoforms
Fragmentation/ clipping	Fragmentation/clipping describes the disruption of a covalent bond in a mAb because of either a spontaneous or enzymatic reaction. Fragmentation can occur during cell culture, is modulated by the purification process and will continue during storage
Aggregation	mAbs may aggregate with each other by association, either without any changes in primary structure (physical aggregation) or by the formation of new covalent bonds (chemical aggregation). Both mechanisms can occur simultaneously and may lead to the formation of either soluble or insoluble aggregates, depending on the protein, environmental condition, and stage of the aggregation process

constant regions of mAbs, potency is impacted, and the Fc-mediated effector function is reduced, respectively. *Oxidation* can significantly alter the biological activity of a mAb [44] and reduce serum half-life [45]. *Deamidation*, by changing the secondary and the tertiary structure of the therapeutic mAb, may reduce biological activity. For example, the induced deamidation of trastuzumab or panitumumab dramatically reduces mAb potency [46, 47]. *Glycosylation* of IgG-Fc is essential for optimal expression of biological activities mediated through Fcγ-RI, Fcγ-RII, Fcγ-RIII, and the C1q component of complement

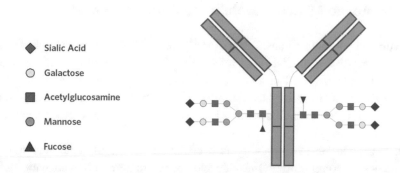

Fig. 12.2 Biantennary glycosylation of mAbs. Monoclonal antibodies are *N*-glycosylated on asparagine 297 forming a biantennary glycan. The structure of the glycan can vary between hosts, the figure shows the schematic representation of an antibody produced in CHO cells. For most of the recombinant mAbs, glycosylation is built on a core fucose with 0, 1, or 2 terminal galactose residues, and various truncated structures with the loss of the core fucose, *N*-acetyl-glucosamine (GlcNAc) or both. This allows for multiple different oligosaccharides structures and potentially more than 400 glycoforms. Glycosylation influences antibody efficacy due to involvement in Fcγ-R binding, pharmacokinetics and is a source of product heterogeneity

(Fig. 12.2). The absence of Asn297-linked glycosylation significantly reduces but does not eliminate effector functions and complement activation of the therapeutic mAb [48]. For instance, the removal of the complete carbohydrate moiety of alemtuzumab abolished complement lysis activity and ADCC but left antigen-binding activity and protein A binding activity intact [49]. Aglycosylated cetuximab did not bind to Fcγ-RI or Fcγ-RIIIa, nor did it have any ADCC activity, whereas the glycosylated molecule showed high receptor-binding affinity and high ADCC activity [50]. The impact of glycosylation on pharmacokinetics, antigen binding, and immunogenicity is further described in the literature [51]. Therefore, to develop a successful therapeutic mAb, it is key to minimize and control the level of mAb variants in the drug product.***

12.14   Protein Engineering to Control mAb Variants

Proper engineering of the mAb sequence is the first approach to minimize potential mAb variants and limit drug product heterogeneity [52]. Typically, amino-terminal glutamine is selected to force the formation of pyroglutamate and reduce the number of charge variants. *N*-glycosylation sites are removed from the V_L and the V_H to prevent the formation of *N*-glycoforms. Mutation of amino acids is introduced to prevent aggregation. Carboxy-terminal lysine is deleted from the sequence by proteolytic cleavage. Methionine in CDRs is avoided to prevent impactful oxidation. The improvement of the mAb sequence can be performed in-silico using dedicated software and/or experimentally by stress studies of mAb candidates [53].

12.15 Improved Manufacturability to Control mAb Variants

Manufacturability is an important aspect of mAb therapeutics development. It is defined as the ability of a mAb drug product to be safe, to maintain pharmaceutical activity over shelf-life, and to be manufactured in a cost-effective and highly reproducible manner. Good manufacturability is mandatory to minimize and control the level of therapeutic mAb variants in the drug product. It is achieved via extensive small-scale process development experiments.

Achieving good manufacturability is a challenging task as mAbs are intrinsically heterogeneous in composition. Often, development programs of therapeutic mAbs are abandoned because of poor manufacturability. Knowing that drug substance and drug product quality attributes are affected by minimal changes in the manufacturing process, everything that happens in the process of synthesizing a mAb in a cell, isolating it from the production system, and purifying it (as well as the choice of materials and methods inherent to the development of the process) may be critical to the quality of that material and patient safety. Therefore, "The process is the product!" is a statement often made in the context of biological therapeutic products [54].

To improve the manufacturability of a mAb, the FDA and other regulatory agencies initiated the Quality by Design (QbD) approach in the early 2000s. The QbD initiative provides guidance on pharmaceutical development to facilitate the design of products and processes that maximizes the product's efficacy and safety profile while enhancing product manufacturability. This risk-based concept links product quality with development and manufacturing activities. Nowadays, QbD activities are fully integrated into product development phases from first toxicology studies to clinical phases and post-marketing activities. The main QbD milestones and deliverables are:

- A *Target Product Profile (TPP))* documenting the intent of the product and its desired features (indication, desired efficacy and safety claims, desired drug format, etc.).
- A *Quality Target Product Profile (QTPP))* summarizing the quality characteristics of a pharmaceutical product that will ideally be achieved to ensure the desired quality, taking into account both safety and efficacy. Attributes represented in the QTPP relate to the intended use of the drug product and are those that impact the patient (route of administration, dosage form, bioavailability, strength, stability, etc.).
- A risk assessment of the product quality attributes based on prior knowledge, clinical and non-clinical studies (including toxicology) to identify the *Critical Quality Attributes (CQA)*.
- A *Process Risk Assessment (PRA)* identifying the manufacturing parameters (inputs) for each process step, such as cell density and integrated cell viability for an upstream process or load temperature, load pH, and load conductivity for a column chromatography step that, if varied, have the potential to impact a CQA.
- A *Process Characterization* study that examines the deliberate variation of the parameters identified as potentially critical by PRA to determine the acceptable limits

of variation. Process characterization studies lead to the classification of process parameters as either *critical process parameters (CPP)* or non-critical process parameters. CPPs are defined as a process parameter whose variability impacts a CQA and must be controlled.

- A *Design Space* representing the multidimensional combination and interaction of input variables (e.g., material attributes) and process parameters that have been demonstrated to provide assurance of quality.
- A defined *control strategy* based on the design space and the CPPs describing in detail the Input Material Controls, the In-Process Controls, the Parameter Controls, the specifications, the product characterization, and the process monitoring.
- *Qualification studies* of the facilities, the utilities and the equipment plus a process *performance qualification study (PPQ)* performed at full-scale mimicking final commercial manufacturing operations.
- *Continued process verification studies* to demonstrate that the process validation package remains up to date across the lifecycle of the product.

12.16 mAb Industrial Manufacturing

Manufacturing of a therapeutic mAb is initiated by the thawing of a cryogenic vial containing the cells. This cryogenic vial is part of a bank of the production cell line. To generate the production cell line, the host cells are transfected with the expression vector and cell clones are screened to find antibody-producing cells. The goal of this transfection and screening process is to select a cell line that grows well and is genetically stable in culture, and that produces high levels of the product in its active form (properly assembled, folded, glycosylated and not fragmented or aggregated). Once a cell line is established, it must be fully characterized and banked for use in long-term production. The first mAb products that entered the market were produced in several different mammalian host cell lines including the original hybridoma or the murine myeloma cell lines SP2/0 and NS0. However, as more mAb products were developed, the biopharmaceutical industry focused its efforts on the use of various CHO cell lines for the production of these products. As a result, CHO cells are currently the dominant host cell line for the production of mAb products.

Every bioprocess requires a synthesis stage, a recovery stage, and a purification stage. The synthesis stage is an upstream processing (USP) activity, while the recovery and purification stages are downstream processing (DSP) operations. The downstream process consists of several steps that deliver the active pharmaceutical ingredient (API) (also referred to as bulk drug substance, or just drug substance (DS)) to the final stage of the biomanufacturing process, formulation and filling (Fill and Finish). In this stage, a sterile filtered DS is transferred to a storage buffer and filled into vials. The formulated DS is called a drug product (DP) and is ready to be given to the patient. The typical manufacturing process for recombinant mAb drug substance is described in Fig. 12.3.

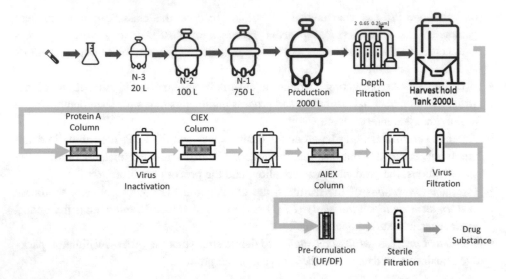

Fig. 12.3 Example of a mAb manufacturing process at 2000 L-scale. Manufacturing is initiated by the thawing of a cryogenic vial containing the cells. The cells are expanded via a series bioreactor to reach a defined cell density to inoculate the production bioreactor. The recovery and purification consist of several process steps that deliver the active pharmaceutical ingredient (API) (also referred to as drug substance (DS)). Between 3 and 6 kg of mAb Drug Substance can be manufactured per batch

All manufacturing operations must be performed under current Good Manufacturing Practices (cGMP). GMP are pharmaceutical regulations into the manufacturing of pharmaceuticals assuring quality, safety, traceability, and reproducibility of the drug product. The GMP is described and audited by several health authorities such as WHO, FDA, or EMA. Basic principle of GMP are listed below:

- Manufacturing facilities must maintain a clean area and a manufacturing area.
- Manufacturing facilities must maintain controlled environmental conditions to prevent cross-contamination from adulterants and allergens that may render the product unsafe for human consumption or use.
- Manufacturing processes must be clearly defined and controlled. All critical processes are validated to ensure consistency and with specifications.
- Manufacturing processes must be controlled, and any changes to the process must be evaluated. Changes that affect the quality of the drug are validated as necessary.
- Instructions and procedures must be written in clear and unambiguous language using good documentation practices.
- Operators must be trained to carry out and document procedures.
- Records must be made, manually or electronically, during manufacture that demonstrate that all the steps required by the defined procedures and instructions were in fact taken

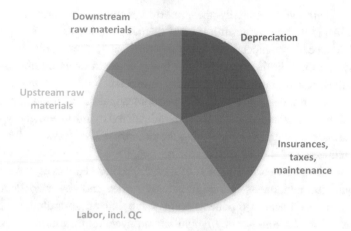

Fig. 12.4 Example of cost categories for mAb manufacturing

and that the quantity and quality of the food or drug were as expected. Deviations must be investigated and documented.

- Records of manufacture (including distribution) that enable the complete history of a batch to be traced must be retained in a comprehensible and accessible form.
- Any distribution of products must minimize any risk to product quality,
- A system must be in place for recalling any batch from sale or supply.
- Complaints about marketed products must be examined, the causes of quality defects must be investigated, and appropriate measures must be taken with respect to the defective products.

The typical cost of manufacturing goods (CoG) for therapeutic mAb drug substance at a commercial scale ranges between 30 and 60$/g. The CoG performance is, however, highly dependent on the overall process performance (yields, productivity, raw material cost), the manufacturing facility utilization, and local taxes or royalties. An example of cost categories for mAbs is described in Fig. 12.4 [55].

12.17 Considerations for Intended Use and mAb Formulation into Drug Product

The intended use for mAb therapeutics is considered very early in the product development program. Typically, the first version of the Target Product Profile (TPP) is proposed during the pre-clinical development stage. This TPP is typically considered the intended shelf-life of the drug product, the mode of administration, and the need for a specific device (for instance an auto-injector).

The complex structure of therapeutic mAb combined with their size is a challenge to develop delivery methods other than parenteral administration. For pharmaceutical products, oral delivery is generally the preferred method. Yet, oral administration of therapeutic mAb is limited by gastrointestinal degradation, low bioavailability, and slow uptake [56]. Some early therapeutic development programs are exploring delivery systems that can pass into the intestinal tract without being digested or an administration via the nasal and pulmonary routes. However, today those are limited by many of the same drawbacks as oral administration, including gastrointestinal degradation and low bioavailability. Therefore, most therapeutic mAb are administered by intravenous (i.v.) or subcutaneous (s.c.) injections, with injections being the primary delivery system for mAbs.

Both lyophilized and liquid forms can be considered when developing a suitable intravenous or subcutaneous formulation of a mAb product. However, the recent trend is to favor liquid formulation as this is frequently easier and faster to develop. Liquid formulation contains fewer components than a lyophilized formulation and does not require as much formulation—or any lyophilization cycle optimization. Moreover, recent years have seen an increased focus on self-administered delivery systems, such as Pre-filled Syringes (PFS) or Pen devices. Self-administration offers the advantages of increased patient compliance, greater ease of use, reduction of pain, and reduced risk of dosing error as compared to other modes of injections. Self-administered delivery systems are easier to develop when containing a liquid formulation.

12.18 Formulation Development

Degradation of a mAb can result from either the chemical degradation of specific amino acid residues in the mAb, physical degradation caused by the loss of tertiary structure, or the covalent or non-covalent aggregation of mAb monomers to form aggregates in solution. Formulation development aims at achieving acceptable long-term stability and high protein concentration in a liquid formulation. This is usually achieved by screening a variety of excipients to provide control of pH and tonicity of the solution, stabilize the mAb structure, prevent surface denaturation or adsorption of the product to container/closure surfaces and reduce degradation of the product (Table 12.2). Excipients used in the formulation of mAb products should ideally meet the appropriate USP, included in the FDA Inactive Ingredient Database or those that comply with the standards for the manufacture of pharmaceutical grade excipients set by the International Pharmaceutical Excipient Council (IPEC). On top of the appropriate buffer to control the pH of the formulation, excipients could be:

- Sodium chloride to control the osmolality
- Surfactants (i.e., polysorbate) to control interaction with the surface of the container closure or the air–liquid interface
- Cryoprotectants (i.e., sucrose or trehalose) to stabilize a mAb during long-term storage as a frozen liquid
- Chelating agents (i.e., EDTA) if the mAb is sensitive to metal ion catalyzed oxidation

Table 12.2 mAb formulation examples

Product name (INN name)	Presentation	Concentration (mg/mL)	Buffer and excipients	Storage condition (°C)	Shelf-life (months)
Actemra (tocilizumab)	Liquid/ syringe	180	Histidine, Arginine, Methionine, Polysorbate 80, pH 6.0	2–8	30
Avastin (bevacizumab)	Liquid	25	Sodium Phosphate, Trehalose, Polysorbate 20, pH 6.2	2–8	24
Enbrel (etanercept)	Liquid/pen/ syringe	50	Sodium Phosphate, Arginine, Sucrose, Sodium Chloride, pH 6.16.5	2–8	30
Humira (adalimumab)	Liquid/pen/ syringe	100	Mannitol, Polysorbate 80, pH 5.2	2–8	24
Orencia (abatacept)	Liquid/pen/ syringe	125	Sodium Phosphate, Sucrose, Poloxamer 188, pH 6.8–7.2	2–8	24
Simponi (golimumab)	Liquid/pen/ syringe	100	Histidine, Sorbitol, Polysorbate 80, pH 5.5	2–8	24
Zinbryta (daclizumab)	Liquid/pen/ Syringe	150	Sodium Chloride, Sodium Succinate, Succinic Acid, Polysorbate 80, pH 6.0	2–8	36

Viscosity, solubility, aggregation, and opalescence are limiting the development of high-concentration antibody formulations (>100 mg/ml). To facilitate the subcutaneous distribution of injection volumes of several milliliters over a greater area and enable painless administration, recombinant human hyaluronidase (rHuPH20) has recently been included in liquid subcutaneous presentations of some mAb (Herceptin SC and MabThera SC). Other approaches, such as the addition of amino acid blends, have been designed to stabilize proteins in high concentration formulations, minimizing aggregation, reducing viscosity, and improving syringeability. Some liquid formulation of mAb can reach concentration up to 200 mg/ml using this approach [57].

12.19 Closing Remarks

In summary, the therapeutic performances of a recombinant mAb does not only depend on its mode of action, its potency, or its affinity to a target. Regulators expect a clear demonstration of controlling and minimizing critical quality attributes in order to prevent safety or efficacy concerns. This is achieved by proper engineering of the molecule and a sound control of manufacturing parameters. The success of a therapeutic mAb is also dependent on its ease of use for the patient or the health care professionals and, obviously, its cost of manufacturing and distribution. Therefore, a good manufacturability and a marketable intended use are key elements to consider.

Take Home Messages

- Therapeutic monoclonal antibodies (mAb) are safe and efficacious injectable drugs; however, some safety issues can occur. Besides side effects that are caused by the activity on target, immunogenicity can play an important role in antibody side effects.
- In general, the more "human like" the structure, the less immunogenic is an antibody. While rodent and chimeric antibodies can cause a vaccination type of immune response involving T cells, humanized and human mAbs usually cause a reversible B cell driven immune response.
- Side effects can also be caused by the Fc-part of mAbs which recruits certain Lymphocytes and elements of the complement system. This can be enhanced by antibody aggregation.
- Several protein engineering technologies exist to reduce the risk of side effects mediated by the host immune system including humanization and Fc engineering.
- MAbs are highly complex and heterogeneous molecules. This heterogenicity is inherent to the biotechnological processes used to manufacture drug products. Some mAb variants are impacting the safety and the efficacy of the product.
- The mAb variants can be minimized by: (1) using proper molecule engineering technology and (2) by a strict understanding and control of every step involved in the manufacturing process.
- The guarantee, consistency, traceability, and quality of the drug substance, the manufacturing process of a therapeutic mAb must comply with the latest process validation guidelines and the good manufacturing practices requested by health authorities.
- The success of a therapeutic mAb is also dependent on its ease of use for the patient or the health care professionals and, obviously, its cost of manufacturing and distribution. Therefore, a good manufacturability and a marketable intended use are key elements to consider.

References

1. Köhler G, Milstein C. Continuous cultures of fused cells secreting antibody of predefined specificity. Nature. 1975;256:495–7.
2. Hansel TT, Kropshofer H, Singer T, et al. The safety and side effects of monoclonal antibodies. Nat Rev Drug Discov. 2010;9:325–38.
3. Force T, Krause DS, Van Etten RA. Molecular mechanisms of cardiotoxicity of tyrosine kinase inhibition. Nat Rev Cancer. 2007;7:332–44.
4. Bianchini D, Jayanth A, Chua YJ, et al. Epidermal growth factor receptor inhibitor-related skin toxicity: mechanisms, treatment, and its potential role as a predictive marker. Clin Colorectal Cancer. 2008;7:33–43.
5. Bohra C, Sokol L, Dalia S. Progressive multifocal leukoencephalopathy and monoclonal antibodies. Cancer Control. 2017;24:1073327481772990.
6. Ransohoff RM. Natalizumab for multiple sclerosis. N Engl J Med. 2007;356:2622–9.
7. Hanke T. Lessons from TGN1412. Lancet (Lond). 2006;368:1569–70; author reply 1570.
8. Kung P, Goldstein G, Reinherz E, et al. Monoclonal antibodies defining distinctive human T cell surface antigens. Science. 1979;206:347–9.
9. Smith S. Ten years of orthoclone OKT3 (muromonab-CD3): a review. J Transpl Coord. 1996;6:109–21.
10. Harding FA, Stickler MM, Razo J, et al. The immunogenicity of humanized and fully human antibodies. MAbs. 2010;2:256–65.
11. Morrison SL, Johnson MJ, Herzenberg LA, et al. Chimeric human antibody molecules: mouse antigen-binding domains with human constant region domains. Proc Natl Acad Sci. 1984;141: c44S. https://doi.org/10.1378/chest.11-2292.
12. Riechmann L, Clark M, Waldmann H, et al. Reshaping human antibodies for therapy. Nature. 1988;332:323–7.
13. Hwang WYK, Foote J. Immunogenicity of engineered antibodies. Methods. 2005;36:3–10.
14. Vidarsson G, Dekkers G, Rispens T. IgG subclasses and allotypes: from structure to effector functions. Front Immunol. 5:520. https://doi.org/10.3389/fimmu.2014.00520.
15. Chung CH, Mirakhur B, Chan E, et al. Cetuximab-induced anaphylaxis and IgE specific for galactose-alpha-1,3-galactose. N Engl J Med. 2008;358:1109–17.
16. Doevendans E, Schellekens H. Immunogenicity of innovative and biosimilar monoclonal antibodies. Antibodies (Basel). 2019;8:21.
17. Barbosa M, Celis E. Immunogenicity of protein therapeutics and the interplay between tolerance and antibody responses. Drug Discov Today. 2007;12:674–81.
18. Squire IB, Lawley W, Fletcher S, et al. Humoral and cellular immune responses up to 7·5 years after administration of streptokinase for acute myocardial infarction. Eur Heart J. 1999;20:1245–52.
19. Bogen B, Ruffini P. Review: to what extent are T cells tolerant to immunoglobulin variable regions? Scand J Immunol. 2009;70:526–30.
20. Pratt K. Anti-drug antibodies: emerging approaches to predict, reduce or reverse biotherapeutic immunogenicity. Antibodies. 2018;7:19.
21. Zhou Q, Qiu H. The mechanistic impact of N-glycosylation on stability, pharmacokinetics, and immunogenicity of therapeutic proteins. J Pharm Sci. 2019;108:1366–77.
22. Carson K, Evens A, Bennett C, et al. Clinical characteristics of erythropoietin-associated pure red cell aplasia. Best Pract Res Clin Haematol. 2005;18:467–72.
23. Li J, Yang C, Xia Y, et al. Thrombocytopenia caused by the development of antibodies to thrombopoietin. Blood. 2001;98:3241–8.

24. Cohen BA, Oger J, Gagnon A, et al. The implications of immunogenicity for protein-based multiple sclerosis therapies. J Neurol Sci. 2008;275:7–17.
25. Dillman RO. Infusion reactions associated with the therapeutic use of monoclonal antibodies in the treatment of malignancy. Cancer Metastasis Rev. 1999;18:465–71.
26. Krishna M, Nadler SG. Immunogenicity to biotherapeutics – the role of anti-drug immune complexes. Front Immunol. 2016;7:21.
27. Baldo B. Adverse events to monoclonal antibodies used for cancer therapy: focus on hypersensitivity responses. Onco Targets Ther. 2013;2:e26333.
28. Jones PT, Dear PH, Foote J, et al. Replacing the complementarity-determining regions in a human antibody with those from a mouse. Nature. 1986;321:522–5.
29. Jespers LS, Roberts A, Mahler SM, et al. Guiding the selection of human antibodies from phage display repertoires to a single epitope of an antigen. Biotechnology (NY). 1994;12:899–903.
30. Kang AS, Jones TM, Burton DR. Antibody redesign by chain shuffling from random combinatorial immunoglobulin libraries. Proc Natl Acad Sci USA. 1991;88:11120–3.
31. Schneider-Merck T, Lammerts van Bueren JJ, Berger S, et al. Human IgG2 antibodies against epidermal growth factor receptor effectively trigger antibody-dependent cellular cytotoxicity but, in contrast to IgG1, only by cells of myeloid lineage. J Immunol. 2010;184:512–20.
32. Wypych J, Li M, Guo A, et al. Human IgG2 antibodies display disulfide-mediated structural isoforms. J Biol Chem. 2008;283:16194–205.
33. Lo K-M, Zhang J, Sun Y, et al. Engineering a pharmacologically superior form of leptin for the treatment of obesity. Protein Eng Des Sel. 2005;18:1–10.
34. Shapiro RI, Plavina T, Schlain BR, et al. Development and validation of immunoassays to quantify the half-antibody exchange of an IgG4 antibody, natalizumab (Tysabri®) with endogenous IgG4. J Pharm Biomed Anal. 2011;55:168–75.
35. Labrijn AF, Buijsse AO, van den Bremer ETJ, et al. Therapeutic IgG4 antibodies engage in Fab-arm exchange with endogenous human IgG4 in vivo. Nat Biotechnol. 2009;27:767–71.
36. Rispens T, Meesters J, den Bleker TH, et al. Fc-Fc interactions of human IgG4 require dissociation of heavy chains and are formed predominantly by the intra-chain hinge isomer. Mol Immunol. 2013;53:35–42.
37. Saxena A, Wu D. Advances in therapeutic fc engineering – modulation of IgG-associated effector functions and serum half-life. Front Immunol. 2016;7:580.
38. Armour KL, Clark MR, Hadley AG, et al. Recombinant human IgG molecules lacking Fcgamma receptor I binding and monocyte triggering activities. Eur J Immunol. 1999;29:2613–24.
39. Xu D, Alegre ML, Varga SS, et al. In vitro characterization of five humanized OKT3 effector function variant antibodies. Cell Immunol. 2000;200:16–26.
40. Saunders KO. Conceptual approaches to modulating antibody effector functions and circulation half-life. Front Immunol. 2019;10:1296.
41. Khetan A, Huang Y, Dolnikova J, et al. Control of misincorporation of serine for asparagine during antibody production using CHO cells. Biotechnol Bioeng. 2010;107:116–23.
42. Cordoba-Rodriguez RV. Aggregates in MAbs and recombinant therapeutic proteins: a regulatory perspective. BioPharm Int. 2008;21:44–53.
43. Vlasak J, Ionescu R. Fragmentation of monoclonal antibodies. MAbs. 2011;3:253–63.
44. Wei Z, Feng J, Lin H-Y, et al. Identification of a single tryptophan residue as critical for binding activity in a humanized monoclonal antibody against respiratory syncytial virus. Anal Chem. 2007;79:2797–805.
45. Wang W, Vlasak J, Li Y, et al. Impact of methionine oxidation in human IgG1 Fc on serum half-life of monoclonal antibodies. Mol Immunol. 2011;48:860–6.

46. Yan B, Steen S, Hambly D, et al. Succinimide formation at Asn 55 in the complementarity determining region of a recombinant monoclonal antibody IgG1 heavy chain. J Pharm Sci. 2009;98:3509–21.

47. Harris RJ, Kabakoff B, Macchi FD, et al. Identification of multiple sources of charge heterogeneity in a recombinant antibody. J Chromatogr B Biomed Sci Appl. 2001;752:233–45.

48. Presta LG. Engineering of therapeutic antibodies to minimize immunogenicity and optimize function. Adv Drug Deliv Rev. 2006;58:640–56.

49. Boyd PN, Lines AC, Patel AK. The effect of the removal of sialic acid, galactose and total carbohydrate on the functional activity of Campath-1H. Mol Immunol. 1995;32:1311–8.

50. Patel D, Guo X, Ng S, et al. IgG isotype, glycosylation, and EGFR expression determine the induction of antibody-dependent cellular cytotoxicity in vitro by cetuximab. Hum Antibodies. 2010;19:89–99.

51. Eon-Duval A, Broly H, Gleixner R. Quality attributes of recombinant therapeutic proteins: an assessment of impact on safety and efficacy as part of a quality by design development approach. Biotechnol Prog. 2012;28:608–22.

52. Beck A, Wurch T, Bailly C, et al. Strategies and challenges for the next generation of therapeutic antibodies. Nat Rev Immunol. 2010;10:345–52.

53. Tesar D, Luoma J, Wyatt EA, et al. Protein engineering to increase the potential of a therapeutic antibody Fab for long-acting delivery to the eye. MAbs. 2017;9:1297–305.

54. Kuehn SE. The process is the product seasoned veteran Dr. Harry Lam shares insights from his 30-year career in CMOs. Pharma Manuf. 2014.

55. Jagschies G. Managing manufacturing economics. In: Jagschies G, Lewine HL, editors. The development of therapeutic monoclonal antibodies. Boston: BioProcess Technology Consultants; 2010. p. 307–32.

56. Ovacik M, Lin K. Tutorial on monoclonal antibody pharmacokinetics and its considerations in early development. Clin Transl Sci. 2018;11:540–52.

57. Kemter K, Altrichter J, Derwand R, et al. Amino acid-based advanced liquid formulation development for highly concentrated therapeutic antibodies balances physical and chemical stability and low viscosity. Biotechnol J. 2018;13:1700523.

Eukaryotic Expression Systems for Upstream Processing of Monoclonal Antibodies

13

Lina Heistinger, David Reinhart, Diethard Mattanovich, and Renate Kunert

Contents

L. Heistinger
Christian Doppler Laboratory for Innovative Immonutherapeutics, Institute of Microbiology and Microbial Biotechnology, Department of Biotechnology, University of Natural Resources and Life Sciences, Vienna (BOKU), Vienna, Austria
e-mail: Lina.heistinger@boku.ac.at

D. Reinhart
Boehringer Ingelheim RCV GmbH and Co KG, Vienna, Austria
e-mail: David.reinhart@boehringer-ingelheim.com

D. Mattanovich
Institute of Microbiology and Microbial Biotechnology, Department of Biotechnology, University of Natural Resources and Life Sciences, Vienna (BOKU), Vienna, Austria
e-mail: Diethard.mattanovich@boku.ac.at

R. Kunert (✉)
Institute of Animal Cell Technology and Systems Biology, Department of Biotechnology, University of Natural Resources and Life Sciences, Vienna (BOKU), Vienna, Austria
e-mail: Renate.kunert@boku.ac.at

© Springer Nature Switzerland AG 2021
F. Rüker, G. Wozniak-Knopp (eds.), *Introduction to Antibody Engineering*,
Learning Materials in Biosciences,
https://doi.org/10.1007/978-3-030-54630-4_13

343

Keywords

Upstream processing · Product quality attributes · Host cell line · Cell culture medium ·
Environmental cultivation parameters · Mode of bioreactor operation · Yeast ·
Filamentous fungi · Glycoengineering · Antibody fragments

What You Will Learn in This Chapter

Monoclonal antibodies (mAbs) represent the largest group of biopharmaceuticals
used for human therapy, diagnostics, and imaging. Full-length mAbs with human-
like posttranslational modifications are almost exclusively expressed in mammalian
cell cultures. Progress in genetic engineering techniques and process development
enables the development of optimized cell lines and manufacturing processes. The
main factors that influence the yield of a production process are the time to accumulate a target amount of biomass, the process duration, and the specific productivity of
the employed cell line. In this chapter, the contribution of the cell line, culture
medium, cultivation parameters, and mode of process operation to product yield
and quality are discussed.

Besides mammalian cell cultures, yeasts and filamentous fungi are widely used
eukaryotic expression systems for recombinant protein production. They are capable
of performing the posttranslational modifications required for mAb production,
although differences in protein glycosylation have to be considered. This chapter
gives an overview of yeast, fungal, and higher eukaryotic species used as production
hosts for full-length mAbs and antibody fragments and highlights critical cell and
process engineering steps required for efficient mAb production.

13.1 Introduction

13.1.1 Antibodies and Different Antibody Formats

Monoclonal antibodies and antibody fragments are important biotechnological products
applied in research, diagnostics, and the treatment of diseases. The most commonly used
class of antibodies is immunoglobulin G (IgG), a heterotetrameric molecule of around
150 kDa consisting of two heavy and two light chains connected by disulfide bonds. Each

of the chains is stabilized by further intramolecular disulfide bonds and consists of a constant part (CH1–CH3 domains in the heavy chain and CL in the light chain) and a variable domain (VH and VL), which determine the binding specificity. The heavy chain of IgG contains one N-glycosylation site in the CH2 domain and the presence and structure of the glycan is important for antibody effector functions [1]. The most widely used antigen-binding antibody fragments are Fab fragments, consisting of the VH-CH1 and VL-CL domains connected by disulfide bonds and single-chain variable fragments (scFV), consisting only of the VH and VL domains connected by a linker peptide. Another important antibody format is single variable domain (VHH) antibodies or "nanobodies," which are derived from the antigen-binding domain of heavy-chain only antibodies as naturally produced by camelids and sharks [2, 3].

13.1.2 Product Quality Attributes

Product quality is an important issue for any biomolecule as it directly affects bioactivity, pharmacokinetics, immunogenicity risk and safety. For mAbs, the main quality attributes typically include variants related to (1) glycosylation, (2) charge distribution, (3) molecular size, (4) oxidation, and (5) structure [4]. Besides process-related impurities (e.g., adventitious agents, host cell protein, DNA). The mentioned quality attributes are highly influenced by the applied expression system (host cell line), cultivation media as well as the conditions present in the bioreactor during the manufacturing process as outlined in the following sections.

13.1.2.1 Glycosylation
The predominant oligosaccharide structure of mAb heavy chains is a biantennary complex-type structure, mostly core-fucosylated partially with a bisecting N-acetylglucosamine, and a small portion may carry one or two sialic acid residues on the antennae [5]. The most significant variation is determined by the degree of galactosylation and a varying portion of truncated glycoforms. Galactosylation mediates complement-dependent cytotoxicity (CDC) due to increased antibody binding to C1q [6], but terminal galactose residues have no impact on serum half-life [7]. Despite fucosylation is naturally occurring in more than 80% of serum IgG, studies have shown that non-fucosylated IgG's increase antibody-dependent cell-mediated cytotoxicity (ADCC) significantly due to enhanced FcγRIIIa receptor binding [8, 9]. The bisecting N-acetyl-glucosamine, bound to the central mannose residue, is another important sugar moiety that enhances ADCC. Sialylation is generally known for its beneficial effect on serum half-life [10], but terminal sialic acid may also reduce ADCC [11]. Finally, non-glycosylated heavy chains as well as high-mannose type glycan structures are also typically discovered.

13.1.2.2 Charge Distribution

Visualization of charge-related mAb variants gives insight into chemical and enzymatic modifications that may have resulted from the bioprocess or the formulation/storage conditions [12]. Charge distribution is typically analyzed through the chromatographic separation of mAb formulations into different peak fractions (main peak, acidic region, and basic region). For instance, acidic charge variants may arise from glyco-variants and/or deamidation reactions, whereas isomerization of aspartic acid, N-terminal pyroglutamic acid, C-terminal lysine, C-terminal proline, or leucine amidation may give rise to basic charge variants [4].

13.1.2.3 Molecular Size

High molecular weight species (HMWS) form due to protein aggregation (i.e., dimers). Low molecular weight species (LMWS) such as protein backbone-truncated fragments expose novel epitopes that can negatively affect immunogenicity or pharmacokinetic properties in vivo along with low or substantially reduced activity relative to the mAb monomer [13]. Both HMWS and LMWS contribute to the size heterogeneity of mAbs and are critical for drug safety and efficacy.

13.1.2.4 Oxidation

Oxidation of methionine may interfere with mAb affinity of binding to its target, thus decreasing its efficacy, especially if the oxidation occurs within the complementarity-determining region [14].

13.1.2.5 Structure

Sequence variants are erroneous amino acid substitutions, either due to DNA mutations or misincorporations of amino acids, which may result in misfolded or aggregated mAbs. Cysteine substitutions can form free sulfhydryl variants, unexpected disulfide bonds, and modified disulfide linkages.

13.2 Antibody Production in Mammalian Cells

13.2.1 Host Cell Lines

CHO, NS0, Sp2/0, HEK293, and PER.C6 cells are the most prominent host cell lines for recombinant mAb expression [15]. Yet, only mAbs produced by CHO, NS0, and Sp2/0 cells were approved for human therapy [16]. All these cell lines are relatively easy to grow in suspension bioreactors, show appreciable production properties, and secrete mAbs with proper posttranslational modifications. Since all immortalized cell lines are aneuploid with inhomogeneous numbers of chromosomes, differences exist even among the cell populations from the same origin. For instance, CHO cells have undergone extensive mutagenesis and clonal selection since their original introduction. Several deletions, translocations, or rearrangements of chromosomal segments have resulted in substantial genetic heterogeneity among the different CHO cell lineages, i.e., CHO-K1, CHO-S, CHO-DG44, and CHO-DXB11 [17–19]. Notably, some CHO cell lines have evolved to favor either protein production or biomass synthesis when the same mAb was expressed under the same conditions [20]. Differing rates of product synthesis among the CHO host cell lines may result from unequal capacities and processing efficiencies of their translational machineries. For instance, a larger ER and higher mitochondrial mass of CHO-K1 were associated with higher productivities compared to CHO-DUXB11 [21]. However, even within a supposedly monoclonal population of the ancestor of CHO-K1 and CHO-DG44 cells substantial heterogeneity exists [22, 23]. Furthermore, cell transfection with the gene of interest results in thousands of clones that differ in their expression levels and growth rates. Therefore, independent of the applied cell line, substantial screening activities have to be performed to select the final production clone that harbors the desired traits. Regarding glycosylation, general differences among the different host cell lines have been identified.

Human cell lines, i.e., PER.C6 and HEK have the obvious benefit of harboring a fully human production machinery delivering correct posttranslational modifications and human identical glycosylation. HEK cells are merely considered for transient mAb expression. HEK cells come with a high degree of transformation, which may impact the glycosylation pattern of the expressed protein [24–26]. In distinct cases, mAbs from HEK cells had lower sialylation compared to CHO cells [27]. Hamster-derived cell lines, including CHO and BHK cells, do not express the human immunogenic glycan epitopes alpha-gal and Neu5Gc (unlike the murine cell lines). In contrast, CHO cells do not express α-2,6-sialyltransferase (α-2,6-SiaT) and β-1,4-N-acetylglucosaminyl-transferase (GnT-III), which add terminally linked α-2,6-sialic acids and bisecting N-acetylglucosamine (GlcNAc) as found in human glycan structures. Instead, CHO cells generate only α-2,3-linked terminal sialylation of N-glycan moieties [28].

13.2.2 Cell Culture Medium

Initial cell culture media contained serum to provide a rich source of minerals, lipids, hormones, metabolites, and beneficial proteins (e.g., growth factors) to cells. To circumvent potential risks regarding lot-to-lot variability, which may negatively affect process robustness, and the introduction of adventitious agents (e.g., viruses, prions, and mycoplasma), serum-free media were developed. In such media, serum is typically substituted with peptones that are water-soluble protein hydrolysates of yeast, plant (e.g., soy or wheat), or casein-derived origin. Peptones are chemically undefined and contain peptides, amino acids, inorganic salts, lipids, vitamins, and sugars [29]. Peptones may exhibit a growth-promoting effect on cells associated with increased recombinant protein expression [30, 31]. However, the chemically undefined nature, lot-to-lot variability, and unspecific effects of peptones in cell cultures reduce the robustness of bioprocesses. Today, most production processes rely on fully chemically defined cell culture media. These media are the most advanced and expensive formulations containing completely defined, animal component-free sources of highly purified recombinant proteins and/or hormones to enable high cell culture performance at a low batch-to-batch variability and thus high process robustness.

Besides other advances such as in cell line development, clone screening and isolation as well as enhanced process monitoring and control, media development has particularly advanced in the last decades to achieve ever-increasing cell and product concentrations. While 30 years ago, old bioprocesses peaked at cell concentrations of 3×10^6 cells/mL and titers of 100 mg/L in simple batch cultures operated for 1 week, two decades later further improvements regarding media formulations and feed strategies enabled considerably longer process durations, higher cell densities, and mAb titers of 1–5 g/L [32]. Today, fed-batches with mAb titers even beyond 10 g/L are becoming reality [15]. Despite numerous media formulations that have become commercially available, it must be considered that each clone has individual demands for the cell culture medium (and environmental cultivation conditions) that further expands in dependence on the expressed product [33]. Therefore, media development is an inherent activity for each newly generated cell line.

Apart from pursuing even higher titers, the focus has shifted toward being able to control product quality and process consistency [34]. Regarding product quality, glycosylation is of particular importance and there are a number of factors that can be varied for modulation. Besides environmental process conditions in the bioreactor (e.g., dissolved oxygen, pH, temperature, pCO_2, and process duration) also the availability of certain metabolites, nutrients, and trace elements influences glycosylation. For instance, high concentrations of ammonia are undesirable as terminal glycosylation is prevented [35]. Glucose and glutamine are the main energy donors to the cell and function as important substrates for the synthesis of intracellular nucleotide-sugars that are essential for N-glycosylation. Glycerol has been shown to enhance sialylation [36]. Galactose, uridine, or the trace element manganese directly affect galactosylation [37]. In general, trace elements can modulate activities of various enzymes and transporters (e.g., glycosyltransferases, mannosidases, and lysosomal hydrolases) and therefore affect glycosylation. In addition, trace elements such as manganese, zinc, and cobalt were shown to decrease the formation of LMWS [38], while copper had the opposite effect [39, 40].

13.2.3 Environmental Cultivation Parameters

Bioprocesses require specific set points that promote the cultivation of cells to reach high cell and product concentrations. Culture operating parameters are frequently targeted during process optimization and include physical as well as chemical parameters. The most important physical parameters include temperature, gas flow rate, and agitation speed. Key chemical parameters comprise dissolved oxygen (DO), carbon dioxide, pH, osmolality, redox potential, and metabolite levels, including substrate, amino acid, and waste byproducts [41]. Care must be taken when modulating these parameters as even small changes can have detrimental effects on yields and the quality of the expressed product.

The temperature of animal cell culture processes is most often set to the physiologic value of 37 °C. In literature, a temperature range of about 25–39 °C has been investigated but there is a clear focus on 30–37 °C [42, 43]. A reduction of the temperature is generally considered to decrease cell-specific growth rates and typically comes with other positive effects such as an extended culture longevity, reduced nutrient consumption, improved viability, and decreased protease activity [44–46]. Additionally, hypothermic conditions may affect the cell cycle and can lead to an increase in cell-specific productivity and therefore improved product yields. However, this effect is cell line dependent. Since changes in the cultivation temperature have also been reported to impact product quality attributes, such as protein glycosylation, this parameter has to be monitored closely [43].

The cultivation pH is another important parameter that greatly affects cell growth and productivity. Typical pH set points for animal cells are neutral to slightly alkaline (7.0–7.2). At a slightly acidic pH (6.8–7.0) cells exhibit a reduced cell metabolism and tend to grow slower [42, 43, 46]. At a slightly alkaline pH ($\geq$7.2), the metabolism of animal cells tends to be more active. The effect of pH on recombinant protein production is cell line-specific. Some reports claim higher specific productivities at more alkaline pH values of 7.6–7.8 [47]. Other groups report fairly constant levels over a broad pH range of 6.8–7.3 [43] or 6.85–7.8 [42]. The setpoint of cultivation pH has to be controlled within a narrow dead band since small deviations can result in changes in product quality attributes, such as protein glycosylation [43]. Setpoints between pH 6.8 and 7.8 may affect galactosylation, sialylation, and microheterogeneity [47, 48].

The DO setpoint typically ranges between 20 and 50% air saturation and is critical for supplying sufficient oxygen to cells. A limitation of DO may result in increased lactate production (acidification), whereas high DO values may cause cytotoxicity [49]. In few cases, DO values exerted cell- and/or product-specific effects on production rates, cell metabolism, and glycosylation [43, 50].

Dissolved CO_2 acidifies the cultivation pH and requires the addition of base if large amounts of CO_2 are trapped in the bioreactor. This is undesired as it increases the osmolality. High CO_2 levels ($\geq$120 mm Hg) may exert inhibitory effects on cell cultures and affect product quality [41, 51].

The osmolality of cell culture media typically ranges between 260 and 320 mOsm/kg. Upon base addition or excessive feeding during fed-batch cultivation, osmolality values increase to even higher numbers. Hyperosmolality has often been associated with increased cell size and reduced cell growth. The reduced cell growth (or no growth) in particular exerts a positive effect on cell-specific production rates and titers for some cell lines [52–55].

Finally, the time point of harvest from the bioreactor can have a critical effect. Despite extensive cultivation typically enables the accumulation of high product concentrations, cell viability naturally decreases toward the end of fermentation and lysed cells release their intracellular content. Thus, degradative enzymes such as glycosidases but also proteases get in contact with products and diminish product quality and quantity [56].

13.2.4 Mode of Bioreactor Operation

For mAb manufacturing, bioreactors can in principle be run in different modes of operation. Historically, batch cultivation was employed for the first production processes and represents the easiest mode of operation. Upon inoculation, maximal cell growth is dictated by the limitation of nutrients (e.g., glucose, glutamine) or waste product inhibition (e.g., ammonia, lactate). Due to relatively low cell and product concentrations, batch cultures are not recommendable for today's bio industrial requirements.

During fed-batch cultivation, nutrients are (semi)continuously supplied to cells in the bioreactor. The controlled addition of nutrients alleviates limiting effects of nutrient depletion and typically increases culture longevity, cell concentrations, cell-specific production rates, and product yield. Limitations to further growth comes from the accumulation of metabolic byproducts or high osmolality due to regular additions of concentrated feed media. Today, fed-batch cultivation is the standard for mAb manufacturing.

Continuous cultivation of animal cells (e.g., chemostat) is characterized by the continuous addition of fresh medium to the bioreactor while the spent medium is continuously harvested. For this type of cultivation, a so-called steady state is desired during which constant conditions can be achieved (e.g., cell-specific growth and production rates, concentrations of nutrients, metabolic byproducts, product and cells, culture volume, dissolved oxygen concentration, pH). This creates a rather stable environment for the cells and the product that is particularly beneficial for nonstable products. However, since rather low cell concentrations are typically reached, chemostat cultures are preferred for studying the effect of different culture variables under steady-state environmental conditions rather than mAb production.

Perfusion cultivation is a type of continuous culture and relies on a constant addition of fresh medium, while the spent medium is continuously removed from the bioreactor. Unlike chemostat cultivation, perfusion cultures employ cell retention devices such as centrifugation, internal and external spin filters, gravity settlers, tangential flow filters, carriers, or hollow fiber modules to maximize biomass. This significantly improves cell concentrations up to 1×10^8 c/mL, and even beyond, as well as space–time yields. In addition, reduced residence times of the product in the bioreactor are beneficial for quality, especially for fragile products. Consequently, perfusion cultures have regained popularity in recent years. Further applications of perfusion technology include the so-called N-1 perfusion. In this case, perfusion assists to accumulate high amounts of biomass at the N-1 stage of the production bioreactor seed train for high cell density seeding of the production bioreactor. As a result, the duration of the growth phase is reduced, thus increasing manufacturing capacity. In an alternative approach, termed hybrid process, the production bioreactor is initially operated in perfusion mode to accumulate biomass. Then, at high cell concentrations, the process is switched to fed-batch cultivation until the end of fermentation. With this strategy, production capacity can be further pushed. A similar approach is termed concentrated fed-batch process. However, during this process type ultrafiltration membranes are employed to retain not only cells but also product in the bioreactor.

The implementation of continuous manufacturing comes with several advantages. For instance, high product-yielding perfusion cultures allow the construction of smaller production bioreactors and therefore also require fewer bioreactors in the seed train. This reduces the initial capital investment. As reviewed by Bielser et al. [57], smaller facilities provide greater manufacturing flexibility, which is of particular importance considering a rapidly changing market because of uncertainties regarding product success, demand variability, competition, clinical trial failures, a growing clinical pipeline including new classes of molecules, the time to the market pressure, or facilitated market adoption. Pressure coming from the payer and the access to developing markets requires the reduction of manufacturing costs and the capitals for new investments.

Xu et al. [58] compared the performance of different process modes using a very similar media to provide a fair comparison in their productivity. Much higher space–time yields were obtained in the perfusion (up to 2.29 g/L/day) and concentrated fed-batch processes (up to 2.04 g/L/day), compared to that in the fed-batch processes with low or high cell density inoculation (ranging from 0.39 to 0.49 g/L/day). Similar results were presented by Bausch et al. [59], a mAb-producing CHO cell line had 3.5-fold higher space–time yields in perfusion culture (1.10 g/L/day) compared to fed-batch mode (0.31 g/L/day). In both studies, high space–time yields were largely due to higher cell concentration maintained as the cell-specific productivities were comparable for all production modes. This demonstrates that continuous technologies can lead to process intensification.

Economic comparisons between fed-batch and continuous perfusion technologies must be done individually for each molecule to address capital expenses (CapEx), operational expenses (OpEx), or cost of goods (CoG). In general, the implementation of perfusion operation may come along with CapEx savings due to the smaller footprint of all unit operations [60, 61]. OpEx requires thorough evaluation as the main driver is media consumption in continuous production. Optimization of perfusion media to maximize space–time yields at low perfusion rates and ideally low media costs will drive down OpEx. In a life cycle and CoG assessment between fed-batch and perfusion processes by Bunnak et al. [62], the authors calculated similar CoGs between both technologies. However, perfusion cultures had higher energy requirements (17%), water consumption (35%), and CO_2 emission (17%). Besides individual comparisons between the implementation of fed-batch or perfusion technology, the full potential of perfusion processes can most likely be realized in a fully continuous end-to-end integrated manufacturing process, where upstream unit operations are directly connected to continuous downstream units [57].

13.3   Antibody Production in Yeast and Filamentous Fungi

13.3.1 General Characteristics of the Expression Systems

Today, mammalian cell lines are the most commonly used expression system for the production of monoclonal antibody biopharmaceuticals [63]. They can perform human-like glycosylations and in a typical optimized process high yields of 3–10 g/L can be reached in 10–16 days [15, 64]. As eukaryotes, yeasts and filamentous fungi generally have the capability to perform all the posttranslational modifications required for the production of complex pharmaceutical proteins. A main advantage is the availability of efficient genetic engineering methods, which allow simple and fast strain development. As microorganisms, yeasts and filamentous fungi exhibit fast growth to high cell densities of more than 100 g/L dry mass on simple and inexpensive substrates and high titers of the desired protein can be obtained in short process times. In contrast to prokaryotic systems like *Escherichia coli*, the recombinant protein can efficiently be targeted into the secretory pathway, where posttranslational modifications like protein glycosylations take place. Also, after secretion from the cells, the correctly folded functional product can easily be purified from the culture supernatant typically containing a very low amount of undesired host cell proteins [65–67].

13.3.2 Glycosylation and Glycoengineered Production Strains

Protein glycosylations are critical for the production of many complex biopharmaceuticals like antibodies. As eukaryotes, yeast and fungi can efficiently perform protein *N*- and *O*-glycosylation. However, they produce high-mannose type glycans with varying numbers of mannose, lacking the additional *N*-acetyl glucosamine, galactose, and sialic acid found in complex mammalian glycans [68, 69]. Due to their differences in structure and composition, fungal glycans are considered as potentially immunogenic in humans. Therefore, substantial efforts have been made to develop yeast strains producing pharmaceutically relevant proteins with human-like glycans. The deletion of the α-1,6-mannosyltransferase Och1 abolished protein hypermannosylation in *Saccharomyces cerevisiae* and *Pichia pastoris* [70, 71]. Different glycoengineering approaches combining the deletion of yeast-like glycosylation with the expression of heterologous glycosyltransferases and glycosidases enabled the production of proteins with human-like *N*-glycans in several yeast species [72–75]. Also, glycoengineered *P. pastoris* strains have been used to successfully produce functional IgG with human-like glycans [76–78]. Another approach is the use of in vitro glycosylation for the precise attachment of desired glycan structures [79, 80]. Fewer studies have been conducted on yeast *O*-glycan engineering. Generally, yeast specific *O*-glycosylation can be partially repressed by the deletion of some (but not all) protein-*O*-mannosyltransferase (PMT) genes or the addition of PMT inhibitors. Both strategies have been shown to improve antibody production and product quality in

P. pastoris and *Ogataea minuta* [81, 82]. For a comprehensive review of yeast glycoengineering please refer to Hamilton et al. or De Wachter et al. [83, 84]. First studies on glycoengineering in the filamentous fungi *Trichoderma reesei* and *Aspergillus oryzae* demonstrated the successful addition of *N*-acetylglucosamine by a mammalian glycosyltransferase, showing the potential for the synthesis of complex glycan structures [85, 86]. Although glycoengineering enables the production of IgG with human-like glycans in yeast and fungi, their main field of application is still the production of different non-glycosylated antibody fragments.

13.3.3 Antibody Production in Yeast

Many different yeast species have been used for the production of recombinant antibodies or antibody fragments, including *S. cerevisiae*, *P. pastoris*, *Ogataea polymorpha*, *O. minuta*, *Kluyveromyces lactis*, *Yarrowia lipolytica*, and *Schizosaccharomyces pombe*. The first successful production of secreted full-length IgG and Fab fragments was done with the well-studied model yeast *S. cerevisiae* [87]. For the first single-chain Fv (scFv) antibody fragments produced in *S. cerevisiae*, the overexpression of the chaperone BiP (Kar2 in yeast) or protein disulfide isomerase (PDI) were found to increase product titers two- to eightfold to up to 20 mg/L in shake flask [88]. Since then, different strain engineering strategies have been developed to further improve antibody folding and secretion in *S. cerevisiae* [89–91]. Also, *S. cerevisiae* has been reported to be well suited for the production of antigen-specific llama single-domain antibodies (VHH) [92].

The most commonly used yeast for the production of antibodies and antibody fragments is the methylotrophic yeast *P. pastoris* (*Komagataella* sp.). A big advantage of *P. pastoris* is its ability to grow on methanol as a sole carbon and energy source, which allows the use of strong methanol inducible promoters (e.g., P_{AOX1}) for recombinant protein production. Also, *P. pastoris* grows to very high cell densities on simple substrates and efficiently secretes proteins to the culture supernatant [93]. The first production of several different scFv antibody fragments in *P. pastoris* with titers of more than 0.1 g/L was reported in 1995 [94]. Applying strain and process optimization, scFv titers beyond 1 g/L can be reached with this yeast [95, 96]. The overexpression of chaperones and folding enzymes of the secretory pathway or the unfolded protein response transcription factor Hac1 have been shown to increase the production of antibody fragments in *P. pastoris* [96, 97]. To optimize the expression of specific products, a number of carbon source dependent and independent promoters are available as alternatives to the strong methanol inducible P_{AOX1} promoter [98–101]. Furthermore, optimized secretion signals have been shown to have a positive impact on recombinant protein production [102–106]. A comparative study on Fab production in *E. coli*, CHO cells, and *P. pastoris* showed that similar amounts of functional products at a slightly lower cost of goods could be produced with the yeast system [107].

Besides *P. pastoris*, also other non-conventional yeasts have been shown to be suitable for the production of antibody fragments. The yeast *K. lactis* can grow on lactate as sole carbon and energy source and is an established host for the production of enzymes in the food industry. A few studies report the soluble production of functional scFv in *K. lactis* [108–110]. Similar levels of scFv could also be obtained with the yeast *Y. lipolytica* [108]. The methylotrophic yeast *O. minuta* has been investigated for the production of full-length antibodies. The use of protease deficient strains, repression of *O*-glycosylation and chaperone overexpression all had a positive impact on the production of functional mAbs with this yeast [81, 111, 112]. Another alternative yeast expression system is the well-studied fission yeast *S. pombe* [113]. Its potential for antibody fragment production has been shown by the successful secretion of an scFv-GFP fusion protein [114]. A recent study comparing different expression hosts also describes the investigation of *Kluyveromyces marxianus, Ogataea polymorpha,* and *Arxula adeninivorans* for the production of antibodies [115].

Not only strain engineering but also process optimization has a great impact on antibody production in yeast. Most commonly, glucose- or methanol-limited fed-batch processes are used for production with *S. cerevisiae* and *P. pastoris*, respectively [95, 116–118]. Mixed-feed fermentations using glycerol and methanol, optimized levels of methanol for induction, and oxygen-limited processes have been described to be beneficial for antibody production with *P. pastoris* [119–122]. Optimization of methanol concentration and cultivation pH allowed high-level production of scFv with more than 4 g/L [95]. A design of experiment approach for the production process of full-length IgG in a glycoengineered *P. pastoris* strain yielded up to 1.6 g/L IgG and the developed process could be successfully scaled-up to 1200 L manufacturing scale [117]. Also, there have been efforts to develop production strains and media for the production of a Fab fragment in a continuous process [123].

13.3.4 Antibody Production in Filamentous Fungi

Filamentous fungi like *Aspergillus niger, A. oryzae,* and *T. reesei* are established hosts for the production of industrial enzymes and other proteins. They have a very high secretion capacity and naturally secrete large amounts of more than 25 g/L of lytic enzymes into the medium. However, yields obtained for recombinant proteins are usually significantly lower [124, 125]. Fusion of the heterologous protein to a native, well-secreted protein has been shown to be beneficial for protein production. Examples for well-characterized sequences include the *T. reesei* cellobiohydrolase (*CBH1*) and the *A. niger* or *Aspergillus awamori* glucoamylase (*glaA*) genes [124]. Fusion of the Fab heavy chain to a truncated form of *CBH1* improved Fab production from 1 to 40 mg/L in *T. reesei* [126]. In the first attempt to produce VHH in *A. oryzae*, 73 mg/L VHH could be produced with fusion to a Taka-amylase A signal sequence (sTAA) and a segment of 28 amino acids from the N-terminal region of *Rhizopus oryzae* lipase (N28) [127]. In a different study, up to 1 g/L VHH in the culture medium could be obtained when a glucoamylase-VHH fusion was used [128]. Other *Aspergillus* species suitable for the production of antibodies are *A. awamori* and *A. niger*. Joosten et al. showed that *A. awamori* can be used for the production of functional VHH and VHH-peroxidase fusion proteins [129, 130]. *A. niger* has been used to produce full-length IgG and Fab fragments. The fusion of the antibody chains to glucoamylase and an optimized KexB cleavage site allowed the production of antibodies with similar properties as mammalian-derived IgG [131]. Recently, also the model fungus *Ustilago maydis* has been investigated as an alternative host for recombinant protein production. The unconventional secretion pathway of the chitinase Cts1 can be used to secrete heterologous proteins while avoiding fungal *N*-glycosylation [132]. It has been shown that this approach is suitable for the production of an scFv and different nanobodies, but the obtained yields are still low [133, 134].

Generally, proteolytic degradation can be an issue for the production of heterologous proteins in filamentous fungi. In many cases, the deletion of host proteases has been shown to improve protein production [135–137]. Also, the addition of BSA has been used as a strategy to reduce the proteolytic degradation of a VHH produced in *A. awamori* [129].

13.4   Conclusion

Today, full-length mAbs with human-like posttranslational modifications are almost exclusively produced in mammalian cell cultures. Despite classical hybridoma cells still being developed for the isolation of individual mAbs, mAbs are often engineered and expressed in standard host cell lines after gene transfer. Progress in molecular biology techniques enabled the efficient genetic manipulation of cells and gene transfer vehicles for optimization of the cell as a production factory [15]. The improvements in cell line and cell culture media development, clone screening and isolation in combination with enhanced process monitoring and control have pathed the way to ever-increasing cell and product concentrations in mammalian cell cultures.

Although there is no yeast or fungi derived therapeutic antibody product on the market yet, yeast and filamentous fungi have great potential as production hosts for antibodies and especially non-glycosylated antibody fragments. A recent comparative study analyzing IgG production in different yeasts and filamentous fungi concluded that the engineering tools available will allow the development of competitive production hosts [115]. Overall, the examples presented here show that a combination of strain and process engineering enables high-level production of functional antibody products in yeast and filamentous fungi, which makes these expression systems attractive alternatives for the development of antibody production processes.

Take Home Messages
- The final application of a mAb product strongly determines the choice of the expression system, process mode, and downstream processing efforts.
- Mammalian cells are the preferred expression system if complex posttranslational modifications are required.
- Host cell engineering and in vitro protein modification approaches make yeasts and filamentous fungi promising alternative expression systems for mAbs products.
- Lower eukaryotes grow faster and reach higher cell densities than animal cell cultures in classical batch and fed-batch processes. The recent development of high cell density continuous processes allowed cell culture technology to catch up with yeast cells.

References

1. Quast I, Peschke B, Lünemann JD. Regulation of antibody effector functions through IgG Fc N-glycosylation. Cell Mol Life Sci. 2017;74:837–47.
2. Liu Y, Huang H. Expression of single-domain antibody in different systems. Appl Microbiol Biotechnol. 2018;102:539–51.
3. Chiu ML, Goulet DR, Teplyakov A, et al. Antibody structure and function: the basis for engineering therapeutics. Antibodies. 2019;8:55.
4. Alt N, Zhang TY, Motchnik P, et al. Determination of critical quality attributes for monoclonal antibodies using quality by design principles. Biologicals. 2016;44:291–305.
5. Raju TS. Terminal sugars of Fc glycans influence antibody effector functions of IgGs. Curr Opin Immunol. 2008;20:471–8.
6. Jefferis R. Glycosylation as a strategy to improve antibody-based therapeutics. Nat Rev Drug Discov. 2009;8:226–34.
7. Jones AJS, Papac DI, Chin EH, et al. Selective clearance of glycoforms of a complex glycoprotein pharmaceutical caused by terminal N-acetylglucosamine is similar in humans and cynomolgus monkeys. Glycobiology. 2007;17:529–40.
8. Miyoshi E, Noda K, Yamaguchi Y, et al. The alpha1-6-fucosyltransferase gene and its biological significance. Biochim Biophys Acta. 1999;1473:9–20.
9. Xue J, Zhu LP, Wei Q. IgG-Fc N-glycosylation at Asn297 and IgA O-glycosylation in the hinge region in health and disease. Glycoconj J. 2013;30:735–45.
10. Raju TS, Lang SE. Diversity in structure and functions of antibody sialylation in the Fc. Curr Opin Biotechnol. 2014;30:147–52.
11. Scallon BJ, Tam SH, McCarthy SG, et al. Higher levels of sialylated Fc glycans in immunoglobulin G molecules can adversely impact functionality. Mol Immunol. 2007;44:1524–34.
12. Lingg N, Berndtsson M, Hintersteiner B, et al. Highly linear pH gradients for analyzing monoclonal antibody charge heterogeneity in the alkaline range: validation of the method parameters. J Chromatogr A. 2014;1373:124–30.
13. Wang S, Liu AP, Yan Y, et al. Characterization of product-related low molecular weight impurities in therapeutic monoclonal antibodies using hydrophilic interaction chromatography coupled with mass spectrometry. J Pharm Biomed Anal. 2018;154:468–75.

14. Sankar K, Hoi KH, Yin Y, et al. Prediction of methionine oxidation risk in monoclonal antibodies using a machine learning method. MAbs. 2018;10:1281–90.
15. Kunert R, Reinhart D. Advances in recombinant antibody manufacturing. Appl Microbiol Biotechnol. 2016;100:3451–61.
16. Dumont J, Euwart D, Mei B, et al. Human cell lines for biopharmaceutical manufacturing: history, status, and future perspectives. Crit Rev Biotechnol. 2016;36:1110–22.
17. Xu X, Nagarajan H, Lewis NE, et al. The genomic sequence of the Chinese hamster ovary (CHO)-K1 cell line. Nat Biotechnol. 2011;29:735–41.
18. Wurm FM, Hacker D. First CHO genome. Nat Biotechnol. 2011;29:718–20.
19. Lewis NE, Liu X, Li Y, et al. Genomic landscapes of Chinese hamster ovary cell lines as revealed by the *Cricetulus griseus* draft genome. Nat Biotechnol. 2013;31:759–65.
20. Reinhart D, Damjanovic L, Kaisermayer C, et al. Bioprocessing of recombinant CHO-K1, CHO-DG44, and CHO-S: CHO expression hosts favor either mAb production or biomass synthesis. Biotechnol J. 2019;14:e1700686.
21. Hu Z, Guo D, Yip SSM, et al. Chinese hamster ovary K1 host cell enables stable cell line development for antibody molecules which are difficult to express in DUXB11-derived dihydrofolate reductase deficient host cell. Biotechnol Prog. 2013;29:980–5.
22. Deaven LL, Petersen DF. The chromosomes of CHO, an aneuploid Chinese hamster cell line: G-band, C-band, and autoradiographic analyses. Chromosoma. 1973;41:129–44.
23. Derouazi M, Martinet D, Besuchet Schmutz N, et al. Genetic characterization of CHO production host DG44 and derivative recombinant cell lines. Biochem Biophys Res Commun. 2006;340:1069–77.
24. Graham FL, Smiley J, Russell WC, et al. Characteristics of a human cell line transformed by DNA from human adenovirus type 5. J Gen Virol. 1977;36:59–72.
25. Varki A. Uniquely human evolution of sialic acid genetics and biology. Proc Natl Acad Sci USA. 2010;107:8939–46.
26. Berger M, Kaup M, Blanchard V. Protein glycosylation and its impact on biotechnology. Adv Biochem Eng Biotechnol. 2012;127:165–85.
27. Croset A, Delafosse L, Gaudry JP, et al. Differences in the glycosylation of recombinant proteins expressed in HEK and CHO cells. J Biotechnol. 2012;161:336–48.
28. Svensson EC, Soreghan B, Paulson JC. Organization of the beta-galactoside alpha 2,6-sialyltransferase gene. Evidence for the transcriptional regulation of terminal glycosylation. J Biol Chem. 1990;265:20863–8.
29. Davami F, Baldi L, Rajendra Y, et al. Peptone supplementation of culture medium has variable effects on the productivity of CHO cells. Int J Mol Cell Med. 2014;3:146–56.
30. Heidemann R, Zhang C, Qi H, et al. The use of peptones as medium additives for the production of a recombinant therapeutic protein in high density perfusion cultures of mammalian cells. Cytotechnology. 2000;32:157–67.
31. Sung YH, Lim SW, Chung JY, et al. Yeast hydrolysate as a low-cost additive to serum-free medium for the production of human thrombopoietin in suspension cultures of Chinese hamster ovary cells. Appl Microbiol Biotechnol. 2004;63:527–36.
32. Jayapal K, Wlaschin KF, Hu WS, et al. Recombinant protein therapeutics from CHO cells – 20 years and counting. Chem Eng Prog. 2007;103:40–7.
33. Reinhart D, Damjanovic L, Kaisermayer C, et al. Benchmarking of commercially available CHO cell culture media for antibody production. Appl Microbiol Biotechnol. 2015;99:4645–57.
34. Kelley B. Industrialization of mAb production technology: the bioprocessing industry at a crossroads. mAbs. 2009;1:443–52.
35. Yang M, Butler M. Effects of ammonia on CHO cell growth, erythropoietin production, and glycosylation. Biotechnol Bioeng. 2000;68:370–80.

36. Rodriguez J, Spearman M, Huzel N, et al. Enhanced production of monomeric interferon-β by CHO cells through the control of culture conditions. Biotechnol Prog. 2005;21:22–30.
37. Liu B, Spearman M, Doering J, et al. The availability of glucose to CHO cells affects the intracellular lipid-linked oligosaccharide distribution, site occupancy and the N-glycosylation profile of a monoclonal antibody. J Biotechnol. 2014;170:17–27.
38. Kao YH, Hewitt DP, Trexler-Schmidt M, et al. Mechanism of antibody reduction in cell culture production processes. Biotechnol Bioeng. 2010;107:622–32.
39. Trexler-Schmidt M, Sargis S, Chiu J, et al. Identification and prevention of antibody disulfide bond reduction during cell culture manufacturing. Biotechnol Bioeng. 2010;106:452–61.
40. Vlasak J, Ionescu R. Fragmentation of monoclonal antibodies. mAbs. 2011;3:253–63.
41. Li F, Vijayasankaran N, Shen A, et al. Cell culture processes for monoclonal antibody production. mAbs. 2010;2:466–79.
42. Yoon SK, Choi SL, Song JY, et al. Effect of culture pH on erythropoietin production by Chinese hamster ovary cells grown in suspension at 32.5 and 37.0°C. Biotechnol Bioeng. 2005;89:345–56.
43. Trummer E, Fauland K, Seidinger S, et al. Process parameter shifting: Part I. Effect of DOT, pH, and temperature on the performance of Epo-Fc expressing CHO cells cultivated in controlled batch bioreactors. Biotechnol Bioeng. 2006;94:1033–44.
44. Ahn WS, Jeon JJ, Jeong YR, et al. Effect of culture temperature on erythropoietin production and glycosylation in a perfusion culture of recombinant CHO cells. Biotechnol Bioeng. 2008;101:1234–44.
45. Kou TC, Fan L, Zhou Y, et al. Increasing the productivity of TNFR-Fc in GS-CHO cells at reduced culture temperatures. Biotechnol Bioprocess Eng. 2011;16:136–43.
46. Kaisermayer C, Reinhart D, Gili A, et al. Biphasic cultivation strategy to avoid Epo-Fc aggregation and optimize protein expression. J Biotechnol. 2016;227:3–9.
47. Borys MC, Linzer DIH, Papoutsakis ET. Culture pH affects expression rates and glycosylation of recombinant mouse placental lactogen proteins by Chinese hamster ovary (CHO) cells. Bio/Technology. 1993;11:720–4.
48. Müthing J, Kemminer SE, Conradt HS, et al. Effects of buffering conditions and culture pH on production rates and glycosylation of clinical phase I anti-melanoma mouse IgG3 monoclonal antibody R24. Biotechnol Bioeng. 2003;83:321–34.
49. Fleischaker RJ, Sinskey AJ. Oxygen demand and supply in cell culture. Eur J Appl Microbiol Biotechnol. 1981;12:193–7.
50. Butler M. Optimisation of the cellular metabolism of glycosylation for recombinant proteins produced by mammalian cell systems. Cytotechnology. 2006;50:57–76.
51. Zanghi JA, Schmelzer AE, Mendoza TP, et al. Bicarbonate concentration and osmolality are key determinants in the inhibition of CHO cell polysialylation under elevated pCO(2) or pH. Biotechnol Bioeng. 1999;65:182–91.
52. Kim TK, Ryu JS, Chung JY, et al. Osmoprotective effect of glycine betaine on thrombopoietin production in hyperosmotic Chinese hamster ovary cell culture: clonal variations. Biotechnol Prog. 2000;16:775–81.
53. Ju HK, Hwang SJ, Jeon CJ, et al. Use of NaCl prevents aggregation of recombinant COMP-angiopoietin-1 in Chinese hamster ovary cells. J Biotechnol. 2009;143:145–50.
54. Shen D, Kiehl TR, Khattak SF, et al. Transcriptomic responses to sodium chloride-induced osmotic stress: a study of industrial fed-batch CHO cell cultures. Biotechnol Prog. 2010;26:1104–15.
55. Reinhart D, Damjanovic L, Castan A, et al. Differential gene expression of a feed-spiked super-producing CHO cell line. J Biotechnol. 2018;285:23–37.
56. Gramer MJ, Goochee CF. Glycosidase activities in Chinese hamster ovary cell lysate and cell culture supernatant. Biotechnol Prog. 1993;9:366–73.

57. Bielser J-M, Wolf M, Souquet J, et al. Perfusion mammalian cell culture for recombinant protein manufacturing – a critical review. Biotechnol Adv. 2018;36:1328–40.
58. Xu S, Gavin J, Jiang R, et al. Bioreactor productivity and media cost comparison for different intensified cell culture processes. Biotechnol Prog. 2017;33:867–78.
59. Bausch M, Schultheiss C, Sieck JB. Recommendations for comparison of productivity between fed-batch and perfusion processes. Biotechnol J. 14:e1700721. https://doi.org/10.1002/biot.201700721.
60. Walther J, Godawat R, Hwang C, et al. The business impact of an integrated continuous biomanufacturing platform for recombinant protein production. J Biotechnol. 2015;213:3–12.
61. Yang O, Prabhu S, Ierapetritou M. Comparison between batch and continuous monoclonal antibody production and economic analysis. Ind Eng Chem Res. 2019;58:5851–63.
62. Bunnak P, Allmendinger R, Ramasamy SV, et al. Life-cycle and cost of goods assessment of fed-batch and perfusion-based manufacturing processes for mAbs. Biotechnol Prog. 2016;32:1324–35.
63. Walsh G. Biopharmaceutical benchmarks 2018. Nat Biotechnol. 2018;36:1136–45.
64. Kelley B, Kiss R, Laird M. A different perspective: how much innovation is really needed for monoclonal antibody production using mammalian cell technology? In: Kiss B, Gottschalk U, Pohlscheidt M, editors. New bioprocessing strategies: development and manufacturing of recombinant antibodies and proteins. Advances in biochemical engineering/biotechnology. New York: Springer; 2018. p. 443–62.
65. Gasser B, Mattanovich D. Antibody production with yeasts and filamentous fungi: on the road to large scale? Biotechnol Lett. 2007;29:201–12.
66. Frenzel A, Hust M, Schirrmann T. Expression of recombinant antibodies. Front Immunol. 2013;4:217.
67. Spadiut O, Capone S, Krainer F, et al. Microbials for the production of monoclonal antibodies and antibody fragments. Trends Biotechnol. 2014;32:54–60.
68. Wildt S, Gerngross TU. The humanization of N-glycosylation pathways in yeast. Nat Rev Microbiol. 2005;3:119–28.
69. Deshpande N, Wilkins MR, Packer N, et al. Protein glycosylation pathways in filamentous fungi. Glycobiology. 2008;18:626–37.
70. Nakanishi-Shindo Y, Nakayama KI, Tanaka A, et al. Structure of the N-linked oligosaccharides that show the complete loss of α-1,6-polymannose outer chain from *och1*, *och1 mnn1*, and *och1 mnn1 alg3* mutants of *Saccharomyces cerevisiae*. J Biol Chem. 1993;268:26338–45.
71. Choi BK, Bobrowicz P, Davidson RC, et al. Use of combinatorial genetic libraries to humanize N-linked glycosylation in the yeast *Pichia pastoris*. Proc Natl Acad Sci USA. 2003;100:5022–7.
72. Hamilton SR, Bobrowicz P, Bobrowicz B, et al. Production of complex human glycoproteins in yeast. Science. 2003;301:1244–6.
73. Oh DB, Park JS, Kim MW, et al. Glycoengineering of the methylotrophic yeast *Hansenula polymorpha* for the production of glycoproteins with trimannosyl core N-glycan by blocking core oligosaccharide assembly. Biotechnol J. 2008;3:659–68.
74. Jacobs PP, Geysens S, Vervecken W, et al. Engineering complex-type N-glycosylation in *Pichia pastoris* using GlycoSwitch technology. Nat Protoc. 2009;4:58–70.
75. Cheon SA, Kim H, Oh DB, et al. Remodeling of the glycosylation pathway in the methylotrophic yeast *Hansenula polymorpha* to produce human hybrid-type N-glycans. J Microbiol. 2012;50:341–8.
76. Li H, Sethuraman N, Stadheim TA, et al. Optimization of humanized IgGs in glycoengineered *Pichia pastoris*. Nat Biotechnol. 2006;24:210–5.
77. Potgieter TI, Cukan M, Drummond JE, et al. Production of monoclonal antibodies by glycoengineered *Pichia pastoris*. J Biotechnol. 2009;139:318–25.

78. Zhang N, Liu L, Dan Dumitru C, et al. Glycoengineered *Pichia* produced anti-HER2 is comparable to trastuzumab in preclinical study. MAbs. 2011;3:289–98.
79. Wei Y, Li C, Huang W, et al. Glycoengineering of human IgG1-Fc through combined yeast expression and in vitro chemoenzymatic glycosylation. Biochemistry. 2008;47:10294–304.
80. Liu CP, Tsai TI, Cheng T, et al. Glycoengineering of antibody (Herceptin) through yeast expression and in vitro enzymatic glycosylation. Proc Natl Acad Sci USA. 2018;115:720–5.
81. Kuroda K, Kobayashi K, Kitagawa Y, et al. Efficient antibody production upon suppression of O mannosylation in the yeast *Ogataea minuta*. Appl Environ Microbiol. 2008;74:446–53.
82. Nylen A, Chen M-T. Production of full-length antibody by *Pichia pastoris*. In: Picanço-Castro V, Swiech K, editors. Recombinant glycoprotein production. Methods in molecular biology. Totowa: Humana Press; 2018. p. 37–48.
83. Hamilton SR, Zha D. Progress in yeast glycosylation engineering. In: Castilho A, editor. Glycoengineering. Methods in molecular biology. New York: Springer; 2015. p. 73–90.
84. De Wachter C, Van Landuyt L, Callewaert N. Engineering of yeast glycoprotein expression. Adv Biochem Eng Biotechnol. 2018. https://doi.org/10.1007/10_2018_69.
85. Maras M, De Bruyn A, Vervecken W, et al. In vivo synthesis of complex N-glycans by expression of human N-acetylglucosaminyltransferase I in the filamentous fungus *Trichoderma reesei*. FEBS Lett. 1999;452:365–70.
86. Kasajima Y, Yamaguchi M, Hirai N, et al. In vivo expression of UDP-N-acetylglucosamine: α-3-D-mannoside β-1,2-N-acetylglucosaminyltransferase I (GnT-1) in *Aspergillus oryzae* and effects on the sugar chain of α-amylase. Biosci Biotechnol Biochem. 2006;70:2662–8.
87. Horwitz AH, Chang CP, Better M, et al. Secretion of functional antibody and fab fragment from yeast cells. Proc Natl Acad Sci USA. 1988;85:8678–82.
88. Shusta EV, Raines RT, Plückthun A, et al. Increasing the secretory capacity of *Saccharomyces cerevisiae* for production of single-chain antibody fragments. Nat Biotechnol. 1998;16:773–7.
89. de Ruijter JC, Koskela EV, Frey AD. Enhancing antibody folding and secretion by tailoring the *Saccharomyces cerevisiae* endoplasmic reticulum. Microb Cell Factories. 2016;15:87.
90. Koskela EV, de Ruijter JC, Frey AD. Following nature's roadmap: folding factors from plasma cells led to improvements in antibody secretion in *S. cerevisiae*. Biotechnol J. 2017;12. https://doi.org/10.1002/biot.201600631.
91. Besada-Lombana PB, Da Silva NA. Engineering the early secretory pathway for increased protein secretion in *Saccharomyces cerevisiae*. Metab Eng. 2019;55:142–51.
92. Frenken LGJ, Van Der Linden RHJ, Hermans PWJJ, et al. Isolation of antigen specific llama V (HH) antibody fragments and their high level secretion by *Saccharomyces cerevisiae*. J Biotechnol. 2000;78:11–21.
93. Zahrl RJ, Peña DA, Mattanovich D, et al. Systems biotechnology for protein production in *Pichia pastoris*. FEMS Yeast Res. 2017;17:fox068. https://doi.org/10.1093/femsyr/fox068.
94. Ridder R, Schmitz R, Legay F, et al. Generation of rabbit monoclonal antibody fragments from a combinatorial phage display library and their production in the yeast *Pichia pastoris*. Bio/Technology. 1995;13:255–60.
95. Damasceno LM, Pla I, Chang HJ, et al. An optimized fermentation process for high-level production of a single-chain Fv antibody fragment in *Pichia pastoris*. Protein Expr Purif. 2004;37:18–26.
96. Damasceno LM, Anderson KA, Ritter G, et al. Cooverexpression of chaperones for enhanced secretion of a single-chain antibody fragment in *Pichia pastoris*. Appl Microbiol Biotechnol. 2007;74:381–9.
97. Gasser B, Maurer M, Gach J, et al. Engineering of *Pichia pastoris* for improved production of antibody fragments. Biotechnol Bioeng. 2006;94:353–61.
98. Prielhofer R, Maurer M, Klein J, et al. Induction without methanol: novel regulated promoters enable high-level expression in *Pichia pastoris*. Microb Cell Factories. 2013;12:5.

99. Gasser B, Steiger MG, Mattanovich D. Methanol regulated yeast promoters: production vehicles and toolbox for synthetic biology. Microb Cell Factories. 2015;14:196.

100. Vogl T, Sturmberger L, Kickenwciz T, et al. A toolbox of diverse promoters related to methanol utilization: functionally verified parts for heterologous pathway expression in *Pichia pastoris*. ACS Synth Biol. 2016;5:172–86.

101. Landes N, Gasser B, Vorauer-Uhl K, et al. The vitamin-sensitive promoter *PTHI11* enables pre-defined autonomous induction of recombinant protein production in *Pichia pastoris*. Biotechnol Bioeng. 2016;113:2633–43.

102. Rakestraw JA, Sazinsky SL, Piatesi A, et al. Directed evolution of a secretory leader for the improved expression of heterologous proteins and full-length antibodies in *Saccharomyces cerevisiae*. Biotechnol Bioeng. 2009;103:1192–201.

103. Govindappa N, Hanumanthappa M, Venkatarangaiah K, et al. A new signal sequence for recombinant protein secretion in *Pichia pastoris*. J Microbiol Biotechnol. 2014;24:337–45.

104. Heiss S, Puxbaum V, Gruber C, et al. Multistep processing of the secretion leader of the extracellular protein Epx1 in *Pichia pastoris* and implications for protein localization. Microbiology. 2015;161:1356–68.

105. Aw R, McKay PF, Shattock RJ, et al. A systematic analysis of the expression of the anti-HIV VRC01 antibody in *Pichia pastoris* through signal peptide optimization. Protein Expr Purif. 2018;149:43–50.

106. Barrero JJ, Casler JC, Valero F, et al. An improved secretion signal enhances the secretion of model proteins from *Pichia pastoris*. Microb Cell Factories. 2018;17:161.

107. Lebozec K, Jandrot-Perrus M, Avenard G, et al. Quality and cost assessment of a recombinant antibody fragment produced from mammalian, yeast and prokaryotic host cells: a case study prior to pharmaceutical development. New Biotechnol. 2018;44:31–40.

108. Swennen D, Paul M-F, Vernis L, et al. Secretion of active anti-Ras single-chain Fv antibody by the yeasts *Yarrowia lipolytica* and *Kluyveromyces lactis*. Microbiology. 2002;148:41–50.

109. Robin S, Petrov K, Dintinger T, et al. Comparison of three microbial hosts for the expression of an active catalytic scFv. Mol Immunol. 2003;39:729–38.

110. Krijger JJ, Baumann J, Wagner M, et al. A novel, lactase-based selection and strain improvement strategy for recombinant protein expression in *Kluyveromyces lactis*. Microb Cell Factories. 2012;11:112.

111. Kuroda K, Kitagawa Y, Kobayashi K, et al. Antibody expression in protease-deficient strains of the methylotrophic yeast *Ogataea minuta*. FEMS Yeast Res. 2007;7:1307–16.

112. Suzuki T, Baba S, Ono M, et al. Efficient antibody production in the methylotrophic yeast *Ogataea minuta* by overexpression of chaperones. J Biosci Bioeng. 2017;124:156–63.

113. Takegawa K, Tohda H, Sasaki M, et al. Production of heterologous proteins using the fission-yeast (*Schizosaccharomyces pombe*) expression system. Biotechnol Appl Biochem. 2009;53:227–35.

114. Naumann JM, Küttner G, Bureik M. Expression and secretion of a CB4-1 scFv-GFP fusion protein by fission yeast. Appl Biochem Biotechnol. 2011;163:80–9.

115. Jiang H, Horwitz AA, Wright C, et al. Challenging the workhorse: comparative analysis of eukaryotic micro-organisms for expressing monoclonal antibodies. Biotechnol Bioeng. 2019;116:1449–62.

116. Thomassen YE, Meijer W, Sierkstra L, et al. Large-scale production of VHH antibody fragments by *Saccharomyces cerevisiae*. Enzym Microb Technol. 2002;20:273–8.

117. Ye J, Ly J, Watts K, et al. Optimization of a glycoengineered *Pichia pastoris* cultivation process for commercial antibody production. Biotechnol Prog. 2011;27:1744–50.

118. Gorlani A, De Haard H, Verrips T. Expression of VHHs in *Saccharomyces cerevisiae*. In: Saerens D, Muyldermans S, editors. Single domain antibodies. Methods in molecular biology. Totowa: Humana Press; 2012. p. 277–86.

119. Hellwig S, Emde F, Raven NPG, et al. Analysis of single-chain antibody production in *Pichia pastoris* using on-line methanol control in fed-batch and mixed-feed fermentations. Biotechnol Bioeng. 2001;74:344–52.
120. Cunha AE, Clemente JJ, Gomes R, et al. Methanol induction optimization for scFv antibody fragment production in *Pichia pastoris*. Biotechnol Bioeng. 2004;86:458–67.
121. Khatri NK, Hoffmann F. Oxygen-limited control of methanol uptake for improved production of a single-chain antibody fragment with recombinant *Pichia pastoris*. Appl Microbiol Biotechnol. 2006;72:492–8.
122. Berdichevsky M, d'Anjou M, Mallem MR, et al. Improved production of monoclonal antibodies through oxygen-limited cultivation of glycoengineered yeast. J Biotechnol. 2011;155:217–24.
123. Cankorur-Cetinkaya A, Narraidoo N, Kasavi C, et al. Process development for the continuous production of heterologous proteins by the industrial yeast, *Komagataella phaffii*. Biotechnol Bioeng. 2018;115:2962–73.
124. Sharma R, Katoch M, Srivastava PS, et al. Approaches for refining heterologous protein production in filamentous fungi. World J Microbiol Biotechnol. 2009;25:2083–94.
125. Sun X, Su X. Harnessing the knowledge of protein secretion for enhanced protein production in filamentous fungi. World J Microbiol Biotechnol. 2019;35:54.
126. Nyyssönen E, Penttilä M, Harkki A, et al. Efficient production of antibody fragments by the filamentous fungus *Trichoderma reesei*. Bio/Technology. 1993;11:591–5.
127. Okazaki F, Aoki JI, Tabuchi S, et al. Efficient heterologous expression and secretion in *Aspergillus oryzae* of a llama variable heavy-chain antibody fragment VHH against EGFR. Appl Microbiol Biotechnol. 2012;96:81–8.
128. Hisada H, Tsutsumi H, Ishida H, et al. High production of llama variable heavy-chain antibody fragment (VHH) fused to various reader proteins by *Aspergillus oryzae*. Appl Microbiol Biotechnol. 2013;97:761–6.
129. Joosten V, Gouka RJ, van den Hondel CAMJJ, et al. Expression and production of llama variable heavy-chain antibody fragments (V(HH)s) by *Aspergillus awamori*. Appl Microbiol Biotechnol. 2005;66:384–92.
130. Joosten V, Roelofs MS, Van Den Dries N, et al. Production of bifunctional proteins by *Aspergillus awamori*: llama variable heavy chain antibody fragment (VHH) R9 coupled to *Arthromyces ramosus* peroxidase (ARP). J Biotechnol. 2005;120:347–59.
131. Ward M, Lin C, Victoria DC, et al. Characterization of humanized antibodies secreted by *Aspergillus niger*. Appl Environ Microbiol. 2004;70:2567–76.
132. Stock J, Sarkari P, Kreibich S, et al. Applying unconventional secretion of the endochitinase Cts1 to export heterologous proteins in *Ustilago maydis*. J Biotechnol. 2012;161:80–91.
133. Sarkari P, Reindl M, Stock J, et al. Improved expression of single-chain antibodies in *Ustilago maydis*. J Biotechnol. 2014;191:165–75.
134. Terfrüchte M, Reindl M, Jankowski S, et al. Applying unconventional secretion in *Ustilago maydis* for the export of functional nanobodies. Int J Mol Sci. 18:937. https://doi.org/10.3390/ijms18050937.
135. van den Hombergh JP, van de Vondervoort PJ, Fraissinet-Tachet L, et al. *Aspergillus* as a host for heterologous protein production: the problem of proteases. Trends Biotechnol. 1997;15:256–63.
136. Yoon J, Maruyama JI, Kitamoto K. Disruption of ten protease genes in the filamentous fungus *Aspergillus oryzae* highly improves production of heterologous proteins. Appl Microbiol Biotechnol. 2011;89:747–59.
137. Landowski CP, Huuskonen A, Wahl R, et al. Enabling low cost biopharmaceuticals: a systematic approach to delete proteases from a well-known protein production host *Trichoderma reesei*. PLoS One. 2015;10:e0134723.

Antibody Validation

14

Gordana Wozniak-Knopp

Contents

G. Wozniak-Knopp (✉)
Christian Doppler Laboratory for Innovative Immunotherapeutics, Institute of Molecular
Biotechnology, Department of Biotechnology, University of Natural Resources and Life Sciences,
Vienna (BOKU), Vienna, Austria
e-mail: gordana.wozniak@boku.ac.at

© Springer Nature Switzerland AG 2021
F. Rüker, G. Wozniak-Knopp (eds.), *Introduction to Antibody Engineering*,
Learning Materials in Biosciences,
https://doi.org/10.1007/978-3-030-54630-4_14

Keywords

Fit-for-purpose antibodies · Research antibody databases · Quality control · MIAPAR
proposal · Human protein atlas

What You Will Learn in This Chapter
This chapter summarizes the approaches for characterization of antibody-based
reagents for research purposes, including their identity and function in different
experimental set-ups, but above all to present the immense base of knowledge
being built by both antibody producers and end users with the purpose to unify the
standards for use of these reagents in research and commercial environments. To
outline the value of difficult-to-describe polyclonal antibodies, the particular aspects
that should be considered when applying animal-derived and other non-recombinant
reagents, are discussed. The standards for reporting on antibodies in scientific
publications and highly structured proposals for characterization of immunoreagents,
with references to databases that contain antibody-based reagent validation as well as
feedback of end users, are outlined. One prominent example is the set of standards for
antibodies presented within the Human Protein Atlas project. The importance of
antibody validation is underlined by the summary of joined efforts of eminent
scientific initiatives that have attempted to define the cornerstones of ontological
characterization of diagnostic antibodies, as well as the criteria of their performance
in particular applications. Finally, a list of essential requirements for the definition of
a characterized reagent poses an orientational landmark for its consideration in
diagnostic and research tests.

14.1   Introduction

14.1.1   State of the Art and Problem Setting

Validation of antibodies and other biological scaffold-based binders remains in the center
of the academic and commercially funded scientific community and of importance to
researchers who apply antibodies as therapeutic and diagnostic reagents. Their essential
properties include an accurate target specificity and adequate affinity. In addition, an
antibody-based reagent must be fit for its intended application and exhibit a sufficient

level of solubility, thermostability, and storage stability. The rise of antibodies as unique targeting molecules was accompanied by the development of standardized procedures that assured the reproducibility and set the standard for testing protocols. The definition of precise requirement for antibody validation was recognized as a task by leaders in antibody engineering and manufacturing, who courageously addressed this complex issue to achieve uniform standards, leading to transparency in the understanding of the properties of a particular antibody-based reagent [1]. The protocols involved are methodologically complementing to derive the integral picture of an antibody's biological, biochemical, and biophysical properties. The resulting consensus is achieved with the contribution both of experts in several various scientific disciplines as well as the end users, whose particular requests and experience can add value to the contents of the characterization reports [2]. These are the basics that guarantee the reproducibility of experiments and aid the advancement of science.

The wide application of antibodies as specific detection reagents has a long history and goes back for nearly a century for the polyclonal antisera that have been used in research, and four decades for monoclonal antibodies [3]. It sounds surprising that at least for animal-derived reagents, the preparations are never described at the molecular level, in sharp contrast to the well-characterized antibody-based therapeutics. Even if monoclonal antibodies are accepted to be reagents of a single defined specificity, hybridomas frequently secrete more than one light and/or heavy chain [4–6], or lose the expression of one of the chains or mutate, and hence their complete characterization at the molecular level requires sequencing of all present genes encoding for the antibody variable domains. In addition, allegedly only about 50% of commercially available antibodies exhibit the specificity to their cognate target [7–9]. A degree of lot-to-lot variability is inherent particularly to polyclonal antibodies [10, 11]. These reagents continue to be widely used; however, their identity remains largely unrevealed and their characterization is limited to the antigens applied for immunization. In contrast, the active ingredient of the novel alternative affinity reagents, such as recombinant antibodies, non-immunoglobulin molecular recognition scaffolds (e.g., DARPINs, anticalins, affibodies, fibronectin domains [12]) and nucleic acid-based affinity reagents (aptamers [13] and SOMAmers [14]) is described on the molecular level, together with concentration, formulation, and the data on its activity [15].

But although monoclonal antibodies react with epitopes in a very defined manner, this renders the antibody–antigen interaction more susceptible to physical and chemical changes of the antigen and antibody structure as well as test conditions. These can affect specificity and the kinetics of the interaction and requires unified reference methods for standardization of a diagnostic test [16].

Questions

1.1.1 What are the requirements for an antibody to be applied as a therapeutic or diagnostic reagent?

1.1.2 What are the obstacles for the molecular definition of polyclonal and monoclonal antibodies?

14.1.2 Advantages of Polyclonal Antibodies

Why do polyclonal antibodies remain so widely used reagents in spite of their biophysical diversity? Two properties are at the forefront of numerous advantages of polyclonals in comparison with monoclonal and recombinant antibody preparations: first, the reactivity with the multiplicity of target antigenic determinants, which translates to a higher sensitivity, and second, the biophysical diversity of their active ingredients, which renders them overall more stable in various environmental conditions [17]. Both properties are beneficial in the research environment: the heterogeneous binding is more likely to function in different specific experimental conditions. It has long been acknowledged that the effective capturing of multiple or variable antigen epitopes in sandwich ELISA assays is more efficiently performed by polyclonal antibodies [18], and the same was established for immunohistochemistry techniques where the effect of tissue fixation and processing may differently affect diverse antigen epitopes [19]. Polyclonal antibodies have proven superior in the detection of low-abundant targets due to the simultaneous interaction of multiple antibodies with various epitopes. Additionally, polyclonal antibodies are excellent secondary reagents as they are less discriminative in their recognition and hence tolerant of small sequence and structural differences in primary antibodies. These properties also render them invaluable reagents when detecting mouse antibodies, which are very diverse in their isotypes, or antibodies from sera of human individuals of different ethnic origin [17]. Polyclonal antibodies can score in recognizing rarely occurring epitopes, as these reagents mostly contain several antibodies reacting with several epitopes. Multi-epitope binding is also the reason for a higher specific reactivity of a polyclonal than a monoclonal antibody in Western blot experiments. To conclude, the challenges encountered in the characterization of polyclonal antibodies are largely set off with their advantageous applicability. The future developments will likely involve a broader use of oligoclonal antibodies that will be compensated for the unique specificity and set of biophysical properties of a monoclonal antibody but offer a higher level of definition as polyclonal reagents [20]. A novel idea suggests the use of polyclonal reagents that can be generated in vitro by combining phage and yeast display, containing hundreds of different antibodies directed toward a target of interest, and which can be reproducibly amplified [21].

Questions

1.2.1 What are the advantages of polyclonal antibody-based reagents?

1.2.2 List examples of methods where polyclonal antibodies are preferentially used.

14.2 Cornerstones for Antibody Validation for Different Scientific Applications

14.2.1 Standards for Reporting the Use of Antibodies in Scientific Publications

Research antibodies are used in a wide range of bioscience disciplines. The commonly agreed standard of reporting of research antibody use enables researchers to reproduce the experiments as well as judge on the conclusions drawn from the presented results. Authors of scientific publications are nowadays requested to demonstrate that every antibody used in their study has been validated for use in each of the specific experiments and species [22]. Each experimental setup requires a separate validation, as the antibody specificity in a certain application, or even in a different matrix, does not promise the same reactivity under different conditions. Even the publicly available information supporting the validation may differ for different applications and species [22].

The first step in the reporting of antibodies' features is their unambiguous identification, implying their unique biochemical composition. It should on one hand always incorporate a unique identifier, but also include all synonymous descriptions. This requires the name of the antibody together with the catalog or clone number and the supplier for commercially available antibodies, and source laboratory and relevant reference for the antibodies from academic labs. Further, the core information on the antibodies includes the host species in which the antibody was raised and whether the antibody is monoclonal or polyclonal. The antibody batch number is relevant due to variability between different antibody production batches particularly with polyclonal antibodies [23], but also to monoclonal antibodies [19]. For the interpretation of the results, it should be known against which antigen the antibody was raised [24], and the pertaining information should be given when the immunization was performed with complex samples, such as cell and tissue lysates. Finally, the details on the purification method, grade of purity, and the content of the active reagent are of importance for the performance of individual experimental techniques.

Questions

2.1.1 Why is it important to standardize reports on the experiments with antibody-based reagents?

2.1.2 What features are included in the unambiguous identification of an antibody?

2.1.3 List other information that aids the validation of an antibody-based reagent.

14.2.2 Publicly Available Antibody Databases

The antibody validation is a complex experimental process, which at a certain level requires rigorous testing with methods such as comparison of antibody's reactivity to wild-type with a knockdown/knockout tissue, and use of a second antibody to a different epitope. Public citations of antibody validation, as well as references to its profile in open databases (e.g., 1degreebio, Antibodypedia, CiteAb, or pAbmAbs) contribute to the experimental reproducibility. Detailed reagent profiling in publications is likely to enhance the impact of such reports and they can be included in antibody search engines and antibody suppliers' data banks. Pivotal Scientific, an international biotechnology consultancy, features an informative list of the 17 major antibody comparison websites (https://pivotalscientific.com/antibody-comparison-websites). The website contains information with description and rating of antibody-based products, their coverage of antibody vendors, as well as their general features, such as functionality and rendering of the website.

In the following paragraph, the portal Antibodypedia (https://www.antibodypedia.com/) is presented in more detail. Antibodypedia was initiated as a publicly available portal, with the main aim to allow the sharing of information on validation of antibodies in which providers can submit their own validation results and reliability scores [25]. Here the validation criteria and submission rules are reported for particular antibody applications, such as Western blots, protein arrays, immunohistochemistry, and immunofluorescence. The contributors are expected to deliver experimental evidence for each antibody, including data on the antigen, and the users can provide feedback and comments on the use of the antibody. The database thus provides a virtual resource of publicly available antibodies toward human proteins with accompanying experimental evidence in an application-specific manner.

Another useful reference for antibody citation is the Research Resource Identifier (RRID), assigned via the Antibody Registry (http://antibodyregistry.org/) [26]. The aim of RRIDs is to enable unambiguous identification of the antibody and other publications that describe the same antibody can be retrieved via an RRID link.

Questions

2.2.1 What is the purpose of publicly available antibody databases?

2.2.2 Which information is included in such antibody databases?

14.2.3 Minimum Information About a Protein Affinity Reagent Proposal

As the collective need for standards to validate antibody specificity and reproducibility of performance has been communicated by the scientific quorums, the extensive discussion of antibody validation has triggered a requirement for standards in reporting practices. A pioneering attempt has been introduced in 2010 by Bourbeillon and colleagues by the minimum information about a protein affinity reagent (MIAPAR) proposal [27]. Its main objective was to formally define how to report information about binders so that for a specific application, one can identify a proper affinity reagent for a particular target. The MIAPAR-compliant reports should deliver the information on how to select the best-suited binder from a catalog, a database or a publication, and offer the possibility of feedback to expand information sharing. For binder producers, MIAPAR files should complement the information from databases and catalogs with research publications, and at the same time allow positioning by binders' comparison. Especially for variable reagents such as polyclonal antibodies, the regularly updated descriptions should refer to the batch of the available material.

The central objective of the MIAPAR is the reliable identification of the affinity reagent–target–application triad. As a binder typically recognizes a target protein or peptide and it may involve the interaction in complex mixtures, the reference has to be made on its intended target and its molecular properties (Table 14.1).

Further, the objective of MIAPAR to achieve convention and clarity proposes three basic criteria (Table 14.2): unambiguity, sufficiency, and practicality. Importantly, the guidelines are expected to be agreed upon within expert quorums and evolve along with technological progress.

MIAPAR was a pioneering incentive to unify the reporting of commonly used antibody reagents; however, experimental approaches best suited to support validation of antibody specificity in particular applications were not yet included as explicit recommendations.

Ignited by the significant concern raised about the identity, quality, and reproducibility of antibodies cited in several scientific publications, expressed by both the vendors as well as the end users, a demand has been posed to use Western blot data as an initial indication of antibody quality as well as distinguish between antibodies allegedly targeting the same antigen. Secondly, best manufacturing practices have been agreed upon to reduce the variability in polyclonal antibodies. Among others, valuable suggestions include preferred use of recombinant antibodies, unique identification of antibody reagents, specification of their immunogens, and their source. Further, a centralized international banking of standard

Table 14.1 An antibody reagent's portfolio contents summarized in MIAPAR

Description of the binding reagent and the immunogen
• Production process may influence the characteristics of the affinity reagent and its target
• Allows the identification of the molecules
Binding properties of the reagent
Delivers the information on:
• Specificity
• Affinity
• Antigen binding kinetics
• Cross-reactivity
Use of the reagent for different techniques
• Information on the compatibility with experimental applications
Links to protocols and experimental results
• Enable the production of the compound
• Help to identify the qualities of the binder
• Support the proposed experimental applications

Table 14.2 The basic criteria of MIAPAR proposal

Criterion	Requirement
Unambiguity	• Standard naming for compounds and processes is used • Database accession numbers included
Sufficiency	Information allows the end user to: • Evaluate experimental result • Interpret the validity of the project outcome • Compare the outcome to similar projects
Practicality	• Guidelines should support further widespread use

antibodies and their ligands has been proposed, along with widely accessible open-source documentation of user experience with antibodies and their certification.

Questions

2.3.1 What was the main goal of the MIAPAR proposal?

2.3.2 Which features were recommended to be included in a reagent's portfolio?

2.3.3 Explain the three basic criteria for validation of an affinity reagent proposed by MIAPAR.

14.3 Scientific Incentives for Antibody Validation

How to set the standards for unambiguous antibody identification and description of properties that make them fit-for-purpose reagents? This was the central task of groups of distinguished scientists, who organized their activities to deliver proposals of standards of best practice to the antibody scientific community. Groups whose solutions can be highlighted included The European Monoclonal Antibody Network (EuroMAbNet) [28], the International Working Group on Antibody Validation (IWGAV), [29] and the Global Biological Standards Institute (GBSI) [30].

EuroMAbNet (www.euromabnet.com) is a network of well-established academic laboratories engaged in monoclonal antibody production. Their practical guide advises to:

1. Acquire the background information on immunogen selection and design; the reactivity with endogenous antigen is to be confirmed.
2. Choose the antibody types in relation to envisaged techniques.
3. Find highly rated suppliers and features of different databases.

In their view, the validation is in the best interest of the user, which is highlighted by pointing out a set of criteria and instructions for researchers to use. These refer to each antibody used in house and take into account the identity of the target antigen and the experimental method to be used.

The IWGAV working group consisted of researchers from institutions, such as Stanford University, Yale University, MIT, UCSD, University of Toronto, National Institutes of Health (NIH), European Molecular Biology Laboratory (EMBL), Niigata University in Japan and Science for Life Laboratory in Sweden. Their recommendations were summarized as "A proposal for validation of antibodies" and published in 2016 in the journal *Nature Methods* [29]. The working group suggested the use of five "pillars" for validation of antibodies with the aim to establish standards for evaluation and quality control of antibodies, for both users and producers of antibodies (Table 14.3).

Table 14.3 The five pillars for antibody validation proposed by the IWGAV working group

Validation strategy		
Goal	Description	Method
Genetic		
Decreased levels of target protein by genetic manipulation	Knock-down (knock-out) of the target protein gene with methods such as CRISPR or siRNA	The staining intensity of the antibody is evaluated before and after the manipulation of the target gene
Recombinant expression		
Increased expression of target protein by recombinant expression	Comparison of the staining with an antibody-independent method that reveals the expression level of the target protein	Two samples that express the target protein at different levels are compared and the level of the target protein is reported by two independent methods
Independent antibodies		
Matched staining of two antibodies reactive with the target protein	Comparison of the reactivity of two independent antibodies with non-overlapping epitopes	The staining pattern of the two antibodies is compared in at least two tissues or cell lines and must show a similar result
Orthogonal		
Target expression pattern is followed with antibody-independent method	Over-expression of the target protein in a target-negative cell line, or expression of a fluorescently tagged target protein in a cell line	Evaluation includes comparing the signal produced by the over-expressed or tagged version of the target protein with the unmodified or endogenous target protein
Capture MS		
Target protein size is determined using mass spectrometry-based size analysis	Staining pattern of the antibody should match the results obtained by the capture MS method	The size revealed by the antibody corresponds to the size of the target protein detected in capture MS

In September 2016, the Global Biological Standards Institute (GBSI) in conjunction with The Antibody Society held a meeting dedicated to "Antibody Validation: Standards, Policies, and Practices," at Asilomar, California (known in the community as the "Asilomar meeting"). Its main goal was to develop standards to certify antibodies. Working groups focused on developing certification standards for specific applications, including Western blot, immunohistochemistry, immunoprecipitation, and sandwich assays, and an extended report on "consensus principles" was the outcome of the meeting [30].

Questions

3.1 Summarize the main points of the practical guide of EuroMAbNet.

3.2 Describe the five pillars for antibody validation listed in "A proposal for antibody validation" by IWGAV working group.

3.3 What was the focus of GBSI "Asilomar meeting"?

14.4 The Human Protein Atlas

14.4.1 Introduction to the Human Protein Atlas

The Human Protein Atlas project aims to evaluate protein expression across cells and tissues [31]. As the performance of antibodies as detection agents in different assays depends on both sensitivity and specificity of epitope binding, antibody validation emerged as a side discipline of the project. All antibodies included in the study have to be compliant with a set of defined criteria and only such can be accepted as a referenced tool. In addition to the standard quality assurance, enhanced antibody validation strategies were defined for specific experimental applications [32]. But to begin with, what is the Human Protein Atlas?

The Human Protein Atlas is a Swedish-based consortium involving several international collaborations, initiated in 2003 and funded by the Knut and Alice Wallenberg Foundation. This project aims to map all human proteins in cells, tissues, and organs using collective information from "–omics" disciplines, including antibody-based imaging, mass spectrometry-based proteomics, transcriptomics, and systems biology. This knowledge resource is open to the scientific community in academia and industry.

The Human Protein Atlas focuses on three aspects of the genome-wide analysis of the human proteins:

- The Tissue Atlas shows the distribution of the proteins across all major tissues and organs in the human body [33].
- The Cell Atlas reveals the localization of proteins at a subcellular level in single cells [34].
- The Pathology Atlas puts the expression level of a protein to relation with the survival of cancer patients [35].

With several thousand citations, European core resource ELIXIR (www.elixir-europe. org) acknowledges the Human Protein Atlas program as a recognized resource in the field of human health and disease [36].

14.4.2 History of Human Protein Atlas

The first version of the Human Protein Atlas website was launched in 2005 and consisted of protein expression data based on approximately 700 antibodies. Since then, each new release has added more data and new functionalities and features to the website (Fig. 14.1). The last, eighteenth major release of the Human Protein Atlas, dated December 2017, covers more than 26,000 antibodies, targeting proteins from almost 17,000 human genes, which amounts to ~87% of the human protein-coding genes [37].

14.4.3 Standard Antibody Validation

To be used for immunohistochemistry and immunocytochemistry or immunofluorescence, all antibodies produced within the Human Protein Atlas project are subject to quality assurance steps (Fig. 14.2): (1) sequencing of plasmid inserts to confirm that the correct epitope signature tag (PrEST) sequence is cloned, (2) mass spectrometry analysis of the size of the resulting recombinant protein to assure that the correct antigen has been produced and purified, and (3) cross-reactivity control of affinity-purified antibodies on protein arrays consisting of glass slides with spotted PrEST fragments [37].

Evaluation of the antibody reliability is performed according to guidelines in the following steps: (4) Western blot for testing of specificity is performed with total protein lysates of cell lines and tissues, (5) immunohistochemical reactivity with normal and cancer tissue is compared with reference data, and (6) confocal microscopy imaging of human cell lines is used to discover the subcellular localizations [37].

The Human Protein Atlas portal (https://www.proteinatlas.org/) features all antibodies that provide a reasonable pattern of immunoreactivity and appreciates the feedback from the research community. Antibodies supplied from sources other than the Human Protein Atlas project have been tested with immunocytochemistry and immunohistochemistry, as well as with Western blot-based methods and the commercially available antibodies are listed along with a link to the antibody provider.

14.4.4 Enhanced Antibody Validation

Antibodies used for Western blot, immunocytochemistry, and immunohistochemistry in the Human Protein Atlas undergo enhanced antibody validation based on the five "pillars" described by the IWGAV working group (see above) [29], and antibodies that fulfill the

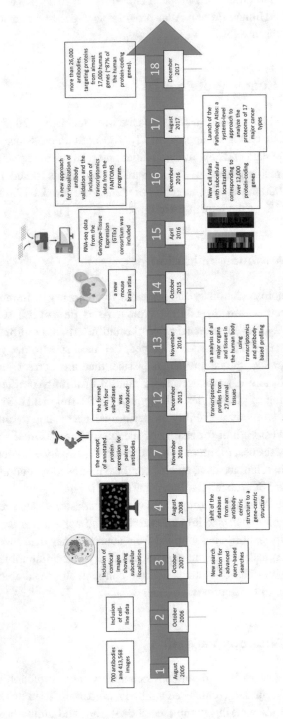

Fig. 14.1 Time arrow illustrating the developments in Human Protein Atlas project

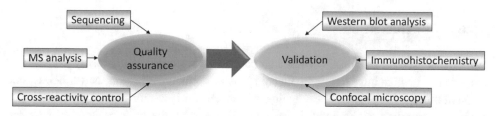

Fig. 14.2 Quality assurance and validation steps required for an antibody-based reagent to be used in the Human Protein Atlas Project

criteria are labeled "Enhanced." The enhanced validation principles are adapted for validation in Western blot, immunocytochemistry, and immunohistochemistry applications. After the application of IWGAV guidelines, over 10,540 antibodies directed to a total of 6787 human protein targets have passed the criteria of enhanced validation [32].

Questions

4.1 Describe the purpose of the Human Protein Atlas Project.

4.2 Why is the Human Protein Atlas important for antibody validation?

4.3 What are the quality assurance steps that an antibody must pass for validation prescribed in Human Protein Atlas and according to which measures is its reliability evaluated?

14.5 Start-off Simple: 10 Validation Rules for Diagnostic and Analytical Antibodies

"Validation is the experimental proof and documentation that a specific antibody is suitable for an intended application or purpose" [38] and hence, it refers to a biochemical compound in connection with a protocol or process. In this view, Weller [38] summarized 10 basic rules applicable to the validation of an antibody-based reagent, together with the "purpose" to explain the orientational guidelines also to novices in the antibody community (Table 14.4).

Questions

5.1 How can an antibody and its antigen be defined in an unambiguous way?

5.2 Comment on the importance of cross-reactivity tests.

5.3 What is "Matrix effect"?

5.4 How can user feedback influence the antibody validation?

14.6 Conclusion

Antibodies used for diagnostic or research purposes have been made broadly available to the scientific community through collaborations or commercialization. Nevertheless, not all went through the process of evaluation at independent laboratories, and not all follow a consistent reporting of the validation results. In this way, the limitations of a particular antibody reagent could initiate efforts to generate new binder candidates with improved activity, specificity, or suitability for a particular experimental procedure. Such well-characterized and validated antibodies will yield reproducible results across independent

Table 14.4 Ten validation rules for diagnostic and analytical antibodies

1. *Definition of the antibody*	
	The binding molecule, most often an antibody, needs to be identifiable in an unambiguous way
Method	Provide a clone number, commercial product number, antibody ID
Purpose	Adds knowledge on the reagent and enhances the reproducibility of scientific work
2. *Definition of the antigen*	
	The target antigen needs to be identifiable in an unambiguous way
Method	• Haptens: chemical formula • Larger molecules: epitope • Cells and tissues for histochemical reagents • Define if native or denatured antigen was used
Purpose	A detailed description of the target is a prerequisite for the assessment of antibody properties
3. *Cross-reactivity testing*	
	Description of binding selectivity of the tested agent
Method	• All relevant substances should be tested for cross-reactivity • Preferred method is immunoprecipitation followed by mass spectrometry
Purpose	Test of several potential cross-reactivity targets indicates a carefully characterized antibody
4. *Concentration of antibodies*	
Method	Determine the concentration of specific and active antibody
Purpose	"Antibody concentration" given on a product label is only a preliminary estimate for assay optimization; the concentration of the active ingredient can be much lower
5. *Documentation*	
Method	Should include information listed in points 1–4
Purpose	Less characterized antibodies require more effort by end user
6. *Determine the binding constant*	
	Relates to the binding strength of antibody/antigen complex
Method	The measurement can be performed with SPR, BLI, ELISA, equilibrium dialysis, etc.
Purpose	• The binding constant defines the sensitivity (detection limit) of the assay • Low-affinity antibodies often lead to weak specific signals

(continued)

Table 14.4 (continued)

7. *Determine the "Matrix effect"*	
	Description of the influence of non-target substances present in the intended assay
Method	Includes description of salts, preservatives, solvents, detergents, and other additives
Purpose	Antibodies, sensitive to matrix compounds, give variable assay results and are less suitable for a practical application [19]
8. *Describe proper storage conditions*	
	Specification of activity under defined storage temperatures
Method	• List recommended storage additives • Test of activity revealing stability under specified storage conditions
Purpose	Very helpful information for end user
9. *Application protocols*	
	Application sheets
Method	Summary of the protocols for a specific experimental method
Purpose	• Such reports reflect practical validation efforts • Better chances of a successful implementation of an assay
10. *Enable user feedback*	
	The exchange of user information in public platforms
Method	Recommendations from users, who already tested the antibody in their specific applications
Purpose	Positive feedback contributes to the acceptance of the antibody in the community

laboratories, leading to significant advances in the field. To achieve this goal, the contribution of antibody producers as well as the end users is required to maximize the availability of results from carefully controlled validation procedures performed before the actual use, as well as the shared experience of end users. This will aid the selection of reagents with a proven track record and also speed up the development of experimental protocols for their optimal use.

Take Home Messages
- Each affinity reagent should be uniquely identifiable by its features. When a definition per molecular structure is not possible, as is often the case for non-recombinant reagents, the standards delineated in the recommendations listed below should be followed.
- In spite of their inherent molecular heterogeneity, polyclonal antibody reagents are valuable in research and diagnostic settings, as they provide a higher sensitivity in many analytical assays, and are less prone to complete inactivation due to specimen preparation and environmental factors.
- There are certain standards to follow when reporting on antibody-based reagents in scientific publications. These have been conveyed to the scientific community

(continued)

as summaries of recommendations, for example, by the MIAPAR proposal, and dedicated EuroMAbNet, IWGAV, and GBSI "Asilomar"—meeting documents.

- Within the Human Protein Atlas project, further guidelines for validation and evaluation of antibodies for a particular application are outlined.
- Definition of the antibody, its target antigen, its affinity binding constant, its stability, an adequate profile of cross-reactivity, the determination of its purity grade and concentration, the potential effect of the matrix in which the assay is performed, accompanied with a comprehensive documentation, is the basis of an adequately validated antibody. Up-to-date experimental data contributed by producers as well as the end users, optimally widely available to the public over internet platforms, should be a part of the portfolio of each antibody used for research and diagnostic purposes.

Answers

1.1.1 An affinity reagent must exhibit accurate target specificity, adequate affinity, a sufficient level of solubility, thermostability, and storage stability.

1.1.2 Polyclonal antibodies contain a mixture of several different antibodies; hybridomas secreting monoclonal antibodies can secret multiple chains, lose the expression of one of the chains or mutate.

1.2.1 In comparison with monoclonal antibodies, polyclonal antibodies are more likely to recognize the desired antigen when it is partially denatured, present in multiple variants or present in low quantities. Their activity is less sensitive to environmental conditions.

1.2.2 Polyclonal antibodies deliver higher sensitivity when used for capturing of multiple antigen variants or epitopes than an assay performed with monoclonal antibody pairs. They offer more sensitivity in assays that include harsh specimen treatment, such as ChIP assays and immunohistochemistry. In Western blot experiments, polyclonal antibodies show a higher specific reactivity due to multi-epitope binding.

2.1.1 The reporting of research antibodies uses a certain standard enabling the researchers to reproduce the experiments as well as to judge on the conclusions drawn from the presented results.

2.1.2 An unambiguous definition of an antibody implies the correct identification of its unique biochemical composition. It should on one hand always incorporate a unique identifier and all synonymous descriptions (name of the antibody, the catalog or clone number and the supplier for commercially available antibodies, and source laboratory and relevant reference for the antibodies from academic labs).

2.1.3 The core information on the antibodies includes the host species in which the antibody was raised, whether it is monoclonal or polyclonal, and the antibody batch number, especially for polyclonal antibodies. It should be known against which antigen the antibody was raised, and the details on the purification method, grade of purity, and the content of the active reagent should be given.

2.2.1 Reports on an antibody that can be accessed through the publicly available databases highlight the experimental reproducibility and such publications are likely to encourage the researchers to access them. The information is available from antibody search engines and linked with antibody suppliers' databases.

2.2.2 Antibody databases include the identification of an antibody, including data on the antigen, and test of its performance for particular antibody applications, such as Western blots, protein arrays, immunohistochemistry, and immunofluorescence. The contributors are expected to deliver experimental evidence for each antibody, and the users can provide feedback and comments on the use of the antibody.

2.3.1 The main goal of MIAPAR proposal was to formally define how to report information about binders so that for a specific application, one can make use of a proper affinity reagent for a particular target.

2.3.2 Features to be included in a reagent's portfolio were the description of the production process of the binding reagent and the immunogen, description of binding properties of the reagent, data on the use of the reagent in different applications, and links to protocols and experimental results.

2.3.3 The main criteria of MIAPAR guidelines were *unambiguity*, implying the use of standard naming conventions for compounds and processes and database accession numbers, *sufficiency*, meaning that the information should be sufficient to evaluate experimental corroboration and interpret the validity of the project outcome, and *practicality*, which should assure that the guidelines allow the continued widespread use of the reagents.

3.1 The practical guide of EuroMAbNet delineates a set of criteria and instructions for researchers to use: to acquire the background information on immunogen selection and design and confirm the reactivity with endogenous antigen, to choose the antibody types in relation to envisaged techniques and to find highly rated suppliers and features of different databases.

3.2 IWGAV working group proposed the five pillars for antibody validation: "genetic," by which expression levels of target protein should be decreased by genetic methods, "recombinant expression," where the increased levels of target protein should be achieved by recombinant expression, "independent antibodies," a method that proposes matched staining of two antibodies binding to the target protein, "orthogonal," where matched target expression patterns should be obtained with antibody-independent methods, and "capture MS," where matched protein size should be determined using mass spectrometry-based size analysis.

3.3 The focus of GBSI "Asilomar meeting" was to develop standards to certify antibodies for specific applications, including Western blot, immunohistochemistry, immunoprecipitation, and sandwich assays.

4.1 Human Protein Atlas Project aims to evaluate protein expression across cells and tissues. This project aims to map all human proteins in cells, tissues, and organs using collective information from antibody-based imaging, mass spectrometry-based proteomics, transcriptomics, and systems biology, and collate this in an open knowledge resource.

4.2 Antibody validation emerged as an extended discipline of the Human Protein Atlas project, as the performance of antibodies as detection agents in different assays depends on both sensitivity and specificity of epitope binding.

4.3 Quality assurance steps for the antibodies validated in Human Protein Atlas project include sequencing of plasmid inserts to assure that the correct epitope signature tag (PrEST) sequence is cloned, mass spectrometry analysis of the size of the resulting recombinant protein to assure that the correct antigen has been produced and purified, and cross-reactivity control of affinity-purified antibodies on protein arrays consisting of glass slides with spotted PrEST fragments. Evaluation of the antibody reliability is further performed: specificity is examined using Western blot in a standardized setup, using total protein lysates from a limited number of tissues, cell lines, and human plasma, immunohistochemical staining of normal and cancer tissue and is compared with available data, and confocal microscopy images of human cell lines stained by indirect immunofluorescence are annotated for subcellular localizations and compared with the available immunohistochemical staining data.

5.1 Identification of the antibody requires a clone number, commercial product number, antibody ID, and preferably information on the production batch. Depending on the class of the antigen, unambiguous identification includes the chemical formula for haptens, epitopes for larger molecules, identification of cells and tissues for histochemical reagents, and the information if native or denatured antigen was used.

5.2 In determining cross-reactivity of an antibody, all relevant substances should be tested, preferably with immunoprecipitation followed by mass spectrometry.

5.3 "Matrix effect" refers to the influence of non-target substances present in the prospective assay. Antibodies, which are very sensitive to matrix compounds, give unstable assay results and are less suitable for a practical application.

5.4 The exchange of user data and experience in "Open Science" platforms can positively influence the acceptance of an antibody in the research community and accelerate the optimization of application-specific protocols.

Acknowledgments The financial support by the Christian Doppler Society (CD Laboratory for innovative Immunotherapeutics), Austrian Federal Ministry for Digital and Economic Affairs and the National Foundation for Research, Technology and Development is gratefully acknowledged.

References

1. Bradbury A, Plückthun A. Reproducibility: standardize antibodies used in research. Nature. 2015;518(7537):27–9.
2. Taussig MJ, Fonseca C, Trimmer JS. Antibody validation: a view from the mountains. New Biotechnol. 2018;45:1–8.
3. Köhler G, Milstein C. Continuous cultures of fused cells secreting antibody of predefined specificity. Nature. 1975;256(5517):495–7.
4. Ruberti F, Cattaneo A, Bradbury A. The use of the RACE method to clone hybridoma cDNA when V region primers fail. J Immunol Methods. 1994;173(1):33–9.
5. Zack DJ, Wong AL, Stempniak M, Weisbart RH. Two kappa immunoglobulin light chains are secreted by an anti-DNA hybridoma: implications for isotypic exclusion. Mol Immunol. 1995;32(17–18):1345–53.
6. Blatt NB, Bill RM, Glick GD. Characterization of a unique anti-DNA hybridoma. Hybridoma. 1998;17(1):33–40.
7. Slaastad H, Wu W, Goullart L, Kanderova V, Tjønnfjord G, Stuchly J, et al. Multiplexed immuno-precipitation with 1725 commercially available antibodies to cellular proteins. Proteomics. 2011;11(23):4578–82.
8. Blow N. Antibodies: the generation game. Nature. 2007;447(7145):741–4.
9. Berglund L, Björling E, Jonasson K, Rockberg J, Fagerberg L, Szigyarto CAK, et al. A whole-genome bioinformatics approach to selection of antigens for systematic antibody generation. Proteomics. 2008;8(14):2832–9.
10. Bordeaux J, Welsh AW, Agarwal S, Killiam E, Baquero MT, Hanna JA, et al. Antibody validation. BioTechniques. 2010;48(3):197–209.
11. Saper CB. A guide to the perplexed on the specificity of antibodies. J Histochem Cytochem. 2009;57(1):1–5.
12. Binz HK, Amstutz P, Plückthun A. Engineering novel binding proteins from nonimmunoglobulin domains. Nat Biotechnol. 2005;23(10):1257–68.
13. Huo Y, Qi L, Lv XJ, Lai T, Zhang J, Zhang ZQ. A sensitive aptasensor for colorimetric detection of adenosine triphosphate based on the protective effect of ATP-aptamer complexes on unmodified gold nanoparticles. Biosens Bioelectron. 2016;78:315–20.
14. Kraemer S, Vaught JD, Bock C, Gold L, Katilius E, Keeney TR, et al. From SOMAmer-based biomarker discovery to diagnostic and clinical applications: a SOMAmer-based, streamlined multiplex proteomic assay. PLoS One. 2011;6(10):e26332.
15. Bradbury ARM, Plückthun A. Getting to reproducible antibodies: the rationale for sequenced recombinant characterized reagents. Protein Eng Des Sel. 2015;28(10):303–5.
16. Albert WHW. The antibody/antiserum as an analytical reagent in quantitative immunoassays. Scand J Clin Lab Invest. 1991;51(S205):79–85.
17. Ascoli CA, Aggeler B. Overlooked benefits of using polyclonal antibodies. BioTechniques. 2018;65(3):127–36.
18. Cox KL, Devanarayan V, Kriauciunas A, Manetta J, Montrose C, Sittampalam S. Immunoassay methods. Assay guidance manual; 2004.
19. Voskuil JLA. Commercial antibodies and their validation. F1000Res. 2014;3:232.
20. Corti D, Kearns JD. Promises and pitfalls for recombinant oligoclonal antibodies-based therapeutics in cancer and infectious disease. Curr Opin Immunol. 2016;40:51–61.
21. Ferrara F, D'Angelo S, Gaiotto T, Naranjo L, Tian H, Gräslund S, et al. Recombinant renewable polyclonal antibodies. MAbs. 2015;7(1):32–41.
22. Chalmers AD, Helsby MA, Fenn JR. Reporting research antibody use: how to increase experimental reproducibility. F1000Res. 2013;2:153.

23. Couchman JR. Commercial antibodies: the good, bad, and really ugly. J Histochem Cytochem. 2009;57(1):7–8.
24. Kalyuzhny AE. The dark side of the immunohistochemical moon: industry. J Histochem Cytochem. 2009;57(12):1099–101.
25. Björling E, Uhlén M. Antibodypedia, a portal for sharing antibody and antigen validation data. Mol Cell Proteomics. 2008;7(10):2028–37.
26. Bandrowski A, Brush M, Grethe JS, Haendel MA, Kennedy DN, Hill S, et al. The resource identification initiative: a cultural shift in publishing. Brain Behav. 2016;6(1):1–14.
27. Bourbeillon J, Orchard S, Benhar I, Borrebaeck C, De Daruvar A, Dübel S, et al. Minimum information about a protein affinity reagent (MIAPAR). Nat Biotechnol. 2010;28(7):650–3.
28. Roncador G, Engel P, Maestre L, Anderson AP, Cordell JL, Cragg MS, et al. The European antibody network's practical guide to finding and validating suitable antibodies for research. MAbs. 2016;8(1):27–36.
29. Uhlen M, Bandrowski A, Carr S, Edwards A, Ellenberg J, Lundberg E, et al. A proposal for validation of antibodies. Nat Methods. 2016;13(10):823–7.
30. Baker M. Biologists plan scoring system for antibodies. Nature. 2016;
31. Uhlen M, Ponten F. Antibody-based proteomics for human tissue profiling. Mol Cell Proteomics. 2005;4(4):384–93.
32. Edfors F, Hober A, Linderbäck K, Maddalo G, Azimi A, Sivertsson Å, et al. Enhanced validation of antibodies for research applications. Nat Commun. 2018;9(1):4130.
33. Fagerberg L, Hallstrom BM, Oksvold P, Kampf C, Djureinovic D, Odeberg J, et al. Analysis of the human tissue-specific expression by genome-wide integration of transcriptomics and antibody-based proteomics. Mol Cell Proteomics. 2014;13(2):397–406.
34. Regev A, Teichmann SA, Lander ES, Amit I, Benoist C, Birney E, et al. The human cell atlas. elife. 2017;6:e27041.
35. Pontén F, Jirström K, Uhlen M. The human protein atlas – a tool for pathology. J Pathol. 2008;216(4):387–93.
36. Durinx C, McEntyre J, Appel R, Apweiler R, Barlow M, Blomberg N, et al. Identifying ELIXIR core data resources. F1000Res. 2017;5:ELIXIR-2422.
37. Thul PJ, Lindskog C. The human protein atlas: a spatial map of the human proteome. Protein Sci. 2018;27(1):233–44.
38. Weller MG. Ten basic rules of antibody validation. Anal Chem Insights. 2018;13:1–5.

Index

© Springer Nature Switzerland AG 2021
F. Rüker, G. Wozniak-Knopp (eds.), *Introduction to Antibody Engineering*,
Learning Materials in Biosciences,
https://doi.org/10.1007/978-3-030-54630-4

Printed in the United States
By Bookmasters